W9-CYV-965

MEDICAL APPLICATIONS OF MICROWAVE IMAGING

OTHER IEEE PRESS BOOKS

MEDICAL APPLICATIONS OF MICROWAVE IMAGING

Edited by
Lawrence E. Larsen, M.D.
John H. Jacobi, M.S.E.E.

Published under the sponsorship of the IEEE Microwave Theory and Techniques Society.

Supported by the U.S. Army Medical Research and Development Command.

IEEE PRESS

The Institute of Electrical and Electronics Engineers, Inc., New York

IEEE PRESS

1985 Editorial Board

M. E. Van Valkenburg, *Editor in Chief*

J. K. Aggarwal, *Editor, Selected Reprint Series*

Glen Wade, *Editor, Special Issue Series*

J. M. Aein	L. H. Fink	E. A. Marcatili
James Aylor	S. K. Ghandhi	J. S. Meditch
J. E. Brittain	Irwin Gray	M. G. Morgan
R. W. Brodersen	H. A. Haus	W. R. Perkins
B. D. Carroll	E. W. Herold	A. C. Schell
R. F. Cotellessa	R. C. Jaeger	Herbert Sherman
M. S. Dresselhaus	J. O. Limb	D. L. Vines
Thelma Estrin	R. W. Lucky	

W. R. Crone, *Managing Editor*

Hans P. Leander, *Technical Editor*

Copyright © 1986 by
THE INSTITUTE OF ELECTRICAL AND ELECTRONICS ENGINEERS, INC
345 East 47th Street, New York, NY 10017-2394
All rights reserved

PRINTED IN THE UNITED STATES OF AMERICA

IEEE Order Number: PC01941

Library of Congress Cataloging-in-Publication Data
Main entry under title:

Medical applications of microwave imaging.

   "Published under the sponsorship of the IEEE Microwave
Theory and Techniques Society. Supported by the U.S. Army
Medical Research Center and Development Command."
   Based on the proceedings of the Symposium on Electromag-
netic Dosimetry Imaging held during the 1980 International
Microwave Symposium in Washington, DC.
   Includes bibliographies and index.
   1. Microwave imaging in medicine—Congresses. I. Larsen,
Lawrence E. II. Jacobi, John H. III. IEEE Microwave Theory
and Techniques Society. IV. Symposium on Electromagnetic
Dosimetry Imaging (1980: Washington, DC.) V. International
Microwave Symposium (1980: Washington, DC.) [DNLM: 1.
Electromagnetics—congresses. 2. Microwaves—diagnostic use
—congresses. 3. Microwaves—therapeutic use—congresses. 4.
Radiation, Non-Ionizing—congresses. WB 141 M4885 1980]
RC78.7.M53M43 1986   616.07'57   85-23923
ISBN 0-87942-196-7

iv

CONTRIBUTORS

AUBIN, JOHN F., Flam and Russell Inc., Horsham, PA

AZIMI, MANI, Michigan State University, East Lansing, MI

BARRETT, ALAN H., Mass. Institute of Technology, Cambridge, MA

BOERNER, WOLFGANG-M., University of Illinois at Chicago, Chicago, IL

BURDETTE, EVERETTE C., University of Illinois, Urbana, IL

CAIN, FRED L., Georgia Tech Research Institute, Atlanta, GA

CHAN, WILSON CHUNG-YEE, Systems and Software Inc., Costa Mesa, CA

FARHAT, NABIL H., University of Pennsylvania, Philadelphia, PA

FLAM, RICHARD P., Flam and Russell, Inc., Horsham, PA

FOTI, STEPHEN J., ERA Technology Limited, Leatherhead, England

GUO, THEODORE C., Catholic University of America, Washington, DC

GUO, WENDY W., Catholic University of America, Washington, DC

JACOBI, JOHN H., Information Development and Applications, Inc., Beltsville, MD

KAK, AVINASH, C., Purdue University, West Lafayette, IN

KIM, YONGMIN, University of Washington, Seattle, WA

LARSEN, LAWRENCE E., Walter Reed Army Institute of Research, Washington, DC*

LIN, JAMES C., University of Illinois at Chicago, Chicago, IL

MYERS, PHILIP C., Mass. Institute of Technology, Cambridge, MA

PETERS, LEON, JR., Ohio State University, Columbus, OH

SEALS, JOSEPH, Georgia Tech Research Institute, Atlanta, GA

SKOLNIK, MERRILL I., Naval Research Laboratory, Washington, DC

SLANEY, MALCOLM G., Schlumberger Palo Alto Research, Palo Alto, CA

WEBSTER, JOHN G., University of Wisconsin, Madison, WI

YOUNG, JONATHAN D., Ohio State University, Columbus, OH

---

* Present address, Medical Microwave Research Corporation, Silver Spring, MD

Table of Contents

Preface

Garrison Rapmund, M.D.
Major General, Medical Corps
Commander, U.S. Army Medical Research Center and Development Command

Often in the past, enthusiasm for the benefits of new technologies has come more quickly than the appreciation of possible associated health hazards. This is certainly true for use of the electromagnetic spectrum. Microwave energy sources have beneficial applications of great importance in both the civilian and military sectors, yet the possibility of hazards has been somewhat ignored. It is very appropriate, therefore, that the United States Army Medical Research and Development Command should sponsor studies to provide the best possible data base for establishment of safety standards, and jointly with the Microwave Theory and Techniques Society of IEEE, sponsor the symposium on which this book is based.

Estimating the amount of exposure to nonionizing radiation, such as microwaves, is far more difficult than that for ionizing radiation. No microwave analog of the x-ray badge dosimeter exists. Indeed, a single microwave exposure event on the skin or clothing would not tell us much about exposure to other areas on the surface nor, more importantly, would it define subcutaneous dose distribution. Although complete quantification of absorbed dose distribution for microwaves remains an elusive goal, the research presented in this book forces us to recognize the weakness of whole-body-average dosimetry. It is clear that tissues and organs differ on a regional and physiological basis in their selective absorptive properties. This fact must be considered as safety standards are developed from experimental dose-response curves based on local dosimetry. The papers of this volume present new information and technologies that should do much to improve dosimetry and safety standards.

The advances reported in the subsequent pages also have important potential to improve existing medical practice in both military and civilian communities. First, there is the potential of microwave radiation for diagnostic imaging. This appears to be especially relevant for early disease detection in soft tissues where the electrical properties of tissue seem to provide a better measure of functional state than that which is possible with conventional x-ray imaging.

Secondly, there are therapeutic applications of microwave power. The original role of diathermy is being enlarged to include use of localized subcutaneous heating for hyperthermia as an adjunct to the treatment of cancer with ionizing radiation.

Lastly, increased use of the electromagnetic spectrum is predictable in both military and civilian work places. In addition to the obvious application of radar to aviation, there are civilian and military uses of microwaves for electronic communication, for industrial heaters in various manufacturing processes, and possibly in the future, as a method of power transmission from space. All these uses are certain to require a close examination of safety standards in the context of risk/benefit analyses. The United States Army Medical Research and Development Command, through sponsorship of symposia such as the one mentioned here, is endeavoring to protect the health of everyone immersed in the modern electromagnetic environment by developing data bases for quantification of hazard and risk estimation through improved dosimetry.

INTRODUCTION

Lawrence E. Larsen

The 1980 International Microwave Symposium

The Microwave Theory and Techniques Society of the Institute of Electrical and Electronics Engineers (IEEE) sponsored a workshop on Microwave Dosimetric Imagery as part of the 1980 International Microwave Symposium in Washington, D.C. The purpose of this workshop was to bring together papers on a variety of techniques for microwave imagery with the goal of enhancing their application to medicine in the context of microwave dosimetry. These original papers have been supplemented by reports of related work (Guo et al., Slaney et al.) that has taken place between the time of the symposium and the publication of this volume. These methods of microwave imagery are pertinent to dosimetry because as a class they represent the best hope for noninvasive dosimetry analysis of biosystems exposed to microwave radiation.

The broad objective of dosimetry is to characterize the spatial distribution of absorbed microwave energy (i.e., the dose distribution) in order to more accurately predict the medical consequences of exposure to microwave electromagnetic fields. The exposures may be environmental, occupational or therapeutic. In some of these cases, the role of the microwave exposure is that of an effector. Hence, we seek a dose-response curve.

Alternatively, microwaves may serve the purpose of a sensor. In this case, the methods may find application in nondestructive evaluation generally, and in medical diagnosis specifically. Although the original intention of the workshop was to support microwave hazards analysis through improved dosimetry, we are not unaware of the additional applications in medicine and industry.

What follows is an overview of the rationale for microwave imagery in dosimetric applications and its treatment in the archival literature. Also, we have attempted to link the book chapters together in their historical context. We have selected various items from the current literature (notably, that published since 1980) to illustrate the technological and scientific evolution in medical microwave imagery since the date of the symposium. Among the references cited, those located in this volume are indicated by bold-faced type.

Material Properties and Microwave Interactions with Biosystems

Microwave propagation in biological media underlies all medical consequences of microwave energy interaction with biosystems. Such propagation implies energy exchange with both reactive and resistive components. The description of material properties of biosystems, therefore, must include both energy dissipation and energy storage. These material properties plus the local electrical field (we ignore magnetic properties for purposes of this volume) determine the basis of electromagnetic interaction between the incident field and biosystems.

Energy exchange between the field and the dielectric in which it propagates may be identified with changes in the biosystem secondary to exposure (Johnson and Guy). These changes are often associated with a dose response curve. The problem of dosimetry is to measure the appropriate dose for prediction of the response.

The physical basis for microwave interaction with biosystems combines properties of both the field and the media within its path of propagation. In realistic settings, the scale of analysis applied to dosimetry must be consistent with the scale of biosystem effectors. Although activation of effectors may have systemic consequences, the effector itself is often discretely localized.

The combination of local electric field and local dielectric properties determines local energy exchange. These combine to form the kernel of a local dosimetry from which target organ correlations may be identified. In this way, local dosimetry may become a means for enhanced generality of experimental results as we try to infer from animal subjects to man.

Not coincidentally, the scattered field is also a product of the local electric field and local dielectric properties. This suggests that measurement of the scattered field may provide a map of the very same tissue and field properties needed for local dosimetric analysis. When the scattered field can be related to the three-dimensional distribution of dielectric properties which characterize the biosystem or target organ, the three-dimensional distribution of absorbed energy is predictable.

Use of the Scattered Field in Dosimetric Analysis

Measurements of the scattered field may be arranged as a spatial series keyed to the locations where the complex field amplitude was probed. This data may be presented in a particularly convenient and informative manner as an image. Dosimetric data displayed as an image offers the additional advantage that the results may be easily correlated with the traditional methods of anatomy, physiology, and pathology.

An imagery-based approach to dosimetry has been frustrated by the poor perceptual quality of microwave images due to the very low spatial resolution of air-coupled systems (Yamaura). Diffraction limitations suggest that increased frequency of operation would increase spatial resolution; but concurrently, the propagation loss is increased. The use of a high dielectric constant liquid as a coupling medium solves this dilemma. The wavelength is decreased in proportion to the square root of the relative dielectric constant, but the propagation loss remains largely a function of frequency and range. As will be shown later in this volume (**Larsen and Jacobi**), the use of water as a coupling medium with an imaging system operating in the radar S-band offers the propagation loss of decimeter waves (ca. 3 to 4 dB/cm) with the resolution (ca. 5 to 7 mm) of millimeter waves (Jacobi, Larsen, and Hast).

The microwave constitutive parameters of biosystems contain the extremely high-contrast information needed for dosimetric imagery. As an illustration of tissue parameters, the range of contrast available in x-ray imagery within soft tissues is less than 2%; whereas in the case of the radar S-band, the range in relative dielectric constant goes from a minimum of about 4 in fat to a maximum of about 80 in cerebral spinal fluid. This range of roughly 20 to 1 in both relative dielectric constant, and dielectric loss factor is also without peer in ultrasound. The range of index of refraction for ultrasound at ca 5 MHz in soft tissue is less than 10%; whereas the comparable range for microwaves in the 2 to 4 GHz band is roughly 4 to 1.

Material Properties of Biosystems as Physiologic Variables

The microwave constitutive properties of biologic dielectrics are also known to be physiologic variables (Burdette et al., 1980). The simplest basis for this physiologic dependency is alteration of the bulk permittivity by changes in local blood flow (**Burdette et al.**). Since blood is a tissue of high dielectric "constant," changes in local perfusion exert a marked effect on bulk permittivity. Local perfusion, in turn, responds to changes in local energy demands, local temperature, local oxygen content, local carbon dioxide level, local hydrogen ion concentration, and local concentration of various metabolites, drugs, hormones etc. Drug and neural action upon arterioles is detectable in the kidney by its effect on the complex permittivity of the renal cortex and medulla at frequencies in the radar S band (Burdette, 1983). Glomerular filtration rate is increased by 9% due to arteriolar dilation secondary to phentolamine administration. In the cortex, the dielectric constant is increased by 5% and the dielectric conductivity is increased by 8% concurrently. Renal nerve stimulation may also be shown to alter renal complex permittivity according to the frequency of nerve stimulation (Burdette, 1983). As the stimulus frequency increases from 2 to 12 Hz, the gomerular filtration rate increases in proportion. This changes both medullary and cortical complex permittivity. Medullary changes, however, have a different slope as a function of stimulus rate than that in the cortex. Similarly, regional variations in complex permittivity may be demonstrated in brain, in normal as well as in induced pathophysiologic states (Burdette and Larsen).

Dielectric properties are also directly temperature dependant (Hasted). In biosystems, temperature changes may be endogenous, due to alterations in local metabolic rate, or exogenous as a result of microwave heating. Either en-

dogenous or exogenous heating provokes local vasodilation. Thus, the temperature change may cause a change in bulk permittivity *in vivo* as well as acting directly upon the individual components of the bulk permittivity through their temperature coefficients. Blood flow also serves to cool the local tissue beds in which vasodilation takes place.

Blood flow serves more functions than simply those attributable to thermal exchange. Additional physiologic correlations may also be expected. This is most evident in the kidney. Blood flow alterations in the renal cortex are related to glomerular filtration. The preservation of glomerular flow rate in the face of variable renal perfusion pressure is accomplished by the process of pressure/flow autoregulation (Burdette, 1983). This physiologic signature of kidney function is mirrored in the complex permittivity of kidney *in vitro* under conditions of perfusion pressure variation. Likewise, regional differences in renal blood flow are correlated with regional differences in complex permittivity. For example, the dielectric constant of cortex is about 10% lower than that of the medulla at comparable perfusion pressures. In the renal medulla, blood flow alterations affect the processes of urine concentration in collecting tubules and the transport of aqueous electrolytes in the Loop of Henle via the vasa recta. Indeed, the changes in conductivity consequent to normal renal physiology may directly become a means for establishing correlations between electric parameters and functional status. For example, there is known to exist a gradient of electrolyte concentration in the medullary pyramids as an inverse function of radial distance from the hilus. This concentration gradient is produced by the so-called counter-current multiplier in the loop of Henle within the renal medulla. The medullary gradient of osmolarity is directly related to the ability of the kidney to produce urine that is concentrated with respect to plasma.

Changes in blood volume obviously are a major concomitant of physiologic activity in the cardiovascular system (**Lin**). The most prominent example is the change in cardiac volume with the contractile cycle; but there are more subtle examples. These include the movement of arterial walls with pulsations in blood pressure, and the partitioning of blood fluids within the intravascular and extra-vascular spaces.

Methods of Microwave Image Formation for Biomedical Applications

Image formation, in broad terms, may be either passive or active. Passive methods employ radiometric measurements of thermal emissions as a method for image formation. The roles of exogenous and endogenous heating can be clarified by the use of microwave thermography (**Barrett and Myers**). Microwave emissions are a function of the first power of the emitter's temperature in degrees Kelvin since the Rayleigh-Jeans approximation is applicable. In broad terms, the sensible depth is inversely related to the center frequency of the microwave receiver's RF section. Microwave thermography is often presented as a longer wavelength analog of the familiar infrared thermography. That is to say, imagery is produced on the basis of photon emission in the microwave region of the spectrum, but the longer wavelength allows sensibility of emissions from greater depth. The mode of operation often compares the microwave flux density from the target with that from a noise source of known temperature and emissivity. The use of active sources is limited to compensation for spatial variations in emissivity (Leudeke et al., Mamouni et al.). Local blood flow influences microwave emissions in two ways: by lowering the thermodynamic temperature, and by altering the bulk permittivity.

Maps of spatial variation in emissivity are an active imaging method. Monostatic and bistatic system configurations are possible.

Active methods of image formation are characterized by the use of controlled microwave sources to interrogate target organ microwave material properties, and thereby to infer the physiologic status of the biosystem. There are a number of possible methods which will be discussed in the following paragraphs. These include scattering parameters, depolarization, dispersion, video pulse, conductivity, and Doppler as well as synthetic aperture methods. In comparison to passive methods, the chief advantages of active techniques are control of the frequency, polarization, and geometry of the source.

Image formation with active sources is possible on the basis of insertion loss (energy dissipation, in broad terms) and phase shift (energy storage) derived from measurement of the scattering parameter S_{21} (Larsen and Jacobi, 1978, 1979). Ultimately, these images are relatable to the spatial distribution of microwave constitutive parameters. Images based on insertion loss presently represent a combination of scattering mechanisms. In addition to tramsmission loss, reflection, diffraction, and refraction all contribute to the failure of transmitted energy to reach the receiving antenna. Nevertheless, methods do exist for separation of reflection coefficients from various layers with time domain data for which transmission coefficients may be estimated under rather restrictive assumptions (Robinson, 1967). Also, methods exist for recovery of the permittivity profile from scattered fields, at least for the one-dimensional case, with relative computational efficiency (Kritikos, et al., 1982).

In the presence of gradual changes in dielectric properties as a function of spatial coordinates and the high propagation loss which characterizes most tissue *in vivo*, insertion loss images may broadly represent energy dissipation or loss factors due to dipole and ionic composition. On the other hand, phase and propagation delay images may broadly represent those properties associated more with energy storage than with energy dissipation. As will be shown later in the book (**Guo et al.**), if the inverse scattering problem is solved, the forward scattered images will be directly relatable to the spatial distribution of complex permittivity, subject only to rather narrow limitations on uniqueness, *videa infra*. Furthermore, conductivity itself may become a basis for image formation (**Kim and Webster**). This is obviously applicable in the global sense of inter-tissue differences. It may be less obvious that conductivity may be applied within an organ, e.g., the gradients within the ca-

nine kidney. The electrolyte concentration gradient in the renal medulla, for example, is closely correlated with physiologic function as previously described *videa supra*.

The properties of biological dielectrics as media for the propagation of microwave energy must also include cognizance of their frequency dependencies. These frequency dependencies or dispersions form another basis for active imagery by allowing estimation of the spatial variations in relaxation frequency. Indeed, it may be possible to estimate the fraction of total water that is bound or vicinal by such a method (Grant et al.). This offers the possibility of significantly improved insight for those diseases characterized by fluid shifts, such as cerebral edema.

Active image formation is also possible on the basis of polarization alteration (Larsen and Jacobi, 1980, 1983). When the wavelength of the microwave radiation in the dielectric becomes comparable to the dimensions of any discontinuity in scattering potential, scalar treatment of the wave equation becomes inappropriate (**Boerner and Chan**). There is a rich background of theory and technique in polarization utilization with radar which may be applied to biomedical microwave imagery (Huynen; Boerner). This is not to say that optimal methods for use of the polarization scattering matrix exist. Simple methods such as incoherent addition of polarization scattering matrix magnitudes are the rule, except for the co-polarization and cross-polarization null methods of Huynen. Larsen and Jacobi (1980) suggested use of the variance/covariance matrix eigenvectors as a statistical description of the polarization scattering matrix by linear combinations of its elements. These linear combinations are optimal in the sense of extracting the greatest fraction of variance due to spatial sampling.

Radar methods for signal processing and data acquisition are broadly applicable to biomedical microwave imagery (**Skolnik**). Examples include chirp and its related pulse compression methods (**Jacobi and Larsen**), (Jacobi and Larsen, 1978, 1980), estimation of the polarization scattering matrix (Larsen and Jacobi, 1983), and estimation of the dielectric tensor. Frequency agility in radar systems is used as a means for target classification. In fact, chirp methods, frequency agility, and microwave holography can be related. An imaging radar technique has been developed upon the idea of filling the Fourier space representation of the target by means of frequency diversity (Chan and Farhat). This is a method of reflection tomography (**Farhat**) which is related to that described by Norton and Linzer (1979) for ultrasonic interrogation of biosystems.

Among the methods of target identification and classification derived from radar, one of the most tantalizing for biomedical application is the video pulse technique (**Young and Peters**). The geometry of the spatial distribution of dielectric properties may be summarized with relatively few numbers by virtue of the intrinsic collapse of dimensionality when complex functions are described by the locations of their poles and zeros (Kennaugh). This is a potentially powerful method of feature extraction that makes automatic classification of imagery possible. Video pulse is also attractive as a high-speed method of data acquisition.

Another area borrowed from radar with special promise

for biomedical application is the use of antenna arrays and synthetic apertures. These are important since they permit electronic beam steering and focusing. Such considerations impact data acquisition rates and system design.

Imaging System Design

The first consideration in system design is the choice of operating frequency. We have previously mentioned the basic dilemma with respect to diffraction limits and penetration depths. Frequency of operation is also related to the choice of forward- or back-scatter system geometry. Back-scatter design affects resolution through the possibility of reduced propagation range for shallow targets, thereby permitting a higher frequency of operation. As will be developed later (**Slaney et al.**), reflection systems permit simpler target illumination for the same Fourier space access with respect to tomographic methods. There are valid arguments for transmission (i.e., forward-scatter systems), however, that relate to better characterization of multi-layered dielectrics with coherent imaging systems and more favorable scattering cross sections (Larsen and Jacobi, 1983).

With respect to system design, rapid data acquisition is needed if microwave imagery is to display fast changes in complex permittivity as a consequence of exogenous heating (e.g., with therapeutic hyperthermia) or other physiologic alteration. Rapid data collection for microwave imagery may be achieved by the use of array technology to avoid electromechanical scanning (Mailloux). Generally, data acquisition is fastest when the receive array elements are polled to sample the forward-scattered fields. The process of beam steering and/or focusing then takes place off-line by data processing. At least two methods of array element polling have been described for medical microwave imagery: the multiplexed receiver (**Foti et al.**) and modulation scattering (Bolomey, 1982a). The multiplexed receiver design depends upon RF switches to connect each array element in turn to one microwave receiver. This technique may be accelerated by the use of PIN diode switches rather than electromechanical switches, but ultimate speed demands the use of parallel receivers. In its full realization, this would require harmonic conversion and reference oscillator phase-locking at each element followed by parallel IF amplifiers and complex ratiometers. In such a setting, image data acquisition would be possible at multi-megahertz rates. The modulation scattering method has a great advantage in terms of low cost, since element-to-element scanning takes place at audio frequencies. Array technology may also make it possible to achieve three-dimensional resolution with reasonable data acquisition times. Three-dimensional resolution has become the norm in medical imagery since the rediscovered Radon (1917) method has been applied to x-ray attenuation tomography.

Methods of Microwave Tomography

Forward-scattered fields in principle represent the radiation pattern of an equivalent current distribution localized over the exit aperture of the target organ (Mittra et al.). The equivalent current distribution, in turn, is the product of the incident electric field and the dielectric properties of the organ. Resolution is available in the transverse plane by sampling the forward-scattered fields in elevation and azimuth. Resolution is available in the axial direction by scanning a sensitive three-dimensional volume (i.e., a three-dimensional beam width) through the target organ (Bolomey et al., 1982b), but interpretability of such images is limited at present (Guo et al. 1984a). Usable axial beam widths are available in the near field of arrays with large numerical aperture (**Guo et al.**); (Guo et al., 1984b); but this can best be achieved in the coupling medium. Volumetric synthesis may also be useful in this application. That is, the receiver becomes effectively a three-dimensional array. This offers some advantages in polarization control and reduction of axial beam width, but data acquisition times are significantly dilated (**Foti et al.**).

Another approach to the problem of resolution recovery is the ray-based algebraic reconstruction technique (ART). This approach has been applied to microwave imagery (Rao et al.), but the images are seriously faulted by inaccurate reconstruction. For example, solid cylinders of uniform loss are reconstructed as nearly tubular in cross section. Similarly, Ermert et al. have investigated both reflection and transmission tomography. Experimental studies were performed with water-loaded and coupled waveguide antennas with low-contrast targets in transmission tomography. The results lacked acceptable accuracy because of the use of the assumption of straight-line propagation.

A modified form of ART has been applied to geotomography by Radcliff and Balanis (1979) as well as by Dines and Lytle (1979). These approaches used numerical damping factors and weighing to ensure convergence of ART solutions in the presence of "noise" modeling of departures from ray optics. Adequate numerical accuracy was difficult to demonstrate. In the case of Radcliff and Balanis, for example, satisfactory results depended upon permitting only deviations within one order of magnitude in conductivity with no variations in the relative dielectric constant. Even so, although the global accuracy was reasonable, the peak error would be unacceptable for most medical systems.

Geotomography is an area where reflection of pulsed RF has had considerable development for mapping of layered media (Wait; Coen et al.). In the seismic exploration industry, the technique of dynamic deconvolution is routinely applied with considerable success (Robinson, 1982). Unfortunately, weak scattering is assumed. This is an approximation required to reach an inverse solution for the scalar wave equation. The total field is represented as a sum of the incident and scattered field wherein the scattered field is assumed to be approximated by the incidence field. In other words, the target must have little effect upon the total field, i.e., the target must be diaphanous.

Time domain methods similar to those in geotomography exist for nonuniform transmission lines. Recently, Lee (1982) developed a new computational method which avoids the recursive polynomial method of Robinson (1967). Lee makes use of the Lagrangian operator to simplify the computational procedure. The method at this point depends upon the boundary conditions of a transmission line and upon layered representation of dielectric gradients with normal incidence of the incident energy. It does have the advantage that lossless media are not required. Profile inversion is possible within an acceptable range of ambiguity when the product of loss and sampling interval is small. At the present time, the range of application of this inverse method to biosystem imagery remains to be explored.

The reason for the inapplicability of ray-based methods such as ART to microwave forward-scatter systems is the fact that a single, linear ray path cannot be reasonably presumed to connect the transmitter and receiver. Neither can a single, curved ray path be assumed to connect the transmitter and receiver. Minor changes in launch angle have a dramatic effect upon the receiver site where the ray is terminated. Thus, the problem of ray linkage (Anderson and Kak) seems to present a barrier to successful application of ray-based methods to microwave systems.

Another class of numerical techniques exist which are customarily stated in radar terminology as inverse scattering methods. This nomenclature is intended to contrast with the so-called forward problem wherein the incident field and scattering object are known but the scattered fields are not. Inverse scattering starts with the scattered field for a known illumination and seeks the characteristics of the scattering object. In that sense, ART is an inverse method. The distinction is more than linguistic, however, since inverse scattering is usually considered as an exact method. Unfortunately, some regularization is necessary to result in a well-posed (numerically approachable) method. It may be argued that regularization is another word for approximation.

In that sense, inverse scattering is related to the methods of diffraction tomography advocated for use in forward-scatter ultrasound systems (Mueller et al.). Multiple views of the target under study are necessary just as in the case of x-ray based methods. The difference is that the Fourier transformation of each projection view, in the microwave case, is not a central section through the two-dimensional Fourier space representation of the object function, but rather, it is an arc in the Fourier space which is tangent to the origin. The radius of the arc is $2k_0$, where k_0 is the wave number of the incident field in the coupling medium. Multiple views fill in the Fourier space in a way analogous to ray-based methods, but important details do differ. The design of the data acquisition system also impacts Fourier space accessibility (*videa infra*).

There does appear to exist a range of applicability for diffraction tomography in microwave systems. Slaney et al. (1984) have shown that first-order diffraction tomography is applicable using the Born approximation when the target is many wavelengths thick, but the total difference in phase shift through the surrounding medium and through the target must be less than ca. 90 degrees for reasonable numerical accuracy (further assuming that the target is lossless). As an example, a 1% deviation of the index of refraction for a cylinder 10 wavelengths in diameter will produce unacceptable errors in both the phase and the amplitude of the reconstruction. In the case of the Rytov ap-

proximation, larger total phase shifts are admissible, but the phase shift per wavelength must be small. In this case, discontinuities in permittivity prevent accurate reconstruction. As an example, a 5% deviation in index of refraction in a cylinder only one wavelength in diameter will ruin the reconstruction. These limitations are consistent with those inferred by Iwata and Nagata (1975); but the severity of the errors for relatively minor deviations is rather surprising. In either the Born or Rytov cases, the problems are consequences of the weak-scattering assumption which underlies much of this work (Wolf).

Slaney et al. also point out that in the case of plane-wave illumination, the range of spatial frequencies is reduced by a factor of two in transmission, as compared to reflection tomography. This loss of access to Fourier space may be restored by the use of a scanned transmitter as well as a scanned receiver (Nahamoo and Kak). In other words, the single-plane wave exposure described by Slaney et al. is replaced by a series of spherical waves from point sources across the same section previously illuminated by the plane wave. This obviously increases data acquisition time and places a premium on high-speed sampling of the forward scattered fields, but the high spatial frequencies are recovered.

Preliminary numerical studies have been performed for the case of lossy media (**Slaney et al.**). Loss in either the coupling medium or the target introduces some problems in the sense of signal-to-noise ratio (especially for higher spatial frequencies), but it offers the advantage that multiple scattering effects are reduced (Azimi and Kak). Furthermore, lossy media confer a distinct advantage to the Born approximation in comparison to the Rytov approximation with respect to the range of variation in the complex index of refraction over which reasonable numerical accuracy may be obtained. These results contradict the optimistic expectation by Greenleaf (1983), but widely held by others working in wave equation tomography, that the Rytov approximation would overcome the limitations imposed by the assumption of weak scattering.

Guo et al. and Guo et al. (1983, 1984b) have proposed a different solution to the problem which is based on a generalization of the Lorentz reciprocity theorem (Carson) to the case of lossy media. This method is proposed to be combined with the method of phase-amplitude conjugation previously described for the self-focusing array. The target is furthermore treated as excess dielectric susceptance with respect to the coupling medium.

The method of profile inversion based on the Gelfand-Levitan solution to the wave equation is a prototype inverse-scattering method with a frustratingly small range of applicability to biosystems. Unlike the first-order approximation in diffraction tomography previously described, the Gelfand-Levitan method is quite general and not limited by the assumption of weak scattering. Until recently, this method has been limited to a few canonical cases of little biomedical interest. However, a new computational procedure (Kritikos et al., 1982) has improved the method's application, at least to one-dimensional cases. The prospect for a similar iterative approach to two-dimensional permittivity distributions must await incorporation of noise sensitivity analysis and proof of convergence—steps that would enhance the results from the one-dimensional case. The method described by Kritikos et al. was applied only to the case of smoothly varying, one-dimensional permittivity distributions. Abrupt or discontinuous changes in permittivity, such as at bone or air junctions, would appear to present a substantial problem.

The inverse scattering method due to Bojarski (1982) is an exact theory which does not share the limitations of the Born and Rytov approximations as applied to the description of the total field. Bojarski deduced a description of the total field that makes no assumptions with respect to weak scattering (Bleistein and Bojarski). Also, only minor assumptions with respect to the uniqueness of the inversion are necessary, namely that the source must be of limited extent and have finite energy, and that all source components must radiate (Bleistein and Cohen). Unfortunately, numerical solutions of the intergral equations derived by Bojarski have been ill-posed and the theory has not enjoyed wide utility in electromagnetic inverse problems of biomedical interest (Stone, 1982a).

Doppler processing is another potential method for microwave tomography. Normally, this technique requires a side-looking antenna to be scanned past the target under study. An equivalent operation is to move the target, often by rotation, within the main beam of a fixed antenna. This is the so-called inverse synthetic aperture radar (**Skolnik**). This method normally depends upon pulse width for range resolution, and Doppler shift for transverse resolution. In dispersive media, RF pulses will be distorted and range resolution will suffer. Unless the dispersion of the media in the propagation path can be compensated by inverse filtering upon reception, similar in practice to pulse compression filters, the loss of range and multipath contamination will limit the utility of this method for biosystem imagery. Doppler processing for CW radars is still applicable for moving reflectors, e.g., blood vessels, but such CW systems lack range discrimination. An alternative method recently described by Mensa et al. (1983) depends upon coherent CW processing and inverse Doppler data acquisition. The problem of lost range resolution in the CW system configuration is ameliorated by multiple views. However, the requirement for uniform propagation constant throughout the target implies sensitivity only to surface reflectivity. This would therefore seem to place the method at a disadvantage. The use of multiple bistatic angles is suggested as a method to avoid multiple frequency irradiations, as in Chan and Farhat. This method also suppresses side lobe levels, but not to any advantage in comparison to frequency diversity.

The similarities of spot light mode synthetic aperture radar (SAR) and tomographic processing in x-ray imagery have recently been clarified (Munson and Sanz; Munson et al.). The formal similarities of air-coupled, spot light mode radar systems to line integral x-ray methods depend upon the fact that spot light mode SAR uses multiple-view angles rather than Doppler processing for lateral resolution. Multiple frequencies are employed in spot light mode SAR by

virtue of the use of a chirped pulse. It seems to fit the category of reflection tomography systems and be subject to the same limitations due to pulse dispersion and multiple media effects.

Lastly, it is pertinent to underline the distinctions between tomography and holography lest the problem be misunderstood. Recently, Stone (1982b) has clarified the differences between holography and inverse scattering. He points out that the field obtained from holographic reconstruction is the volume integral of the product of the source term and the complex conjugate of the free space Green's function, not the source term alone. When the wavelength of the interrogating radiation is very small in comparison to the correlation distances in the target, then the unperturbed far-field samples, when reconstructed near the scatter, will approximate the shape of the target. Under these circumstances, the holographically reconstructed field presents the three-dimensional distribution of surface reflectance, not the three-dimensional complex index of refraction of a penetrable target. Nevertheless, it is a very powerful method, especially Fourier transform holography as described by **Farhat**.

Resolution recovery in the direction of propagation and preservation of high spatial frequencies in the scattered field seem to be the central problems in the further development of forward-scatter dosimetric imagery. Until these problems are solved, only projections of isolated organs may be studied. This is a significant limitation since *in vivo* studies are excluded. However, a vast body of data is available *in vitro* that remains to be explored. Central to this is the study of physiologic and pathophysiologic correlations in isolated organs maintained by extracorporal circulation.

Microwave Imagery and Dosimetric Analysis

Finally, the relationship of these methods to the problem of dosimetry needs to be underlined. Dosimetry is the measurement of absorbed dose. This is to be distinguished from the measurement of incident energy density per unit time (milliwatts/cm^2) which is properly described as densitometry. Once dosimetry is understood to be absorption measurement with the units of watts/kilogram, the corollary of spatial distribution must be addressed. Of course, it is possible (if not informative) to dispense with the issue of spatial distribution by the use of whole-body averaged dosimetry. This is perilous inasmuch as the conditions for large variation in the local dose from the average dose are well established and present in biologic systems such as experimental animals and man. Classic work with simple geometries of homogeneous dielectric composition has shown ratios of peak to average energy absorption of ca. 10 to 1 in the case of a sphere with a radius comparable to one-half the wavelength incident upon it (Kritikos and Schwan). In adult humans, this occurs in the UHF band. More complicated geometries are treated by moment methods in block models with uniform dielectric properties (Hagmann et al., 1979a). In this case, the peak-to-peak variation in SAR is ca 10 to 1 even with the relatively crude 180 cell models used. Finer resolution is available when smaller body regions are modeled. In the case of head models, the peak SAR was shown to be 4 times the average SAR with a peak-to-peak variation of ca 20 to 1 (Hagmann et al., 1979b). Heterogeneous dielectrics present still more complications. Barber et al. (1979) have shown that in a head model composed of 6 layers, the SAR in the so-called supraresonant region is increased from an absorption efficiency of 60% to 140%.

More recently, the limitations of these absorption models has become a matter of concern. The models are reasonably accurate for whole body average SAR, but even for simple models there are significant errors in estimation of the SAR distribution (Massoudi et al.). For example, a homogeneous cube when subdivided into cells may introduce an error of 2 to 1. Experimental studies of scale models represent a more accurate technique than the present numerical methods for SAR distribution analysis. Thermographic studies of dielectric phantoms (Guy et al.) has shown a peak to average SAR ratio of 13 to 1. When an exposure to a highly directive source takes place, such as in the case of centimeter or millimeter wavelengths radiated by antennas of moderate aperture dimensions, the ratio of peak to average absorption becomes even higher, perhaps approaching 100 to 1. Under such circumstances, whole-body averaged dosimetry is nearly meaningless.

As a result of the nonuniformity of the spatial distribution of either incident or absorbed energy, some method of local dosimetry becomes essential. The available techniques for local or (to borrow an expression from the famous 19th century physiologist Claude Bernard) target organ dosimetry may be divided into two broad categories: invasive and noninvasive. The invasive methods are represented by a variety of thermometric methods with various suitabilities for use in microwave fields. Unsuitable methods include the use of metallic conductors or sensors. These are not suitable for use during microwave exposure because of lead coupling to the incident field. The problem is crudely analogous to the measurement of air temperature with a thermometer in direct sunlight.

Metallic leads or sensors also are not suitable for studies of incremental temperature elevation since the high thermal conductivity of metals disrupts the spatial distribution of temperatures. Shielded or allegedly cross polarized metallic structures are also unsuitable (Cetas). In the former case, the use of a shield distorts the spatial distribution of absorbed energy in comparison to the unperturbed dielectric. In the case of the latter, cross polarization to the incident field depends upon linear polarization at the site of temperature measurement. This is manifestly impossible since the target must depolarize the incident field for any wavelength less than ca 20 meters.

Suitable thermometric methods include the use of dielectric probes such as optical etalons (Christensen), temperature transducers in optic fiber jackets (Giallorenzi et al.), or decoupled microwave integrated circuit electrodes (Larsen et al.; Bowman). Liquid crystal probes with optic fiber leads are less satisfactory due to hysteresis and drift in the calibration curves which relate sensor reflectance to temperature.

In any case, all invasive methods have unyielding prob-

lems with sampling error. Since the spatial gradients in absorbed dose may be steep and the locations of their maxima or minima are not easily predicted, multiple sites of measurement are necessary. This approach has obvious practical limitations when the number of sample points exceeds two or three, not to mention unnecessary trauma. The problem of high sampling density without trauma to the target organ requires a noninvasive method of dosimetry.

The leading technological candidate for noninvasive dosimetry is microwave imagery. The absorbed dose may be estimated from the spatial distribution of heating subsequent to irradiation (under standard assumptions regarding thermal diffusion, conduction and radiation) by radiometric methods. The problem of estimation of brightness temperature in depth for simple dielectric models has been approached from the perspective of microwave remote sensing by satellite surveys of earth resources (Coen et al.). Since brightness temperature is the product of emissivity and thermodynamic temperature, some form of emissivity compensation is necessary. Nevertheless, multifrequency information and the assumption of smooth spatial gradients of complex permittivity distribution limit the range of applicability (Njoku and Kong). Correlation radiometry offers another approach to the problem of estimation of brightness temperature at depth. This is a result of increased signal levels from regions of overlap between two receiving antenna patterns (Fujimoto). The method depends upon forming the product of separate antenna patterns to peak the response within the Fraunhofer zone of the separate antennas. At this moment, it is unclear how well this technique will work in lossy media with dielectric heterogeneities in the volumes subsumed by either or both antennas. In any event, the limitations inferred from Njoku and Kong will still apply, but correlation processing may provide some relief from decreased signal with increasing range (Hill & Goldner).

Active methods may also address the problem by use of the temperature coefficient of complex permittivity (Hasted). Indeed, a 1°C sensitivity with an S_{21} imaging system operating at 3 GHz has recently been demonstrated in vitro (Bolomey et al., 1982c, 1983). In this case, the temperature of a renal perfusate was shown to influence kidney images ways that were not simply the result of desreased loss at elevated temperature. Active methods offer the additional promise that the spatial variation in the real and imaginary parts of the complex permittivity may be separately estimated. Given this information, the spatial distribution of absorbed dose may be estimated without the need for exogenous heating.

Should the tomography problem reach satisfactory resolution (pun intended), the scattered field may be represented as a superposition of discrete scattering centers. Each of these, in turn, may be associated with an equivalent current distribution which is the result of the local incident field and the local complex permittivity (Mittra et al.). At that point dosimetric imagery will have realized its potential for providing three-dimensional maps of the spatial distribution of complex permittivity as a function of frequency (Ghodgaonkar et al.). Subject to the limitations of interpretation due to discrete representation, the absorbed dose will be available for plane wave and spherical wave illumination of the target under study. By the use of the plane wave spectrum method (Tai), more complex exposure geometries may be approached.

References

A. H. Anderson and A. C. Kak, "Digital ray tracing in two-dimensional refractive fields," *J. Acoust. Soc. Am.*, Vol. 72, pp. 1593–1606, 1982.

M. Azimi and A. C. Kak, "Distortion in diffraction imaging caused by multiple scattering," *IEEE Trans. Medical Imaging*, Vol. MI-2, pp. 176–195, 1983.

P. W. Barber, O. P. Gandhi, M. J. Hagmann, and I. Chatterjee, "Electromagnetic absorption in a multilayered model of man," *IEEE Trans. Biomed. Eng*, Vol. BME-26, pp. 400–405, 1979.

A. Barrett and P. Myers, this volume, p. 41.

C. Bernard, *An Introduction to the Experimental Study of Medicine*, translated by H. C. Green, reprinted as Dover T400, New York, 1957.

N. Bleistein and N. N. Bojarski, "Recently developed formulations of the inverse problem in acoustics and electromagnetics," Report MS-R-7501, Denver Research Institute, University of Denver, Colorado, 1974.

N. Bleistein and J. Cohen, "Nonuniqueness in the inverse source problem in acoustics and electromagnetics," *J. Math Phys.*, Vol. 18, pp. 194–201, 1977.

W-M. Boerner, "Polarization utilization in electromagnetic inverse scattering," in *Topics in Current Physics*, H. P. Baltes, Ed., Chap. 7, Springer-Verlag, Berlin, 1980.

W.-M. Boerner and C-Y. Chan, this volume, p. 213.

N. N. Bojarski, "Inverse scattering, inverse field, and inverse source theory," in the *11th International Symposium on Acoustic Imaging and Acoustic Holography*, pp. 399–408, J. P. Powers, Ed., Plenum, New York, 1982.

J. Ch. Bolomey, "La methode de diffusion modulee: une approche au releve des cartes de champs microondes en temps reel," *L'onde electrique*, Vol. 62, pp. 73–78, 1982a.

J. Ch. Bolomey, A. Izadnegahdar, L. Jofre, Ch. Pichot, G. Peronnet, and M. Solaimani, "Microwave diffraction tomography for biomedical applications," *IEEE Trans. Microwave Theory Tech.*, Vol. MTT-30, No. 11, pp. 1998–2000, 1982b.

J. Ch. Bolomey, G. Peronnet, Ch. Pichot, L. Jofre, M. Gautherie, and J. L. Guerquin-Kern, "L'Imagerie Micro-onde Active en Genie Biomedical," Ecole Superieure d'Electricite, Oct. 1982c.

J. Ch. Bolomey, L. Jofre, and G. Peronnet, "On the possible use of microwave active imaging for remote thermal sensing," *IEEE Trans. Microwave Theory Tech.*, Vol. MTT-31, No. 9, pp. 777–781, 1983.

R. R. Bowman, "A probe for measuring temperature in radiofrequency heated materials," *IEEE Trans. Microwave Theory Tech.*, Vol. MTT-24, No. 1, pp. 43–45, 1976.

E. C. Burdette, F. L. Cain, and J. Seals, "*In vivo* probe measurement technique for determining dielectric properties at VHF through microwave frequencies," *IEEE Trans. Microwave Theory Tech.*, Vol. MTT-28, No. 4, pp. 414–427, 1980.

E. C. Burdette, "Dielectric properties of tissue measured with an *in situ* probe technique," Contract Report, DAMD17-8044, U.S. Army Medical Research and Development Command, April, 1983.

E. C. Burdette, P. G. Friederich, R. L. Seaman, and L. E. Larsen, "*In situ* permittivity of canine brain: regional variations and postmortem changes," *IEEE Trans. Microwave Theory Tech.*, Vol. MTT-34, No. 1, pp. 38–50, 1986.

E. C. Burdette, E. L. Cain, and J. Seals, this volume.

J. R. Carson, "Reciprocal theorems in radio communication," *Proc. IRE*, Vol. 17, pp. 952–956, 1929.

I. C. Cetas, "Temperature measurement in microwave diathermy fields: principles and probes," *Proc. Int. Symp. Cancer Therapy* by *Hyperthermia and Radiation*, J. E. Robinson, Ed., Am. College Radiology, Bethesda, Md., p. 193, 1976.

C. K. Chan, and N. H. Farhat, "Frequency swept tomographic imaging of three-dimensional perfectly conducting objects," *IEEE Trans. Antennas and Propagation*, Vol. AP-29, pp. 312–319, 1981.

D. A. Christensen, "A new nonperturbing temperature probe using semiconductor band edge shift," *J. Bioengineering*, Vol. 1, pp. 541–545, 1977.

S. Coen, K. Mei, and D. J. Angelakos, "Inverse scattering techniques applied to remote sensing of layered media," *IEEE Trans. Antennas and Propagation*, Vol. AP-29, pp. 298–306, 1981.

K. A. Dines and R. J. Lytle, "Computerized geophysical tomography," *Proc. IEEE*, Vol. 67, No. 7, pp. 1065–1073, 1979.

H. Ermert, G. Fulle, and D. Hiller, "Microwave computerized tomography," *11th European Microwave Conf.*, pp. 421–426, 1981.

N. H. Farhat, this volume, p. 66.

S. J. Foti, R. P. Flam, J. F. Aubin, L. E. Larsen, J. H. Jacobi, this volume.

K. Fujimoto, "On the correlation radiometer-technique," *IEEE Trans. Mircowave Theory Tech.*, Vol. MTT-12, No. 2, pp. 203–212, 1964.

D. K. Ghodgaonkar, O. P. Gandhi, M. J. Hagmann, "Estimation of complex permittivities of three-dimensional inhomogeneous biological bodies," *IEEE Trans. Microwave Theory Tech.*, Vol. MTT-31, No. 6, pp. 442–446, 1983.

T. G. Giallorenzi, J. A. Bucaro, A. Dandridge, G. H. Sigel, Jr., J. H. Cole, S. C. Rashleigh, and R. G. Priest, "Optical fiber sensor technology," *IEEE Trans. Quantum Electronics*, Vol. QE-18, No. 4, pp. 626–665, 1982.

E. H. Grant, R. J. Sheppard, and G. P. South, "The importance of bound water studies in the determination of energy absorption in biological tissue," *5th European Microwave Conf.*, pp. 366–370, Hamburg, 1975.

J. F. Greenleaf, "Computerized tomography with ultrasound," *Proc. IEEE*, Vol. 71, No. 3, pp. 330–337, 1983.

T. C. Guo, W. W. Guo, and L. E. Larsen, "Microwave imagery: an inverse scattering approach," in *Proc. IEEE 8th Conf. on Infrared Millimeter Waves*, 1983.

T. C. Guo, W. W. Guo, and L. E. Larsen, "Comment on Microwave Diffraction Tomography for Biomedical Applications," *IEEE Trans. Microwave Theory Tech.*, Vol. MTT-32, No. 4, pp. 473–474, 1984a.

T. C. Guo, W. W. Guo, and L. E. Larsen, "A local field study of a water-immersed microwave antenna array for medical imagery and therapy," *IEEE Trans. Microwave Theory Tech.*, Vol. MTT-32, No. 8, pp. 844–854, 1984b.

T. C. Guo, W. W. Guo, and L. E. Larsen, this volume, p. 167.

A. W. Guy, C-K. Chou, and B. Neuhaus, "Average SAR and SAR distribution in man exposed to 450 MHz radiofrequency radiation," *IEEE Trans. Microwave Theory Tech.*, Vol. MTT-32, No. 8, pp. 752–763, 1984.

M. J. Hagmann, O. P. Gandhi, and C. H. Durney, "Numerical calculation of electromagnetic energy deposition for a realistic model of man," *IEEE Trans. Microwave Theory Tech.*, Vol. 27, No. 9, pp. 804–809, 1979a.

M. J. Hagmann, O. P. Gandhi, J. A. D'Andrea, and I. Chatterjee, "Head resonance: Numerical solutions and experimental results," *IEEE Trans. Microwave Theory Tech.*, Vol. MTT-27, No. 9, pp. 809–813, 1979b.

J. B. Hasted, *Aqueous Dielectrics*, Chapman and Hall, London, 1972.

J. C. Hill and R. B. Goldner, "The thermal and spatial resolution of a broad band correlation radiometer with application to medical microwave thermography," *IEEE Trans. Microwave Theory & Tech.*, MTT 33, No. 8, 718–722, 1985.

J. R. Huynen, *Phenomenological Theory of Radar Targets*, Drukkerij-Bronder, Rotterdam, 1970.

K. Iwata and R. Nagata, "Calculation of refractive index distribution from interferograms using the Born and Rytov approximations," *Japan J. Applied Phys.*, Vol. 14, pp. 1921–1927, 1975.

J. H. Jacobi and L. E. Larsen, "Microwave interrogation of dielectric targets: Part II, by microwave time delay spectroscopic methods," *Med. Phys.*, Vol. 5, No. 6, pp. 509–513, 1978.

J. H. Jacobi, L. E. Larsen, and C. T. Hast, "Water immersed microwave antennas and their application to microwave interrogation of biological targets," *IEEE Trans. Microwave Theory Tech.*, Vol. MTT-27, No. 1, pp. 70–78, 1979.

J. H. Jacobi and L. E. Larsen, "Microwave time delay spectroscopic imagery of isolated canine kidney," *Med. Phys.*, Vol. 7, pp. 1–7, 1980.

J. H. Jacobi and L. E. Larsen, this volume, p. 138.

C. C. Johnson and A. W. Guy, "Nonionizing electromagnetic wave effects in biological materials and systems," *Proc. IEEE*, Vol. 60, pp. 692–718, 1972.

E. M. Kennaugh, "Transient and impulse approximations," *Proc. IEEE*, Vol. 53, No. 8, pp. 893–901, 1965.

Y. Kim and J. G. Webster, this volume, p. 106.

H. N. Kritikos and H. P. Schwan, "Hot spots generated in conducting spheres by electromagnetic waves and biological implications," *IEEE Trans. Biomed. Eng.*, Vol. BME-19, No. 1, pp. 53–58, 1972.

H. N. Kritikos, D. L. Jaggard, and D. B. Ge, "Numeric reconstruction of smooth dielectric profiles," *Proc. IEEE*, Vol. 70, No. 3, pp. 295–297, 1982.

L. E. Larsen, R. A. Moore, J. H. Jacobi, F. A. Halgas, and P. V. Brown, "A microwave compatable MIC temperature electrode for use in biologic dielectrics," *IEEE Trans. Microwave Theory Tech.*, Vol. MTT-27, No. 7, pp. 673–679, 1979.

L. E. Larsen and J. H. Jacobi, "Microwave interrogation of dielectric targets, Part I: by scattering parameters," *Med. Phys.* Vol. 5, No. 6, pp. 500–508, 1978.

L. E. Larsen and J. H. Jacobi, "Microwave scattering parameter imagery of isolated canine kidney," *Med. Phys.*, Vol. 6, No. 5, pp. 394–403, 1979.

L. E. Larsen and J. H. Jacobi, "The use of orthogonal polarization in microwave imagery of isolated canine kidney," *IEEE Trans. Nucl. Sci.*, Vol. NS-27, No. 3, pp. 1184–1191, 1980.

L. E. Larsen and J. H. Jacobi, "Methods of Microwave Imagery for Medical Applications," in *NATO Advanced Studies Institute Series No. 61: Diagnostic Imaging*, R. C. Reba, Ed., pp. 61–131, Martinus-Nijoff, The Hauge, 1983.

L. E. Larsen and J. H. Jacobi, this volume, p. 118.

C. Q. Lee, "Wave propagation and profile inversion in lossy inhomogeneous media," *Proc. IEEE*, Vol. 70, No. 3, pp. 219–228, 1982.

K. M. Leudeke, B. Schiek, and J. Kohler, "Radiation balance microwave thermography for industrial and medical applications," *Electron. Lett.*, Vol. 14, pp. 194–196, 1978.

J. C. Lin, this volume, p. 47.

R. J. Mailloux, "Phased array theory and technology," *Proc. IEEE*, Vol. 70, No. 1, pp. 246–292, 1982.

A. Mamouni, F. Bliot, Y. Leroy, and Y. Moschetto, "A modified radiometer for temperature and microwave properties measurement of biological substances," in *Proc. 7th European Microwave Cont.*, Copenhagen, pp. 713–717, 1977.

H. Massoudi, C. H. Durney, and M. F. Iskander, "Limitations of the cubical block model of man in calculating SAR distribution," *IEEE Trans. Microwave Theory Tech.*, Vol. MTT-32, No. 8, pp. 746–752, 1984.

D. L. Mensa, S. Halevy, and G. Wade, "Coherent Doppler tomography for microwave imaging," *Proc. IEEE*, Vol. 71, No. 2, pp. 254–261, 1983.

R. Mittra, W. L. Ko, and Y. Rahmat-Samii, "Transform approach to electromagnetic scattering," *Proc. IEEE*, Vol. 67, No. 11, pp. 1486–1503, 1979.

R. K. Mueller, M. Kaveh, and G. Wade, "Reconstructive tomography and applications to ultrasonics," *Proc. IEEE*, Vol. 67, No. 4, pp. 567–587, 1979.

D. C. Munson and J. L. C. Sanz, "Image reconstruction from frequency-offset Fourier data," *Proc. IEEE*, Vol. 72, No. 6, pp. 661–669, 1984.

D. C. Munson, J. D. O'Brien, and W. K. Jenkins, "A tomographic formulation of spotlight-mode synthetic aperture radar," *Proc. IEEE*, Vol. 71, No. 8, pp. 917–925, 1983.

D. Nahamoo and A. C. Kak, "Ultrasonic Diffraction Imaging," School of Electrical Engineering, Purdue University, TR-EE, pp. 82–20, 1982.

E. G. Njoku and J-A Kong, "Theory for remote sensing of near-surface soil moisture," *J. Geophys. Res.*, Vol. 82, pp. 3108–3118, 1977.

S. J. Norton and M. Linzer, "Ultrasonic reflectivity tomography: reconstruction with circular transducer arrays," *Ultrasonic Imaging*, Vol. 1, pp. 154–184, 1979.

R. D. Radcliff and C. A. Balanis, "Reconstruction algorithms for geophysical applications in noisy environments," *Proc. IEEE*, Vol. 67, No. 3, pp. 1060–1064, 1979.

J. Radon, "On the determination of functions from their integrals along certain manifolds," *Berichte Saechsische Akademie der Wissenschaften*, Vol. 69, pp. 262–277, 1917.

P. S. Rao, K. Santosh, and E. G. Gregg, "Computed tomography with microwaves," *Radiol.*, Vol. 135, pp. 769–770, 1980.

E. A. Robinson, "Wave Propagation in Layered Media," Chap. 3 in *Multichannel Time Series Analysis*, Holden Day, San Francisco, 1967.

E. A. Robinson, "Spectral approach to geophysical inversion by Lorentz, Fourier, and Radon transforms," *Proc. IEEE*, Vol. 70, No. 2, pp. 1039–1055, 1982.

M. I. Skolnik, this volume, p. 59.

M. Slaney, A. C. Kak, and L. E. Larsen, "Limitations of imaging with first-order diffraction tomography," *IEEE Trans. Microwave Theory Tech.*, Vol. MTT-32, No. 8, pp. 860–874, 1984.

M. Slaney, M. Azimi, A. C. Kak, and L. E. Larsen, this volume.

W. R. Stone, "An exact theory for coherent acoustic probing," in the *11th International Symposium on Acoustic Imaging and Acoustic Holography*, J. P. Powers, Ed., pp. 365–384, Plenum, New York, 1982a.

W. R. Stone, "Acoustical holography is, at best, only a partial solution to the inverse scattering problem," *Ibid.*, pp. 385–398.

C. T. Tai, *Dyadic Green's Functions in Electromagnetic Theory*, Intereducational Publishers, Scranton, Pa., 1971.

J. R. Wait, *Electromagnetic waves in Stratified Media*, Pergamon, New York, 1970.

E. Wolf, "Three-dimensional structure determination of semitransparent objects from holographic data," *Opt. Commun.*, Vol. 1, No. 4, pp. 153–156, 1969.

L. Yamaura, "Mapping of microwave power transmitted through the human thorax," *Proc. IEEE*, Vol. 67, No. 8, pp. 1170–1171, 1979.

J. D. Young and L. Peters, this volume, p. 82.

In-Situ Tissue Permittivity at Microwave Frequencies: Perspective, Techniques, Results

E. C. Burdette, F. L. Cain, and J. Seals

A review of published tissue permittivity data is presented, and the significance of *in-situ* versus excised tissue permittivity measurements is discussed. Shortcomings of conventional tissue permittivity measurement techniques are briefly described, and various *in-situ* measurement methods which have been investigated are discussed. A novel *in-situ* probe measurement technique is presented together with recent results of *in-situ* living tissue permittivity measurements. The effects of induced physiological changes on measured *in-situ* permittivity are presented and discussed with reference to EM imagery for dosimetric and diagnostic purposes.

1. INTRODUCTION

In this paper, electrical properties of tissues will be reviewed, and the significance *in-situ* dielectric permittivity information will be presented over the frequency spectrum from radio frequencies (RF) through microwave frequencies. Recent *in-situ* data, some of which is previously unpublished, will be presented. These data will be compared to previously published and newly published data on excised tissues.

The interaction of an electromagnetic (EM) field with a biological system is highly dependent upon the electrical properties of that system. Thus, in beneficial biomedical applications of EM radiation and in the determination of possible EM radiation hazards with respect to personnel safety, an accurate knowledge of the respective tissue electrical properties is essential. Areas of biomedical research which would benefit from *in-situ* electrical property information include EM thawing of cryopreserved organs and tissues, EM-induced hyperthermia in cancer treatment, detection of pathological conditions in tissues, and diagnostic monitoring applications such as lung water content. In potential EM radiation hazards research, the magnitude and distribution of the absorbed energy are dependent upon the interaction of the incident EM field with the various biological constituents of the exposed organism. These interactions are in turn dependent upon the local geometry and electrical properties of those constituents. Thus, an accurate knowledge of the various tissue electrical properties is crucial to an accurate determination of absorbed power.

The determination of the electrical properties of biological tissues has in the past been limited to measurements over restricted frequency ranges on separate excised tissue samples. These previous results served to validate electrical impedance models of tissue structure, but were not capable of yielding information on the actual living *in-situ* tissue electrical properties. In addition, the influence of surrounding tissue organization, blood perfusion, and the development of pathological conditions could not be ascertained from these *in-vitro* measurements. The recent development of a small (2mm diameter) probe that operates over a wide frequency range (1 MHz to 10,000 MHz) provides the opportunity for obtaining dielectric data over a range of physiological conditions and offers the potential of using dielectric measurements for the detection of local tissue pathology. It is also possible to correlate the time rate of change of the *in-vitro* tissue electrical properties with tissue blood flow, in a manner similar to that of impedance plethysmography.

Presented in this paper is a brief discussion of an *in-situ* dielectric tissue probe, its theory of operation, and an examination of its validity through measurements of the electrical properties of known materials. The design of the probe will be illustrated, and a description of the required impedance measurement system provided. New data on the *in-situ* properties of living tissues over a wide range of frequencies, *in-situ* brain data measured as a function of time postmortem, and correlation of measured dielectric changes in kidney tissues with changes in renal blood flow will be presented.

2. FUNDAMENTALS

The electrical properties to be presented and discussed include the conductivity, σ, dielectric permittivity relative to

Biomedical Research Branch, Electronics Technology Laboratory, Engineering Experiment Station, Georgia Institute of Technology, Atlanta, Georgia 30332. April 1982.

free space, ϵ_r', and the loss tangent, tanδ. The conductivity is defined as the conductance of a unit volume of matter. The dielectric permittivity is the capacitance of a unit volume of matter, divided by the permittivity of free space ϵ_o. One frequently introduces a parameter known as the complex permittivity, $\epsilon^* = \epsilon' - j\epsilon''$. This parameter is a complex quantity whose real part ϵ' is conventionally expressed relative to the permittivity of free space by the ratio ϵ'/ϵ_o, which is the relative dielectric constant, expressed as ϵ_r'. The imaginary part of the complex permittivity is called the loss factor, which can be expressed relative to the free space permittivity as $\epsilon_r'' = \epsilon''/\epsilon_o$. The loss tangent is simply the ratio $\epsilon''/\epsilon' = $ tanδ, where tanδ is the loss tangent of the material. The conductivity σ is just $\omega\epsilon''$ where $\omega = 2\pi f$ is the angular frequency and f represents the frequency in hertz of the EM field. Magnetic properties shall not be considered since the magnetic permeability of most biological material is equal to that of free space and magnetic losses are essentially non-existent.

A material's electrical properties are essentially a measure of its ability to interact with EM energy. Since this interaction results from the presence of components within the material that can be affected by the electric and magnetic forces generated by the EM fields, a material's electrical properties are a direct consequence of its composition and structure. In non-magnetic materials such as tissues, an EM field will primarily act upon components within the material that possess a net electrical charge and/or an electric dipole moment [1]. The motion imparted to these components results in an electric current flow within the material. In biological tissues, components possessing a net electrical charge are mainly ions (e.g., Na^+, Cl^-, K^+, Ca^{2+}, etc.). Polar molecules (such as water) are the main source of electric dipole moments in tissues. Protein structures, muscle, fat, bulk, etc., are additional sources of electric dipole moments [2]. Because the electrical properties of a tissue are determined by such a wide variety of components, as reflected in different dielectric dispersions, these properties exhibit significant variations as a function of parameters such as frequency, temperature, tissue type, and vascularization. In turn, these variations in tissue electrical properties can serve as a measurement of different physiological conditions.

For most materials, the complex permittivity and the conductivity properties are frequency dependent or dispersive. Therefore, their time-dependent characteristics are dominant in response to a step input of current or voltage. Considering the exponential response to a step input, one obtains the Debye equation [3]

$$\epsilon_r^* = \epsilon_{r\infty}' + \frac{\epsilon_{rs}' - \epsilon_{r\infty}'}{1 + j\omega\tau} \tag{1}$$

where ϵ_r^* is the complex permittivity relative to that of free space, τ is the relaxation time constant, and ϵ_{rs}' and $\epsilon_{r\infty}'$ are the relative static dielectric constant (at DC) and the relative "optical dielectric constant" (at millimeter wave frequencies), respectively. The Debye equation may be written in several forms. By separating it into its real and imaginary parts, the standard version is obtained as

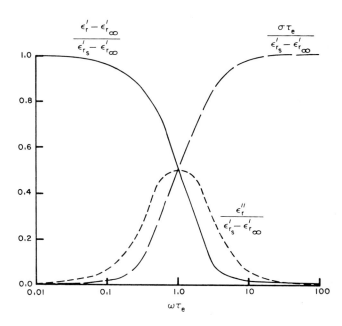

Fig. 1.   Dispersion of a pure dielectric showing dielectric constant, loss factor, and conductivity of a simple relaxation spectrum in normalized form.

$$\epsilon_r' = \epsilon_{rs}' + \frac{\epsilon_{rs}' - \epsilon_{r\infty}'}{1 + (\omega\tau)^2} \tag{2}$$

and

$$\epsilon_r'' = \frac{\epsilon_{rs}' - \epsilon_\infty'\,\omega\tau}{1 + (\omega\tau)^2} \tag{3}$$

A normalized logarithmic plot of dielectric dispersion, loss factor, and conductivity of a simple relaxation spectrum is shown in Fig. 1. Finally, from Eqs. (2) and (3) the following linear equations in $\omega\epsilon_r'$ and ϵ_r'/ω are obtained as

$$\epsilon_r' = \epsilon_{rs}' - \omega\epsilon_r''\tau \tag{4}$$

and

$$\epsilon_r' = \epsilon_{r\infty}' + \frac{\epsilon_r''}{\omega\tau} \tag{5}$$

The Kronig-Kramer relationships [4] show that the frequency response of the permittivity entirely determines the response of the conductivity and vice versa for linear systems. Thus, it follows

$$\epsilon_r' = \epsilon_{rs}' - \sigma\tau. \tag{6}$$

The behavior of the dispersion over a broad frequency range where the relaxation is clustered about a central value can be usefully represented in graphical form by the Cole-Cole function [5]

$$\epsilon_r^* = \epsilon_{r\infty}' + \frac{\epsilon_{rs}' - \epsilon_{r\infty}'}{1 + (j\omega\tau)^{1-\sigma}}. \tag{7}$$

where ϵ_r' is plotted against ϵ_r'' in the complex plane. Points obeying Eq. (7) fall on a semicircle with its center at ϵ_{rs}' +

$\epsilon_{r_\infty}'/2$. The optical and static dielectric constants and the relaxation time are easily obtained from the Cole-Cole plot [5]. The influence of physiological factors (cellular membranes, proteins) are reflected as a depression of the center of the semicircle below the real axis (abscissa).

The above formulation holds for the case of polar molecules in a liquid dielectric, exhibiting a single relaxation time. If a specimen of material composed of structural components of different dielectric constants and conductivities is exposed to a step potential, charge accumulation at dielectric boundaries occurs until a constant current density exists. Since this charge buildup requires time which is dependent upon conduction magnitude on either side of the dielectric boundary, relaxation for each dielectric is established. The Maxwell-Wagner structural formulation for heterogeneous dielectric materials involves a summation of the Debye dispersion for each relaxation. Thus the second term of Eq. (2) becomes a summation over the different relaxation times. The Debye dispersion for tissue, summing over three major relaxation regions, is given by Eq. (8) in Part IV. The dispersion phenomena for tissue exhibits multiple relaxation times whose behavior is not dominated by water at all frequencies. Furthermore, even at microwave frequencies, where the bulk tissue electrical properties are dominated by the presence of water, the influence of protein molecules has been shown to contribute to these characteristics in a way not accounted for by the simple Debye expression [6]. These effects and others (such as blood perfusion) are even more apparent when the measurements are performed *in-situ* on living tissues.

3. PREVIOUS TISSUE ELECTRICAL PROPERTY INVESTIGATIONS

Several investigations of *in-vitro* tissue dielectric properties were conducted by Schwan [2]. Short-circuited transmission line and waveguide techniques were utilized to perform these measurements. By combining these measured data with the low frequency tissue data earlier measured by Rajewsky [7] and Osswald [8], and with the microwave frequency data from Herrick [9], a useful base of *in-vitro* dielectric data was established for various normal tissues over an extremely wide frequency range. Examination of these data reveals the existence of three dispersion regions in tissue dielectric properties in the frequency range from 1 Hz to 30 GHz. These dispersion regions are important because they are characterized by significant variations in dielectric property values as a function of frequency. Schwan entitled these regions the α, β and γ dispersion regions and related the dispersion phenomena to various tissue components [2]. The α-dispersion region (0.1–100 kHz) results from the frequency dependence of the apparent outer-cell membrane impedance. This membrane impedance frequency dependence may result from a number of factors: a frequency-dependent access to inner membrane systems (mitochondrial membrane), a frequency-dependent outer membrane conductance, a boundary potential-related series capacitance element, and a frequency-dependent membrane capacitance due to ionic gating currents. The β dispersion, which occurs at approximately 1–20 MHz, is due to the presence of cellular membranes (enclosing bound water) which act as insulating structures. Bound water is that water bound to macromolecules, typically proteins. As the EM frequency is increased, the cellular membranes are effectively short-circuited. The cellular membrane/bound water effects decrease with increasing frequency and, at microwave frequencies, the tissue electrical properties are largely influenced by the water and electrolyte content of the tissue. The magnitudes of the dielectric dispersions $\epsilon_{r_s}' - \epsilon_{r_\infty}'$ for ice, bound water, and free water are similar, but the relaxation frequencies are different. Ice relaxes at audio frequencies, bound water at 100–1000 MHz, and free water near 20 GHz. Because of its water content, tissue exhibits another dispersion phenomenon (γ) in the microwave region near 20 GHz (25 GHz at body temperature). Between the β and γ dispersions is a small dispersion known as the δ dispersion [10], which is characterized by a fairly broad spectrum of characteristic frequencies extending from a few hundred MHz to several GHz. Its magnitude is considerably smaller than the other dispersions, and it is probably caused by a partial rotation of polar molecules together with a dispersion of protein bound water. Heterogeneous dielectrics are characterized by relaxation described by the Maxwell-Wagner formulation which provides volume weighting based on the constituent contributing complex permittivities [11]. The β dispersion results from a Maxwell-Wagner type structural relaxation.

Investigations of tissue electrical properties were also conducted by Cook [12] and by Roberts and Cook [13]. Cook employed a short-circuited coaxial transmission line to obtain electrical property data for muscle, skin, and fatty tissue over a limited frequency range (1.8 to 5.0 GHz). Roberts and Cook investigated the electrical properties of blood, muscle, and fatty tissue in the frequency range Schwan described as the γ dispersion region (approximately 1 to 30 GHz). Their investigations verified the dispersion phenomenon exhibited by the electrical properties of biological tissues in this frequency range. By including a term to account for ionic conductivity and using the relaxation times of pure water, Debye dispersion equations could be utilized to describe the electrical properties of muscle and blood over the UHF/microwave frequency range [13]. This fact verified that the electrical properties of these tissues in this frequency range are determined primarily by their water and electrolyte content. However, recent investigations on excised tissues by Foster, et al. [14] indicate an effect on the tissue dielectric properties at microwave frequencies due in part to dielectric relaxation of the tissue proteins themselves. Recent previously unpublished data from *in-situ* measurements by Burdette, et al., also indicate that protein relaxation plays an important role and that blood flow changes significantly influence tissue dielectric characteristics. Larsen, Jacobi, and Krey [15] reported results of changes in permittivity dispersion over the 3–30 MHz frequency range associated with physiological and drug-induced patho-physiological states of the cell membrane in living cell suspensions. Their results illustrate the sensitivity of dielectric dispersion to membrane alteration at MHz frequencies.

Several investigations have been directed at obtaining tissue electrical property data for various normal animal and human tissues. These investigations include the surveys by Tinga and Nelson [16], Johnson and Guy [17], Geddes and Baker [18], and Schwan and Foster [19]. Johnson and Guy classified tissues into two categories: (1) muscle, skin, and other tissues with high water content, and (2) fat, bone, and other tissues with low water content. However, by classifying tissues into only two categories, no accounting is made for individual tissue types (e.g., brain, muscle) or for possible physiological influences. These previous investigations have resulted in data that have proven to be extremely useful to studies involving the interaction of biological tissues with EM energy. However, because these results were obtained from *in-vitro* measurements, none of these data reflect actual physiological conditions of living tissues and none of these data represent measurements of changes in electrical properties due to fluid shifts or blood flow. *In-vitro* measurements performed at microwave frequencies as a function of time post-excision [2, pp. 165–174] indicate substantial dielectric changes due to cellular degeneration (which are reflected in the Cole-Cole plot). The physiological significance of these magnitude changes was not addressed.

4. TEMPERATURE COEFFICIENTS

It is possible to determine the temperature coefficients for the dielectric properties of tissues from the data presented by Osswald [8], Geddes and Baker [18], Schwan and Li [20], and Foster et al. [14]. Rewriting Eq. (1) for the dielectric constant in terms of the sum of exponentials as

$$\epsilon_r' = \epsilon_{r\infty}' + \sum_{n=1}^{3} \frac{\Delta\epsilon_{r_n}'}{1 + (j\omega\tau_n)^\alpha}, \tag{8}$$

the dielectric characteristics can be modeled as a sum of relaxation processes where the index n is different for the α-, β-, and γ-dispersion ranges. If the three dispersion ranges are well separated such as in the case of the α-, β-, and γ-dispersion regions, the analysis is straightforward. The critical frequency f_c is that frequency within each dispersion region for which the temperature coefficient is largest. For each dispersion, $\epsilon_{r\infty}'$ and the relaxation $r = \frac{1}{2}f_c$ are reflected in the temperature coefficient. The temperature coefficient of the α dispersion is dependent on the ionic conductance of the tissue electrolyte, which is approximately 2% per °C at frequencies away from f_c. In the β-dispersion region, the insulating cellular membranes are charged via the electrolytes bound by the membrane. In that case, the charging time constant of the membrane is inversely proportional to the electrolyte ionic conductivity, the temperature coefficient of which is 2% per °C. In the γ-dispersion region the relaxation frequency of tissue is close to that of unbound water. The temperature dependence due to water is again approximately 2% per °C at frequencies not near f_c. Thus, if possible impedance changes in the cell membrane are ig-

nored (their contribution to the temperature coefficients is small), f_c then increases with temperature at about 2% per °C in all dispersion regions (over the temperature range from room temperature to physiological temperature) for f not near f_c.

5. DIELECTRIC PROPERTY MEASUREMENT TECHNIQUES

Measurement techniques that may be utilized to determine tissue dielectric properties are based on measuring the effects of the interaction of tissue with an EM field at a specific frequency. Several standard measurement techniques are in existence that may be utilized to measure tissue dielectric properties. These standard techniques include impedance bridges, resonant circuits, transmission lines/waveguides, and resonant cavities. A detailed analysis of these techniques can be found in Von Hippel [1], and from those discussions, it can be seen that only the transmission line/waveguide techniques can be utilized over an appreciable frequency range. However, it should be noted that waveguide measurement methods are bandwidth-limited by the low-frequency cut-off and high-frequency multimoding. Multimoding especially becomes a problem when measuring materials having high dielectric constants (e.g., water). Because of a desire to characterize the dielectric properties of tissue over the radio frequency (RF) through microwave frequency ranges, only the transmission line/waveguide techniques have found widespread utilization for dielectric property measurements of tissue [2], [12].

Usually, the transmission line or waveguide is terminated in a short circuit, and the tissue being measured fills a portion of the line or waveguide (usually adjacent to the short-circuiting plane). The presence of the tissue will affect the field distributions (standing wave pattern) usually present in the line or waveguide, and these effects can then be measured and utilized to calculate the tissue electrical properties.

Although short-circuited transmission lines and waveguides have been widely utilized for tissue measurements, several characteristics of the technique make it unattractive. First, excision of tissue samples is required. Thus, *in-situ* measurements cannot be made. Other characteristics of this technique are that tissue samples must conform to the exact dimensions of the transmission line or waveguide and that the interface between tissue and air must be smooth and normal to the direction of propagation of the incident fields. When a short-circuited waveguide is used, significant quantity of tissue is required. Since uniform tissue specimens large enough to fill a section of waveguide at UHF/microwave frequencies below approximately 10 GHz are often not available from a single tissue and are difficult to prepare, smaller tissue samples are usually diced, ground, chopped, etc., and then poured much like a liquid into the waveguide sample holder [2], [9]. This makes sample preparation somewhat easier, however, variations in surface smoothness due to dicing or chopping, or alteration of cellular/tissue membrane structure from grinding of the sample, could significantly affect the accuracy of the measured dielectric properties relative to *in-situ* values.

When short-circuited transmission lines are employed, smaller tissue samples may be utilized since the dimensions of a transmission line are usually smaller than the dimensions of a waveguide [1]. By using a small enough coaxial transmission line, a sample suitable for measurement may be prepared from a single small sample. But even measurements with a short-circuited coaxial line are limited by the accuracy with which the sample thickness can be measured and by the fact that the measurements are performed *in-vitro* on an excised tissue sample.

6. *IN-SITU* VERSUS EXCISED TISSUE PERMITTIVITY MEASUREMENTS

Questions addressing the importance of *in-situ* tissue electrical property data as opposed to data obtained from measurements of excised tissues have long been of concern to investigators. As early as 1922, Osterhout described significant changes in the dielectric characteristics of biological materials following death due to the loss of membrane function and the breakdown to cellular structure [21]. Rajewsky's low frequency measurements showed a deterioration of the dielectric properties concommitant with a significant decrease in metabolic rate [7]. However, the reported changes did not begin until approximately one day following death of the organism. Recent *in-situ* low frequency measurements by Burdette, et al. [22] revealed changes in electrical conductivity of the liver within an hour following death, while one day was required for the observation of changes in the dielectric permittivity. The observed changes in low frequency dielectric properties are caused by a breakdown of cellular membranes, which are primarily responsible for the low frequency characteristics properties (i.e., α and β dispersions). Schwan and Foster [19] state that the high frequency data are relatively unaffected by death of the tissue because at microwave frequencies, the dielectric characteristics of tissue are predominantly due to the water and protein contents of the tissue. However, some of their data [2], [19] do show magnitude changes in dielectric properties following death. They also indicate that changes in the tissue do occur upon excision which may affect its dielectric properties. Factors which may play a role in dielectric changes due to excision include blood loss and moisture loss.

It is true that at high frequencies the tissue dielectric characteristics largely reflect those of water. However, in recent measurements performed *in-situ* by Burdette, et al. (and discussed later in this paper) changes in the dielectric characteristics of brain tissue at microwave frequencies were seen immediately upon termination of the experimental animal. These initial changes were followed by a slower, gradual reduction in permittivity and conductivity over a longer period of time (two hours). The initial change was attributed to blood loss, while the slower changes were attributed to a combination of tissue water loss and autolysis. From those *in-situ* studies of permittivity, there appear to be significant influences on tissue dielectric characteristics due to the fact that the tissue is living and *in-situ*, even at microwave frequencies.

7. *IN-SITU* PERMITTIVITY MEASUREMENTS

7.1 Introduction

Efforts in the area of *in-situ* permittivity measurements at RF and microwave frequencies began a decade ago with preliminary work performed by Magin and Burns [23]. This technique, based on an antenna modeling theorem, utilized a monopole probe that behaved electrically as a short antenna which could be inserted into tissue. Although available instrumentation limited the frequency range of their technique to 10–100 MHz, preliminary dielectric property data for tissues in mice, rats, and dogs were measured. More recently, Toler and Seals [24] improved the accuracy of the monopole antenna approach from 10–100 MHz. Hahn [25] modified the instrumentation and probe configuration and performed dielectric property measurements on various animal tissues. However, the frequency range of this technique was only 3–100 MHz. A second *in-situ* dielectric property technique utilizing a probe described as an "open transmission line resonator" has also been recently investigated [26]. This technique utilizes a section of open-ended coaxial cable that is placed in contact with a sample of the material being measured. Because of instrumentation problems, this technique was accurate over only a limited frequency range from 1.0–4.0 GHz. Joines, Tanabe and U [27] utilized this technique to perform preliminary dielectric property measurements on normal and neoplastic tissues in human subjects. Their limited results indicated possible differences in the dielectric properties of normal and neoplastic tissues. Guy [28] developed and used a "four-electrode" technique to measure the electrical conductivity of normal canine muscle tissues. This technique was also evaluated by Toler and Seals [24]. They found that large sample volumes were required in order to obtain valid results and that measurement accuracy degraded rapidly as the frequency was increased above 40 MHz. *In-situ* measurements of rat thigh muscle were reported by Edrich and Hardee [29] using an open-ended waveguide technique at 40–54 GHz and 85–90 GHz. However, the probable error in their data was not reported, suggesting a need for further developmental efforts of *in-situ* permittivity measurement methods they utilized at millimeter wave frequencies.

These various *in-situ* measurement techniques represented an improvement over conventional *in-vitro* techniques; however the frequency range and suitability for implantation is limited. A broadband "*in-vivo* probe" technique was developed by Burdette, Cain, and Seals [30]. The *in-vivo* probe technique operates over a frequency range from 1 MHz to 10 GHz. The accuracy of the technique over this frequency range has been verified by measurements of reference materials such as water, methanol, ethylene glycol, etc., which have well documented dielectric properties. The technique has been successfully employed to perform dielectric property measurements on a variety of animal tissues under *in-vivo* conditions and on samples of muscle-equivalent phantom modeling materials [24], [30]–[32].

7.2 Theoretical Basis

Before discussing physical probe design and developmental efforts, it is appropriate to mention the theoretical basis underlying the *in-vivo* dielectric measurement probe. Detailed analyses of theoretical aspects of the probe have been previously presented [31], [32], but only recently were the various theoretical considerations presented in a single report [30]. Because the details of these analyses have been previously reported, only a brief synopsis of the theory is given here.

The theoretical basis of the *in-vivo* measurement probe stems from the application of an antenna modeling theorem to the characterization of unknown dielectric media [30], [33]. Simply stated, the antenna modeling theorem equates the terminal impedance of an antenna operating at frequency ω in a dielectric material to its terminal impedance in free space at frequency $n\omega$, where n is the complex index of refraction of the dielectric material. In a non-magnetic material ($\mu = \mu_o$), the theorem is expressed mathematically as

$$\frac{Z(\omega, \epsilon^*)}{\eta} = \frac{Z(n\omega, \epsilon_o)}{\eta_o}, \qquad (9)$$

where ω, ϵ^* and ϵ_o are as previously defined,

$\eta = \sqrt{\mu_o/\epsilon^*}$ = the complex intrinsic impedance of the dielectric medium,

$\eta_o = \sqrt{\mu_o/\epsilon_o}$ = the intrinsic impedance of free space, and

$n = \sqrt{\epsilon^*/\epsilon_o}$ = the complex index of refraction of the dielectric medium relative to that of air.

This theorem is applicable for any probe provided an analytical expression for the terminal impedance of the antenna is known both in free space and in the dielectric medium under study. The theorem as stated in Eq. (9) assumes that the medium surrounding the antenna is infinite in extent, or conversely, the theorem is valid provided the probe's radiation field is contained completely within the medium.

For probe antennas one-tenth wavelength or less in length, the terminal impedance is given by

$$Z(\omega, \epsilon_o) = A\omega^2 + \frac{1}{jC\omega}, \qquad (10)$$

where A and C are constants determined by the physical dimensions of the antenna [34]. From a knowledge of the above constants and the complex terminal impedance Z(ω,

ϵ^*) of the probe antenna in a dielectric material, the complex permittivity, and thus the relative dielectric constant, conductivity, and loss tangent values can be obtained from the theorem of Eq. (9) by substitution. Utilizing the form of antenna impedance given in Eq. (10), one obtains

$$Z(\omega, \epsilon^*) = A\omega^2 \sqrt{\frac{\epsilon^*}{\epsilon_o}} + \frac{1}{jC\omega \frac{\epsilon^*}{\epsilon_o}} \qquad (11)$$

Equation (11) relates the complex impedance of a lossy dielectric medium at a frequency ω to the relative complex dielectric constant of the medium. In terms of dielectric constant and loss tangent, Eq. (11) becomes

$$Z(\omega, \epsilon^*) = A^2 \sqrt{\epsilon_r'(1 - j\tan\delta)} + \frac{1}{jC\omega[\epsilon_r'(1 - j\tan\delta)]}, \qquad (12)$$

which is a restatement of the theorem of Eq. (9) for a short monopole. This equation can be placed in the form Z = R + jX which reduces to two real equations to give

$$R = \frac{\sin 2\delta}{\epsilon_r'\omega C} + A\sqrt{\epsilon_r'}\,\omega^2\,\frac{\sqrt{\sec\delta + 1}}{2} \qquad (13)$$

and

$$X = \frac{\cos^2\delta}{\epsilon_r'\omega C} + A\sqrt{\epsilon_r'}\,\omega^2\,\frac{\sqrt{\sec\delta - 1}}{2}. \qquad (14)$$

The parameters R and X are the real and imaginary components of the measured impedance, A and C are the physical constants of the probe, and all other parameters are known except ϵ_x' and δ. Because of the inverse pair of equations corresponding to Eqs. (13) and (14) cannot be easily obtained, an iterative method of solution is utilized. The second term in both Eqs. (13) and (14) is small at low frequencies. Neglecting these high frequency terms, one obtains equations involving only the first term of Eqs. (13) and (14). The solution of these resulting low frequency equations are readily obtained by noting that $\tan\delta = R/X$. Utilizing the low frequency solution for δ, Eq. (14) can be solved for ϵ_r'. The validity of the solutions for δ and ϵ_r' may be determined by substituting these values into Eq. (14). The correct values for δ and ϵ_r' will satisfy both Eqs. (13) and (14) exactly, where ϵ_r' is the relative dielectric constant and $\tan\delta$ is the loss tangent.

For very short (or infinitesimal) monopole probes whose center conductor approaches zero length, the probe impedance in free space is totally reactive. In this case, the probe is essentially an open-circuit transmission line having only a fringing field. Thus, the minimum sample volume necessary to obtain accurate measurement results is primarily dependent upon the distance between the tightly-coupled center and outer conductors of the probe. The free-space impedance of the open-circuited probe is simply

$$Z(\omega, \epsilon_o) = \frac{1}{jC\omega}, \qquad (15)$$

where C is as indicated in Eq. (10). Expanding this imped-

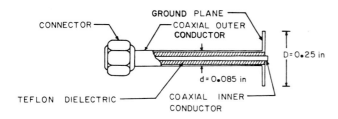

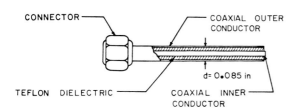

(a) PROBE WITH GROUND PLANE

(b) PROBE WITHOUT GROUND PLANE

Fig. 2. *In-situ* dielectric measurement probe configurations (from [30]).

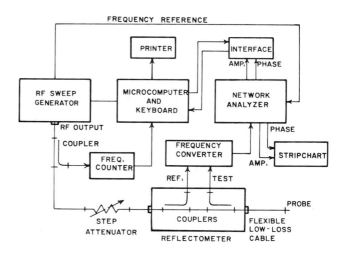

Fig. 4. Block diagram of microcomputer-controlled impedance measurement instrumentation used with the *in-vivo* dielectric measurement probe.

ance expression in the antenna modeling theorem yields the result

$$Z(\omega,\epsilon^*) = \frac{1}{jC\omega[\epsilon_r'(1 - j\tan\delta)]} \quad (16)$$

which is the same expression as the imaginary part of Eq. (12) above. Therefore, for the case of an open-circuit transmission line, the antenna modeling theorem reduces to Eq. (16). A more extensive treatment of the open-circuit transmission line probe is presented in References 30 and 31.

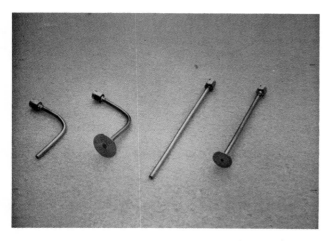

Fig. 3. Photograph of *in-situ* probes employed most extensively in dielectric measurements.

7.3   *In-Situ* Measurement Probes and System

A number of probe configurations suitable for *in-situ* dielectric measurements have been investigated [30]. The lengths of the extended center conductor of the probes have ranged from infinitesimal lengths of less than 0.02 inch to lengths of 0.4 inch, and probe outside diameters have ranged from the size of an 18-gauge hypodermic needle (approximately 0.042 inch) to 0.141 inch. The configurations used most extensively are diagrammed in Fig. 2 and four probes are pictured in Fig. 3.

The *in-situ* measurement probes are each fabricated from a section of open-minded semi-rigid 0.085-inch diameter coaxial cable with a slightly extended center conductor. An SMA connector is attached to the probe by first removing the center conductor and teflon dielectric material, then soldering the connector to the outer conductor and reassembling the probe using the center conductor as the center pin of the connector, thus avoiding additional soldering. Thus, the SMA connector is attached without heating the teflon dielectric. While disassembled, the probe conductors are first flashed with nickel plating and then gold plated. Gold plating of the probe greatly reduces chemical reactions between the probe and the electrolyte in the tissue. This process virtually eliminates oxidation of the probe's metallic surfaces and helps to minimize electrode polarization effects at lower frequencies (1 MHz–100 MHz) [24], [30].

The impedance measurement instrumentation employed to measure the terminal impedance of the probe is schematically illustrated in Fig. 4 and a photograph of one system configuration is shown as Fig. 5. The key components of the measurement system are the probe and a network analyzer. The relative amplitude and phase difference between the

Fig. 5. Photograph of *in-situ* probe measurement system showing microcomputer-controlled data acquisition/data processing system.

semi-automated data acquisition/data processing system whose key components are an analog/digital converter and microprocessor is used to increase the rate of acquisition and processing of *in-situ* permittivity measurement data.

When a network analyzer system is used for performing microwave measurements, there exist certain inherent measurement errors which can be separated into two categories: instrument errors and test set/connection errors. Instrument errors are measurement variations due to noise, imperfect conversions in such equipment as the frequency converter, cross-talk, inaccurate logarithmic conversion, non-linearity in displays, and overall drift of the system. Test set/connection errors are due to the directional couplers in the reflectometer, imperfect cables, and the use of connector adapters. The instrument errors exhibited by modern network analyzers are very small.

The primary source of measurement uncertainty is due to test set/connector errors at UHF and microwave frequencies. These uncertainties are quantified as directivity, source match, and frequency tracking errors. An analytical model to account for test set/connection errors has been developed by Hewlett-Packard for correcting reflectivity measurements on their semi-automatic network analyzer system [35]. This model has been implemented for use with the *in-vivo* measurement probe and equations which correct for the open-circuit fringing capacitance of the probe have been added to the algorithm. Error-corrected data for

reference and reflected signal channels is measured by the network analyzer, which yields the terminal impedance of the probe in terms of the magnitude and phase angle of the reflection coefficient. These data are used as input data to a computer algorithm which corrects systemic measurement errors and computes the dielectric property information. A

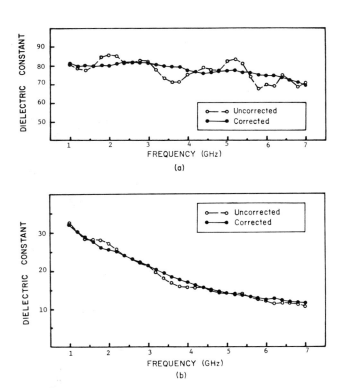

Fig. 6. Relative dielectric constant of (a) water and (b) methanol measured at a temperature of 20°C with and without systemic error correction.

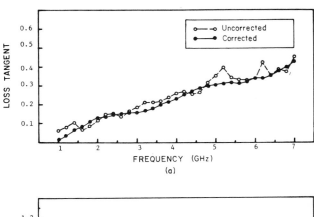

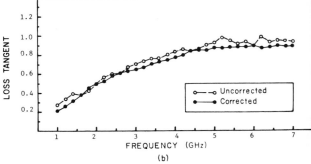

Fig. 7. Loss tangent of (a) water and (b) methanol measured at a temperature of 20°C with and without systemic error correction.

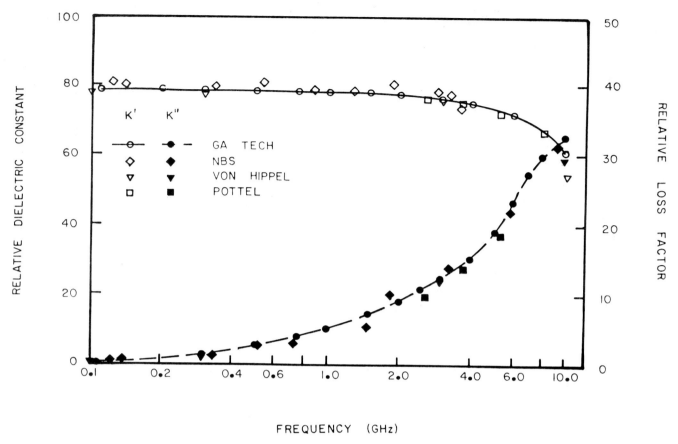

Fig. 8. Probe-measured relative dielectric constant K′ and relative loss factor K″ of dionized water at 23°C compared to results from References [1,16,36,37] and Ref. [30].

7.4 Accuracy Studies

In any measurement technique, the first thing one must do is validate the technique. This was done in this case by measuring standard dielectric materials. Results using the probe measurement method on water, methanol, and ethylene glycol are presented in Figs. 8, 9, and 10 for relative dielectric constant and relative loss factor. Results for water and methanol are also presented in tabular format at selected frequencies in Tables I and II. The accuracy of this dielectric measurement technique is approximately ±5% when compared to reference data [1], [36]–[38]. An interesting observation is that for a specific material measured at the same temperature, the variability of data from different references is often greater than the variability of the data obtained using the *in-vivo* measurement probe. The accuracy of probe-measured dielectric data on pure liquids

is within ±3% when compared to values predicted by the Debye theory [6].

8. *IN-SITU* PERMITTIVITY MEASUREMENTS—RECENT RESULTS

In 1976, efforts to extend the *in-situ* probe measurement technique to microwave frequencies were reported [39]. During the past several years, the *in-situ* probe technique described above and by Burdette et al. [30–32] has been developed to the extent that the probe measurement system operates under microcomputer control. This system has been utilized for measurements of living tissues as a function of frequency over the 3-MHz to 10-GHz range and as a function of physiological state at microwave frequencies [40]. In 1979, Athey [41] described preliminary efforts to develop impedance calibration standards which would attach directly to the end of the type of probe developed by Burdette, Cain, and Seals [30]. However, the probes which we used were approximately twice the diameter of those used by Burdette, et al.

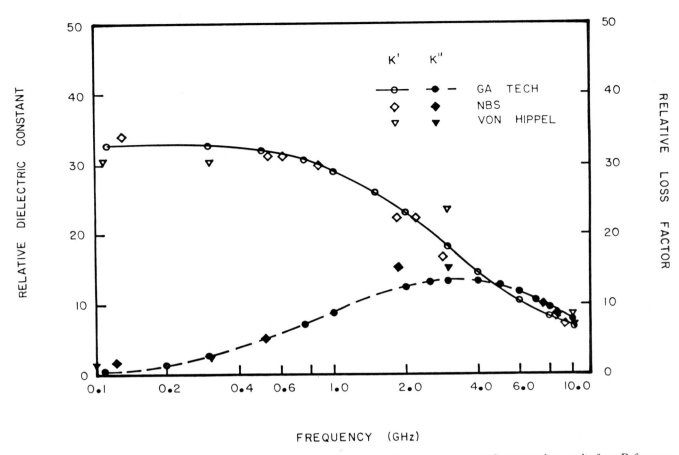

Fig. 9. Probe-measured relative dielectric constant K′ and relative loss factor K″ of methanol at 23°C compared to results from References [1,16,36] and Ref. [30].

8.1 *In-Situ* Permittivity Measurements as a Function of Frequency

The *in-situ* probe measurement technique described in Reference 30 has provided useful information about the dielectric characteristics of tissue for electromagnetic applications in biomedicine, including hyperthermia, organ preservation and thawing, radiometric and/or dielectric imaging, etc. *In-situ* permittivity measurements have been made as a function of frequency for muscle, kidney, fat, and brain. The results of the *in-situ* canine tissue measurements are summarized in Figs. 11 through 14. The relative dielectric constant and conductivity of *in-situ* canine muscle tissue are compared to *in-vitro* muscle data from Schwan [2] in Fig. 11. The uncertainty in the measured data is expressed as the standard error of the mean (SEM). The differences in the results of the *in-vitro* measurements of human autopsy muscle and the *in-situ* measurements of animal muscle are attributed primarily to blood flow and temperature. The maximum dielectric difference is approximately 15%, of which 6% is attributed to sample temperature difference. The remainder of the dielectric difference is attributed to

blood flow and to changes in δ-dispersion protein relaxation due to significant changes in protein metabolism. Further, the actual physiological differences which exist between *in-situ* and *in-vitro* tissues due to cellular membrane changes could also contribute to these differences. Measurement results of *in-situ* and *in-vitro* canine kidney cortical tissue are presented in Fig. 12. Again, these results are compared to other *in-vitro* data. Note that although the *in-vitro* K′(K′ = ϵ_r′) data obtained from probe measurements of kidneys are in close agreement with earlier published *in-vitro* data, the *in-situ* data are different from both sets of *in-vitro* data. The results of measurements performed on canine fat and brain are shown in Figs. 13 and 14, respectively. When compared to *in-vitro* data [18], the *in-situ* canine fat tissue measurement results (Fig. 13) exhibited relative dielectric constant values a factor of approximately 1.5–2 times greater than reported *in-vitro* results at frequencies above 100 MHz. These differences in K′ are primarily attributed to the likely difference in water content between the *in-situ* and *in-vitro* measurement conditions; explicit water content was not reported *in-vitro* and no measurement of *in-situ* water content in fat was

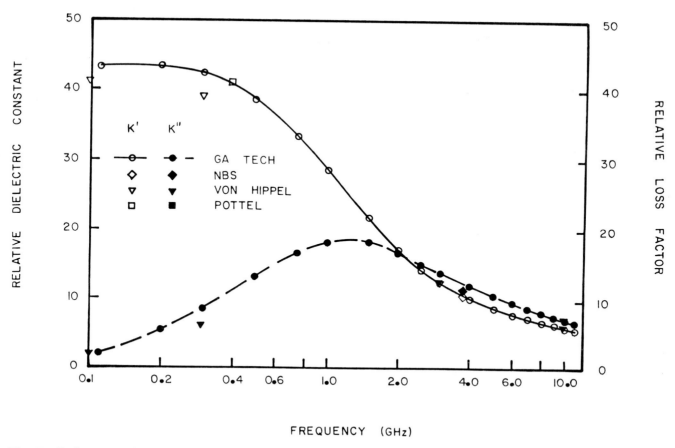

Fig. 10. Probe-measured relative dielectric constant K' and relative loss factor K'' of ethylene glycol at 23°C compared to results from References [1,36] and Ref. [30].

performed. *In-situ* and reported *in-vitro* [2], [19] conductivity values compared very favorably. At all frequencies, the conductivity measured *in-situ* was within the range of reported values nearer the smaller value [18]. The reported *in-vitro* values were not included in Fig. 13 because of their wide range. The *in-situ* brain data presented in Fig. 14 are from probe measurements of the pial surface of dog brain where the probe was positioned on the left ectosylvian gyrus. Under one condition, the probe was placed over a pial vessel (approximately 0.5-mm diameter) and in the second condition, the probe was directly on pia with no macroscopic vessels. These data are compared to recent *in-vitro* dielectric measurements of gray matter separated from dog brains [42].

8.2 Physiological Effects on Dielectric Parameters

In this subsection, the results of measurements performed to examine possible changes in permittivity brought about by physiological changes in canine brain and kidney tissues are presented [40,43]. The principal topics addressed are the effects of the following parameters on dielectric properties at a frequency at 2450 MHz:

- Changes in regional blood flow,
- physiological changes due to nerve stimulation and drugs, and
- postmortem changes.

Results were obtained from twenty-four dogs. All measurements were performed as a function of time at 2450 MHz using a 0.085-in diameter probe. The dogs used in these acute experiments weighed 12–19 kg and were maintained NPO overnight prior to surgery. All surgical procedures and subsequent dielectric measurements took place under phenobarbital anesthesia, 50 mg/kg administered IP.

During each experiment, numerous variables were measured simultaneously. Some of these were recorded continuously; others were recorded at discrete time increments. The physiological parameters recorded were arterial and venous blood pressures, electrocardiogram (EKG), blood flow (renal), and blood pH, P_{O_2}, and P_{CO_2}. The blood pressures and EKG were recorded continuously on a physiograph stripchart. Values of pH, P_{O_2}, and P_{O_2}, and P_{CO_2} were also recorded at periodic intervals. Dielectric data recorded digitally at different intervals consisted of the frequency, the

TABLE I

Complex Permittivity (K′ − jK″) of Water

Frequency MHz Data Source		100	110	122	138.9	300	334.1	500	514.6	731.7	1271.2	1500	1851.9	2500	2610
NBS	K′	*	*	*	*	*	*	*	*	*	*	*	*	*	*
(Hasted)	K″	*	*	*	*	*	*	*	*	*	*	*	*	*	*
NBS	K′	*	*	*	*	*	*	*	*	*	*	*	*	*	*
(Collie)	K″	*	*	*	*	*	*	*	*	*	*	*	*	*	*
NBS	K′	*	*	*	80.4	*	80.0	*	*	79.0	78.7	*	*	*	*
(Burdon)	K″	*	*	*	0.63	*	1.44	*	*	2.8	5.3	*	*	*	*
NBS	K′	*	*	80.8	*	*	*	*	80.8	*	*	*	80.7	*	*
(Slevogt)	K″	*	*	0.60	*	*	*	*	2.8	*	*	*	10.2	*	*
Von	K′	78.0	*	*	*	77.5	*	*	*	*	*	*	*	*	*
Hipple	K″	0.39	*	*	*	1.24	*	*	*	*	*	*	*	*	*
Pottel	K′	*	*	*	*	*	*	*	*	*	*	*	*	*	76.9
	K″	*	*	*	*	*	*	*	*	*	*	*	*	*	9.77
Cook	K′	*	*	*	*	*	*	*	*	*	*	*	*	*	*
	K″	*	*	*	*	*	*	*	*	*	*	*	*	*	*
Burdette	K′	80.0	79.9	79.8	79.7	79.3	79.2	78.8	78.8	78.4	78.1	78.0	77.8	77.1	77.0
	K″	0	0	0.10	0.30	1.20	1.40	2.34	2.40	3.70	6.45	7.40	8.80	11.15	11.6

Frequency MHz Data Source		2873.6	3000	3253.8	3571.4	3750	4630	5300	5882.4	7950	9140	9345.8	9375	9390	10000
NBS	K′	*	*	77.8	*	*	*	*	*	*	63.0	*	*	*	*
(Hasted)	K″	*	*	13.9	*	*	*	*	*	*	31.5	*	*	*	*
NBS	K′	*	77.42	*	*	*	*	*	*	*	*	61.41	*	*	*
(Collie)	K″	13.1	*	*	*	*	*	*	*	*	*	31.8	*	*	*
NBS	K′	*	*	*	74.0	*	*	*	67.0	*	*	*	56.7	*	*
(Burdon)	K″	*	*	*	14.3	*	*	*	22.0	*	*	*	34.0	*	*
NBS	K′	78.6	*	*	*	*	*	*	*	*	*	*	*	*	*
(Slevogt)	K″	12.1	*	*	*	*	*	*	*	*	*	*	*	*	*
Von	K′	*	76.7	*	*	*	*	*	*	*	*	*	*	*	55.0
Hipple	K″	*	12.04	*	*	*	*	*	*	*	*	*	*	*	29.7
Pottel	K′	*	*	*	75.6	*	72.9	*	67.6	*	*	*	*	*	*
	K″	*	*	*	13.74	*	18.58	*	25.74	*	*	*	*	*	*
Cook	K′	*	77.7	*	*	*	74.0	*	*	*	*	*	*	61.5	*
	K″	*	13.0	*	*	*	18.8	*	*	*	*	*	*	31.6	*
Burdette	K′	76.6	76.4	76.1	75.6	75.3	74.1	73.1	72.3	67.8	64.3	63.2	53.2	63.1	59.9
	K″	12.35	12.8	13.65	14.7	15.25	17.85	20.1	22.25	29.2	31.75	32.8	32.85	32.9	33.

* Indicates data not reported.

amplitude and phase of the complex reflection coefficient, and the computed dielectric properties—relative dielectric constant, conductivity, and loss tangent—of the tissue contacted by the probe. The quantities of drugs administered to the experimental animal were also recorded.

a. In-Situ Brain Permittivity— Regional Changes

There were four types of *in-situ* measurements made according to the probe antenna location:

- Dural,
- Pial,
- Brain (shallow), and
- Brain (deep),

Dural measurements were made on the surface of the dura mater, the outermost membrane covering the brain; pial measurements were made on the pia mater, the innermost covering. The dura, a thick, tough membrane, is separated from the thin pia by cerebrospinal fluid (CSF) in the areas where measurements were performed. The pia is very thin, elastic, and conforms to the surfaces of the brain. Pial measurements may thus be considered as taken on the brain's surface. Measurements were also performed with the probe at two relative depths in brain tissue. The probe tip was less than 1.3 cm below the surface for shallow brain measurements and between 1.3 and 2.5 cm for deep brain measurements. The shallow and deep measurements correspond roughly to gray and white matter, respectively. Data from the four probe locations noted above are presented in Table III, with most of the data being from dural and pial locations. There are not as many data for locations beneath the surface (the shallow- and deep-brain locations) as there are for surface measurements because emphasis was placed on noninvasive (surface) measurements.

For the dural measurements, the mean of multiple measurements in one animal for dielectric constant ranged from 45.3 to 53.9; and for conductivity, the mean ranged from 18.2 to 23.5 mmho/cm. The differences in dielectric properties for the dura are probably due to different thicknesses of the

TABLE II

Complex Permittivity (K′ − jK″) of Methanol

Data Source		100.0	123.5	300.0	514.6	535.7	612.2	877.2	1840.5	2230.5	2873.6
NBS	K′	*	*	*	*	*	*	*	*	*	*
(Lane)	K″	*	*	*	*	*	*	*	*	*	*
NBS	K′	*	*	*	*	*	*	*	*	*	*
(Poley)	K″	*	*	*	*	*	*	*	*	*	*
NBS	K′	*	34.6	*	34.3	*	*	*	22.6	*	17.0
(Slevogt)	K″	*	1.55	*	5.15	*	*	*	15.4	*	17.4
NBS	K′	*	*	*	*	*	*	*	*	*	*
(Baz)	K″	*	*	*	*	*	*	*	*	*	*
NBS	K′	*	*	*	*	31.4	31.6	30.0	*	22.7	*
(V. Ardenne)	K″	*	*	*	*	*	*	*	*	*	*
Von	K′	31.0	*	30.9	*	*	*	*	*	*	*
Hipple	K″	1.17	*	2.47	*	*	*	*	*	*	*
Burdette	K′	33.5	33.5	33.5	32.2	31.8	31.6	30.0	24.3	22.3	19.2
	K″	0.50	0.80	2.90	5.20	5.45	6.05	5.20	12.4	13.2	13.7

Data Source		3000	3333.3	4838.7	7518.8	7894.7	8547.0	9345.8	9375.0	10000	10714.3
NBS	K′	*	*	*	*	*	*	8.33	*	*	*
(Lane)	K″	*	*	*	*	*	*	8.16	*	*	*
NBS	K′	*	*	*	9.72	*	8.68	*	7.78	*	*
(Poley)	K″	*	*	*	10.20	*	9.14	*	7.69	*	*
NBS	K′	*	*	*	*	*	*	*	*	*	*
(Slevogt)	K″	*	*	*	*	*	*	*	*	*	*
NBS	K′	*	15.8	7.57	*	4.44	*	*	*	*	3.5
(Baz)	K″	*	11.0	7.90	*	5.52	*	*	*	*	4.24
NBS	K′	*	*	*	*	*	*	*	*	*	*
(V. Ardenne)	K″	*	*	*	*	*	*	*	*	*	*
Von	K′	23.9	*	*	*	*	*	*	*	8.9	*
Hipple	K″	15.3	*	*	*	*	*	*	*	7.21	*
Burdette	K′	18.5	17.2	12.8	8.9	8.6	8.0	7.4	7.4	7.2	6.8
	K″	13.8	13.7	12.75	10.2	9.8	9.3	8.5	8.45	7.9	7.35

* Indicates data not reported.

dura itself rather than due to interanimal differences since measured values were distinctly different for two locations (anterior and posterior) in the same animal when this variability was studied. The measured properties of the dura are also influenced to a certain extent by the underlying CSF. This was demonstrated in one set of measurements in which the CSF was drained from beneath the measurement site through a small dural hole located a few millimeters away. For this condition, the measured dielectric constant was 44.0 ± 6.1 (mean ± S.D.) reduced from 49.9 ± 4.1 measured with

TABLE III

In-Situ Dielectric Measurements—Living Canine Brain
Mean Values ± SEM
(n = 5)

Dura mater	
K′ = 49.9 ± 1.9	σ = 20.7 ± 1.3
Pia mater	
K′ = 57.2 ± 0.9	σ = 19.3 ± 0.7
*Shallow brain (gray matter)	
K′ = 43.0	σ = 17.4
Deep brain (white matter)	
K′ = 32.3 ± 1.7	σ = 12.1 ± 0.8

* N = 1 Dog.

the CSF present. Only small differences in conductivity, with overlapping sample values, were observed. Based on these results, dielectric properties of only dura would probably best be measured in a sample of dura alone. However, the results of such measurements would not necessarily reflect *in-situ* conditions.

For the pial measurements, the mean of multiple measurements performed on each animal for dielectric constant ranged from 55.0 to 59.2; for conductivity, the mean ranged from 17.8 mmho/cm to 21.8 mmho/cm. The pial measurements are not as patterned among animals as the dural measurements. Generally, the values for typical measurements are slightly greater than those obtained for the dural measurements. Additional pial data are discussed subsequently.

For the shallow-brain measurements, the mean relative dielectric constant, 43.0, and the mean conductivity, 17.4 mmho/cm, are smaller than values of the respective properties for both dural and pial measurements. For the deep-brain measurement cases, the mean value ranges of relative dielectric constant 33.9 to 30.6, and of conductivity, 12.8 mmho/cm to 11.3 mmho/cm, are even smaller. Thus, there seems to be a pattern of smaller dielectric constant and smaller conductivity with increasing distance below the surface of the brain. A similar trend was observed by Foster

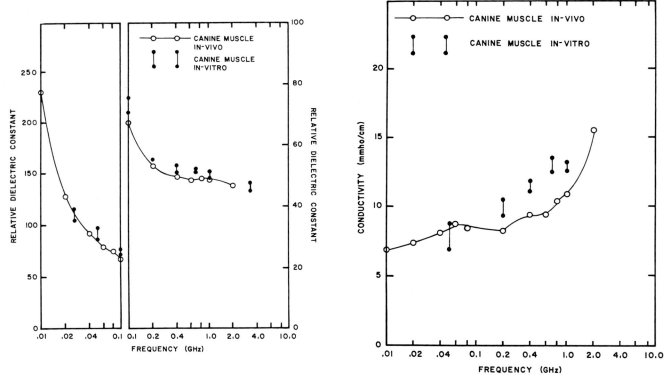

Fig. 11.   (a) Relative dielectric constant of *in-situ* canine muscle at 34°C [30] compared to 37°C *in-vitro* muscle data from [19]. (b) Conductivity of *in-situ* canine muscle at 34°C [30] compared to 37°C *in-vitro* muscle data from [19].

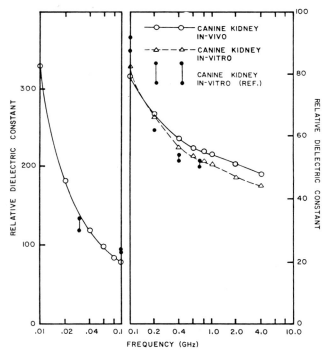

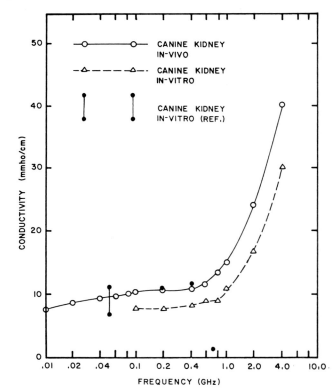

Fig. 12.   (a) Relative dielectric constant of probe-measured *in-situ* and *in-vitro* canine kidney from [30] compared to *in-vitro* kidney data from [19]. (b) Conductivity of probe-measured *in-situ* and *in-vitro* canine kidney from [30] compared to *in-vitro* kidney data from [19].

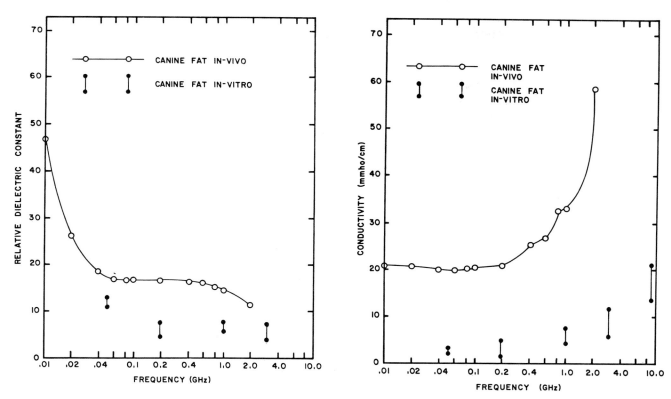

Fig. 13. (a) Relative dielectric constant of *in-situ* canine fat at 35°C [30] compared to 37°C *in-vitro* fat data from [19]. (b) Conductivity of *in-situ* canine fat at 35°C [30] compared to 37°C *in-vitro* fat data from [19].

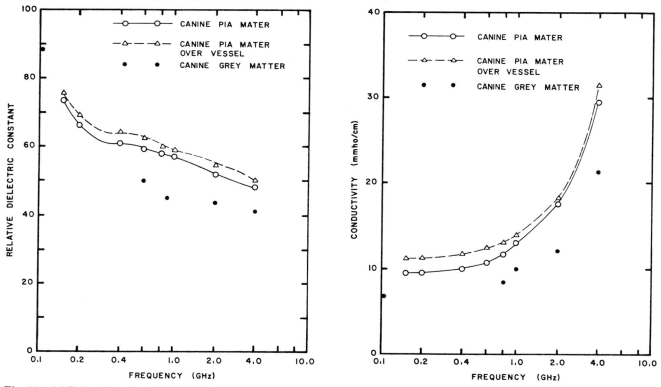

Fig. 14. (a) Relative dielectric constant of *in-situ* canine brain (pia mater) at 36°C compared to 37°C *in-vitro* canine brain (gray matter) data from [42]. (b) Conductivity of *in-situ* canine brain (pia mater) at 36°C compared to 37°C *in-vitro* canine brain (gray matter) data from [42].

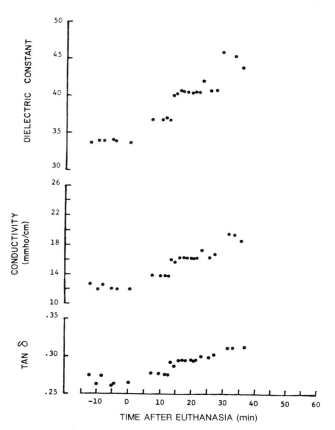

Fig. 15. Measured dielectric properties of *in-situ* deep canine brain (white matter) at 37.5 ± 1.0°C as a function of time postmortem. Lethal injection: pentobarbital overdose (from [40]).

[42] for measurements of excised dog brain. This pattern may be related to our finding that after the cessation of blood flow to the head, the dielectric properties at deeper locations changed more slowly than at shallow or surface locations. The relationship may be caused by the greater blood perfusion of gray matter located near the surface of the brain [44]. It should be noted that our shallow- and deep-brain dielectric data agree well with recently published results [42] for gray and white matter, respectively, for dog brain.

To study differences in dielectric properties between antemortem and postmortem brain tissue, measurements were made before, during and after injections of pentobarbital, calcium chloride (CaCl₂) or potassium chloride (KCl). In two animals euthanized with pentobarbital overdoses, measurements were made with the probe in brain tissue. Figures 15 and 16 show the results from these two experiments. The data for one animal were obtained with the probe in a deep-brain position (Fig. 15), and in the other animal data were obtained with the probe in a shallow-brain position (Fig. 16). Each data point for the deep-brain position is the average of three to six measurements taken within one minute. Data points for the shallow-brain position represent single measurements taken at one-minute intervals.

The time at which blood flow to the head stopped in these

two experiments was taken as that time when the arterial blood pressure went to zero. In both experiments, the dielectric constant and conductivity increased after the blood flow stopped. For the deep position (Fig. 15), the increase in dielectric constant occurred between seven and ten minutes after the cessation of blood flow. The values of dielectric constant and conductivity are lower for the deep position. This can be expected if the myelin of the deeper white matter is assumed to contribute to the overall dielectric constant of the tissue at 2450 MHz. For the deep position, all the values increased monotonically until the end of the experiment at 36 minutes after the cessation of blood flow. For the shallow position (Fig. 16), increases in all dielectric property values began within one-minute of the cessation of blood flow. The values increased for the first seven to eight minutes, leveled off for about six minutes, and then decreased until the end of the experiment at 53 minutes after cessation of blood flow.

The differences in the two time courses could be due to the different types of brain tissue measured. The type of neural tissue may determine not only the static dielectric properties, but also the changes in these properties due to physiological states. Alternatively, the differences in time courses may be

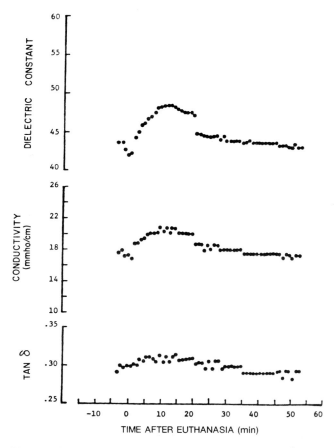

Fig. 16. Measured dielectric properties of *in-situ* shallow canine brain (gray matter) as a function of time postmortem. Lethal injection: pentobarbital overdose (from [40]).

TABLE IV

Summary of Antemortem/Postmortem *In-Situ* Dielectric Studies with KCl Euthanasia
(N = 8)

Time after Death (min.)	Dielectric Constant ± SE	Conductivity ± SE (mmho/cm)	Brain Temperature (°C) ± SD	Colonic Temperature (°C) ± SD
−10	57.8 ± 0.5	20.4 ± 0.2	36.6 ± 1.5	36.1 ± 2.1
−.05	57.8 ± 0.5	20.3 ± 0.2	36.4 ± 1.3	36.1 ± 2.0
+.1	60.9 ± 0.6	21.1 ± 0.2	36.2 ± 1.2	36.1 ± 1.0
10	58.1 ± 0.7	19.2 ± 0.2	36.5 ± 1.5	36.0 ± 2.1
40	55.1 ± 0.8	17.6 ± 0.3	35.6 ± 1.7	35.6 ± 2.3
60	52.8 ± 0.7	17.0 ± 0.2	35.7 ± 1.6	34.9 ± 2.6
75	52.3 ± 0.7	17.4 ± 0.2	35.9 ± 1.7	34.0 ± 2.6
*90	52.3 ± 1.1	17.6 ± 0.4	35.4 ± 1.7	33.6 ± 2.2

* N = 5 for 90 minutes postmortem.

related to blood flow changes in the tissue [44], [45]. Blood flow through the gray matter (shallow) is known to be several times larger than the blood flow through white matter (deep). Thus, the effect of a cessation of blood flow to the brain would be expected to have a more rapid and larger effect on gray matter than on white matter. Changes in dielectric properties could be the direct result of the lack of blood flow or, because of the greater metabolic activity of gray matter, they could be the result of changes in metabolic processes caused by the lack of nutrients in the gray matter [45]. Because of the differences in blood flow through gray and white matter, a comparison of the changes in dielectric properties of gray and white brain matter may be a sensitive test for analyzing the effects of blood flow on dielectric properties.

b.   *In-Situ Pial Measurements— Antemortem/Postmortem Results*

In each of the antemortem/postmortem experiments involving euthanasia by either KCl or $CaCl_2$, measurement of the *in-situ* dielectric properties of dog brain were performed with the dura mater and arachnoid removed and the probe placed directly on the epipial layer of the pia mater over the ectosylvian gyrus. The procedure used in each of these acute experiments is summarized below.

Following administration of pentobarbital anesthesia, tracheotomy, and stabilization of the animal, arterial and venous femoral cannulations were performed and transducers connected for monitoring and recording systemic blood pressures. Femoral cannulations were also performed for extracorporeal monitoring of blood gases, EKG electrodes were connected for monitoring heart rate, and a temperature probe was inserted rectally to the level of the colon for monitoring systemic temperature. The dog's head was mounted in a stereotaxic frame, and a three-fourths inch diameter circular opening was made in the skull over the ectosylvian gyrus. Bone wax was used to stop bleeding, and a dura hook and fine scissors were used to remove the dura covering the gyrus. Before placing the probe on the pia, the cerebrospinal fluid at the measurement site was drained and

the arachnoid was removed. At the conclusion of the above procedures, a miniature thermistor temperature sensor was placed on the pial surface at the edge of the circular opening of the skull and the *in-situ* measurement probe was positioned on the pia.

In-situ permittivity measurements were performed over a two-hour period. Both dielectric property data and physiological data were recorded for a period sufficiently long to ensure stability of all measured parameters prior to euthanasia performed by injecting a 30 cc bolus of saturated KCl or $CaCl_2$ into the femoral vein. All electrical and physiological parameters were monitored for 90 minutes postmortem, based on the time at which all systemic pressures are zero. The results of these experiments are summarized in Tables IV and V. Before intravenous administration of the injection of either KCl or $CaCl_2$, the mean arterial blood pressure was 174/130 mmHg with a SEM of 3.2/3.6 in eleven dogs. Mean venous blood pressure was 11.4/4.2 mmHg ± 0.8/0.7 SEM and mean heart rate was 162 beats/min ± 22 SD.

When a KCl injection was used to euthanize the animal, the heart went into fibrillation immediately following administration and all systemic pressures dropped to zero. This was followed by cessation of all myocardial electrical activity. The data summary in Table IV illustrates the measured electrical property changes which occur immediately following KCl injection. The dielectric constant increased rapidly from a mean of 57.8 to 60.9, followed by a slow decrease to 52.3 over a 90-minute period postmortem. In a similar fashion, the conductivity rapidly increased from 20.3 mmho/cm to 21.1 mmho/cm upon KCl injection and slowly decreased to 17 mmho/cm.

The results of dielectric measurements performed with a concentrated aqueous $CaCl_2$ solution used for euthanasia are presented in Table V. The arterial blood pressure increased immediately upon this injection and then decreased to zero, all within 20 seconds after the injection. This drop of blood pressure to zero was more rapid than seen in the experiments using overdoses of pentobarbital, but less rapid than the rate observed following administration of KCl. In this respect, the results of experiments performed using KCl or $CaCl_2$ euthanasia represent the effects of an infarct upon cerebral blood flow. As indicated by the data in Table V,

TABLE V

Summary of Antemortem/Portmortem *In-Situ* Dielectric Studies with CaCl₂ Euthanasia
(N = 3)

Time after Death (min.)	Dielectric Constant ± SE	Conductivity ± SE (mmho/cm)	Brain Temperature (°C) ± SD	Colonic Temperature (°C) ± SD
−10	57.3 ± 2.7	25.0 ± 2.2	38.6 ± 2.5	37.2 ± 1.2
−.5	58.5 ± 2.3	25.6 ± 2.0	40.0 ± 1.3	37.9 ± 0.8
−.1	67.6 ± 2.2	31.5 ± 2.1	40.0 ± 1.3	37.9 ± 0.8
−.05	66.1 ± 0.1	31.3 ± 0.7	40.0 ± 1.3	37.9 ± 0.8
+.05	65.7 ± 0.3	31.1 ± 0.5	40.2 ± 0.9	37.7 ± 0.7
.1	57.9 ± 4.5	26.8 ± 2.4	38.7 ± 2.5	37.7 ± 0.7
.5	59.2 ± 2.9	26.5 ± 2.1	38.7 ± 2.5	37.7 ± 0.7
2	57.3 ± 2.7	25.2 ± 1.9	37.5 ± 2.1	37.7 ± 0.7
10	56.6 ± 2.9	24.1 ± 1.9	35.5 ± 3.5	37.4 ± 0.2
40	56.7 ± 2.1	23.1 ± 1.4	37.9 ± 0.4	37.1 ± 0.1
90	57.7 ± 0.7	23.3 ± 1.6	36.4 ± 1.7	36.3 ± 0.1

there was a rapid, very large increase in the relative dielectric constant and conductivity to 67.6 and 31.5, respectively, following injection. The brain expanded during this period, corresponding to the increase in blood pressure. Subsequent to the blood pressure increase, a rapid return to values similar to those prevailing before the injection was observed, following by a gradual decrease in both properties. In cases of CaCl₂ euthanasia, the values peak sooner and appear to change more than in the KCl cases.

While nominal values for the dielectric constant are approximately equal, the CaCl₂ euthanasia results in a change of approximately 17% in measured dielectric constant. This peak occurs 6–10 seconds before systemic pressures reach zero. KCl euthanasia on the other hand produces a total change of about 5%, and the peak occurs 6–10 seconds after systemic pressures reach zero. This indicates that the CaCl₂ injection causes a stronger and more abrupt change in blood flow to the brain, as would be expected from the different mechanisms by which the K^+ and Ca^{2+} ions effect myocardial depolarization. Note that the measured dielectric property changes were due to blood flow rather than the Ca^{2+} and K^+ ions *per se*, because any change in the systemic concentration of these ions resulting from their injection was too small to significantly affect the tissue's dielectric characteristics.

Following the postmortem measurement period, the craniotomy was enlarged and one cerebral hemisphere was excised. The tissue was thoroughly homogenized and a small sample measured under controlled temperature conditions. Data were measured at 1°C intervals from 26°C to 40°C at a frequency of 2450 MHz. The permittivity profile (from four dogs) is nearly flat over this temperature range; the mean relative dielectric constant is 43.3 ± 0.57 (mean ± SEM) and the mean conductivity is 15.2 ± 0.23 mmho/cm. These mean dielectric property values for the 26° − 40°C temperature range are very near their mean values at 37°C (K′ = 43.5; σ = 15.4). The above described *in-situ* and *in-vitro* measurements were performed using the probe dielectric measurement technique.

There exist little published experimental data with which

to compare the data obtained in the majority of the experiments just described. Using the probe technique developed in our laboratory, we have measured live rat brain in a location equivalent to the pial measurements described here [31]. At 2450 MHz, a dielectric constant of 54 and a conductivity of 20 mmho/cm were measured, both of which are in good agreement with the pial values shown in Table III. Measurements using a slotted line with homogenized brain tissue 2 to 24 hours after specimen collection have been made. The reported average dielectric constant was 30–35 and the average conductivity ranged from 15 to 20 mmho/cm with a large amount of scatter in the data. This average dielectric constant is similar to the value for antemortem deep brain and postmortem pia, but smaller than our data for homogenized brain. The average conductivity is comparable in antemortem pia and shallow brain and to our homogenized brain data, but is larger than antemortem deep brain and postmortem pia. The differences between those data [46] and the data measured in our laboratory could be attributed to the difference between antemortem and postmortem brain tissue or to the averaging effect of homogenizing the tissue (considering that the homogenized brain used in those measurements could have had a larger volume percentage of white matter than our homogenized brain tissue).

c. Correlation of Dielectric Property Changes with Radioactive Tracer Measurements of Renal Blood Flow

Experiments were performed in eight separate isolated kidneys in which dielectric measurements and radioactive tracer flow measurements were performed simultaneously. The objective of these experiments was to provide not only a comparison of each technique with the measured flow/ pressure data for a functioning autoregulating kidney, but also to provide direct comparison of dielectric property measurement results and independent flow rate measurements.

The kidneys were approached through a midline abdominal incision under pentobarbital anesthesia (30 mg/kg).

The renal blood vessels and ureter were isolated and the renal artery(ies) traced from the renal pelvis to its junction with the aorta. Ligatures were tied loosely around the vessels and the ureter was severed. The renal artery was then clamped, cannulated distal to the clamp, and cut between the clamp and the cannula. The renal arterial cannula was connected to an IV drip of heparinized saline which enabled initial flushing of the kidney immediately upon cannulation. Next, the renal vein was tied and cut distal to the tie, and the kidney placed on the external perfusion circuit illustrated in Figure 17. The period between the time at which arterial flow was first stopped and the time when perfusion on the external circuit was begun for each kidney was two to four minutes, with some renal flow always maintained by the IV drip. Similar surgical procedures were followed when isolating the other kidney.

The perfusion circuit included an in-line flowmeter and an on-line variable resistance located above the site of pressure measurement. The in-line variable resistance made possible the use of perfusion pressure as the forcing function. For each kidney, the total RBF was recorded as a function of perfusion pressure.

After a kidney was set up on the flow-controlled supply from the perfusion circuit, it was allowed to stabilize for 15–20 minutes while the network analyzer impedance measurement system and recorders used in conjunction with the dielectric measurement probe were calibrated. Once initial flow/pressure/weight information had been gathered, a small section of the renal capsule was separated and cut away and the dielectric measurement probe (Figure 2) was placed in contact with the renal cortex. The probe's contact pressure against the kidney was maintained relatively constant through use of a spring-loaded probe holder, which is diagrammed in Figure 18. Total renal flow rate was increased in sequential steps by increasing the speed of the perfusion pump, allowing both pressure and dielectric properties to reach a steady state condition before incrementing flow rate. Data were recorded for each of six different pressure-flow conditions. Following the cortex measurements, the kidney was punctured using an 18 ga. needle and the probe was in-

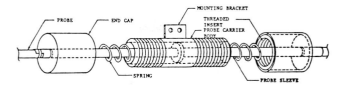

Fig. 18. Diagram of the spring-loaded probe holder used in conjunction with the dielectric measurement probe for the purpose of maintaining a uniform probe contact force.

serted into the medullary region. The measurement sites in the kidney are indicated in Figure 19. Care was taken to place the probe at the same measurement sites in each experiment. Data collection in medulla tissue was performed at the same flow rates used for cortex measurements. Two additional measurement sites (resulting in a total of four), one in the inner cortical region and one deep in the medulla, were included in the combined radioactive tracer/dielectric property renal flow studies.

In order to determine if the relationship between renal flow and measured dielectric properties which was observed *in-vitro* also existed *in-vivo*, experiments in six animals were also performed wherein the dielectric measurements were made *in-vivo*. The surgical procedure for these investigations was similar to that used in the *in-vitro* experiments.

The femoral vessels were cannulated and the left (and subsequently, the right) kidney and renal vessels located and isolated. Heparin was administered intravenously, 500

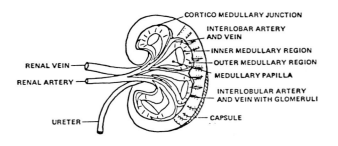

(a) Renal anatomy in longitudinal cross-section

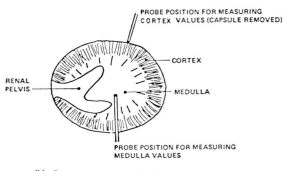

(b) Transverse cross-sectional view of dielectric property measurement sites

Fig. 19. Two anatomical views of kidney. Sites where dielectric property measurements were performed are shown on transverse view of kidney.

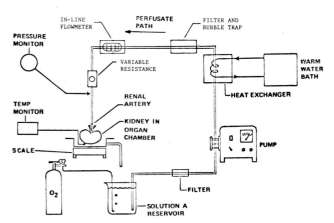

Fig. 17. Diagrammatic illustration of experimental setup used for perfusion during *in-vitro* renal dielectric measurements.

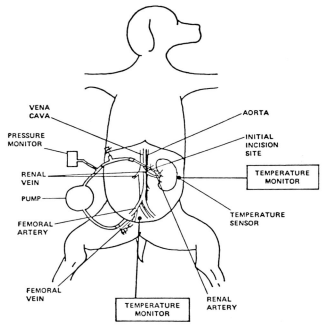

Fig. 20. Diagram of extracorporeal perfusion circuit used for control of renal blood flow during *in-situ* renal dielectric property measurements. Note site of initial surgical incision.

units/kg followed by 5 cc anti-coagulant citrate dextrose (ACD). The renal artery was ligated at the base of the aorta and cannulated and the cannula was connected to two branches of a Y-shaped tube, of which the third branch was clamped off. The left femoral arterial cannula was connected through a Holter pump which drove the extracorporeal perfusion circuit illustrated in Figure 20. The blood delivered by the circuit perfused the kidney through the renal artery.

The results of renal dielectric property vs flow rate experiments are shown in Figures 21–24. Relative dielectric constant and conductivity (averaged from four zones within the kidney) are plotted in Figures 21/22, as a function of pressure and as a function of flow in Figures 23/24. Note that the complex permittivity versus perfusion pressure results have the same sigmoidal relationship observed for flow versus pressure in the normal autoregulating kidney. If indeed the changes in complex permittivity versus pressure (expressed as percentages changes from values at baseline pressure) are indicative of flow changes within the kidney, then the complex permittivity changes plotted as a function of flow rate should exhibit a linear or nearly linear relationship. The correlation of total renal flow rate with renal dielectric properties from the cases shown in Figures 21 and 22 is presented in Figures 23 and 24. In each case, very good

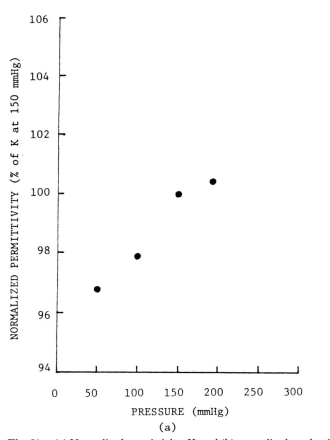

(a)

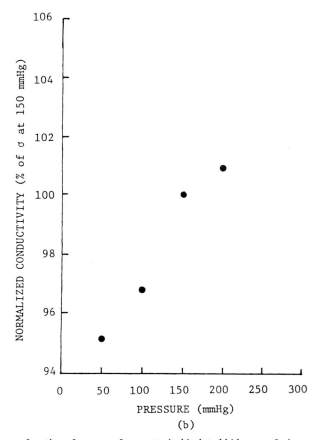

(b)

Fig. 21. (a) Normalized permittivity, K, and (b) normalized conductivity, σ, as a function of pressure for one typical isolated kidney perfusion experiment.

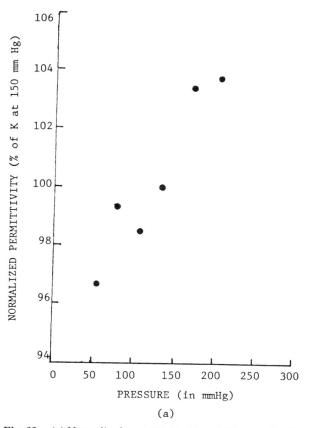

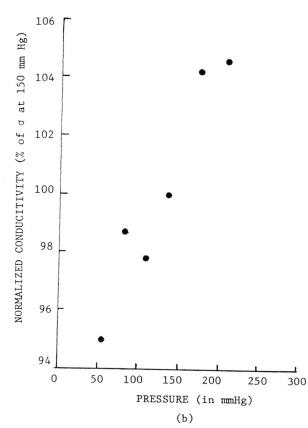

Fig. 22. (a) Normalized permittivity, K, and (b) normalized conductivity, σ, as a function of pressure for a second typical isolated kidney perfusion experiment.

correlation between changes in renal flow and dielectric properties was exhibited. In cases where flow was autoregulated, dielectric constant and conductivity changes corresponded directly with renal pressure-flow autoregulation.

d. Distribution of Renal Flow Measured via Dielectric Property Changes

Measurements of radioactive tracer uptake and dielectric properties were made simultaneously in eight kidneys. Effective Tc99m infusion period was 8 to 12 seconds, measured from first appearance on the gamma camera's scope monitor to initial clearance through the renal vein. Mean arterial blood pressure was 134 mm Hg with a standard error of 4.2.

The local perfusate flow in each of the four zones described in Section 7 (outer cortex, inner cortex, outer medulla, inner medulla) was determined from corresponding regional Tc99m time-activity curves (obtained using light pen techniques). These results were compared to relative zonal differences in measured dielectric properties measured using multiple implantable dielectric probes. The zonal fractions of total renal perfusate flow determined by Tc99m activity averaged for all eight experiments are shown in Table VI.

A direct comparison of perfusate flow measured in the same kidney and same zonal regions within each kidney

using the Tc99m activity curves and relative dielectric property changes was also performed. Table VII shows the zonal flow fractions determined using radioactive tracers and the relative differences in permittivity and conductivity for each of the zones. As evident from Tables VII and VIII, the average zonal flow fractions between zones obtained with Tc99m were quite similar to relative percentage differences in dielectric property fractions between zones. Zonal conductivity fractions corresponded more closely with Tc99m results than did the permittivity zonal fractions. Fractional

TABLE VI

Percentage of Total Renal Perfusate Flow for Each Zone Determined from Tc 99m Activity[1]
(n = 8 kidneys)

Zone[2]	Fractional Flow (%)
	(Mean ± S.D.)
Outer Cortex (OC)	61.4 ± 5.1
Inner Cortex (IC)	28.5 ± 2.6
Outer Medulla (OM)	7.3 ± 2.4
Inner Medulla (IM)	1.2 ± 0.5

[1] Fractional Flow = Zonal Flow (ml/g − min) × $\dfrac{\text{Zonal Volume}}{\text{Total Kidney Volume}}$

[2] Zonal volumes expressed as percentages of total kidney volume are OC = 41, IC = 29, OM = 20, IM = 10.

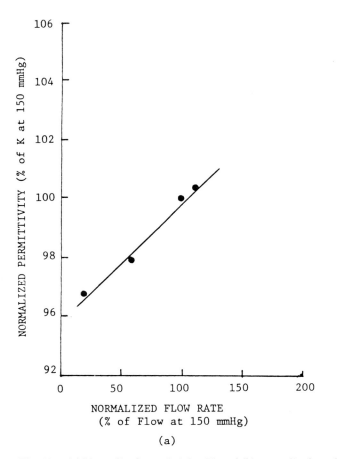
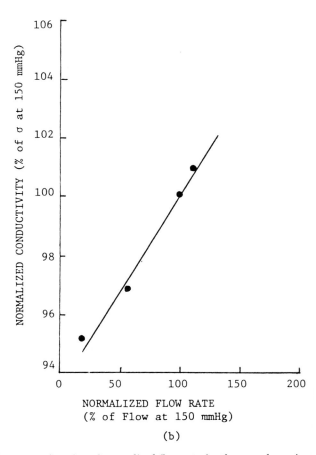

(a)   (b)

Fig. 23. (a) Normalized permittivity, K, and (b) normalized conductivity, σ, as a function of normalized flow rate for the case shown in Figure 3.

changes measured between inner cortex and outer medulla were slightly greater for K and σ (3.07 and 3.04, respectively) than for Tc99m (2.92). However, differences between outer medulla and inner medulla were greater for Tc99m (5.64) than for K (5.00), but were smaller than measured changes in σ (6.50). Percentage differences between the two methods were generally small, averaging 6.5 ± 3.7% (mean ± S.D.).

TABLE VII

Zonal Fraction of Total Renal Perfusate Flow Determined from Tc 99m Activity and Compared with Percentage Changes in Relative Permittivity and Electrical Conductivity at 2450 MHz. Mean ± S.D. (n = 8 Kidneys)

Zone	Tc 99m Fractional Flow (%)	* Change in Permittivity, K (%)	* Change in Conductivity, σ(%)
Outer Cortex	61.4 ± 5.1	26.1 ± 3.4	38.9 ± 3.6
Inner Cortex	28.6 ± 2.6	12.2 ± 2.1	18.2 ± 2.9
Outer Medulla	7.3 ± 2.4	3.0 ± 0.5	4.5 ± 0.4
Inner Medulla	1.1 ± 0.5	0.5 ± 0.2	0.6 ± 0.2

* Percentage of the sum total of the changes in relative permittivity and conductivity from baseline values measured prior to beginning perfusion.

e. Effects of Nerve Stimulation and Pharmacological Agents on Renal Dielectric Properties and Flow

A series of experiments designed to examine sympathetic nerve stimulation effects on renal flow were conducted. The decentralized renal nerve of dog kidneys was stimulated at frequencies of 2, 6, and 12 Hz while monitoring changes in total renal flow, glomerular filtration rate (GFR), and renal dielectric properties. The responses in three kidneys perfused at a constant renal perfusion pressure (RPP) of 150

TABLE VIII

Fractional Differences in Renal Flow and Dielectric Properties Between Zones

Zones	Fractional Changes				
	Tc 99m	I-Ap*	THO*	K	σ
OC-IC	1.146	1.131	0.876	1.139	1.137
IC-OM	2.92	3.13	2.97	3.07	3.04
OM-IM	5.64	16.25	24.66	5.00	6.50
Cortex-Medulla	9.71	11.23	9.98	9.94	10.19

* I-Ap and THO data from Clausen, et al. (1979).

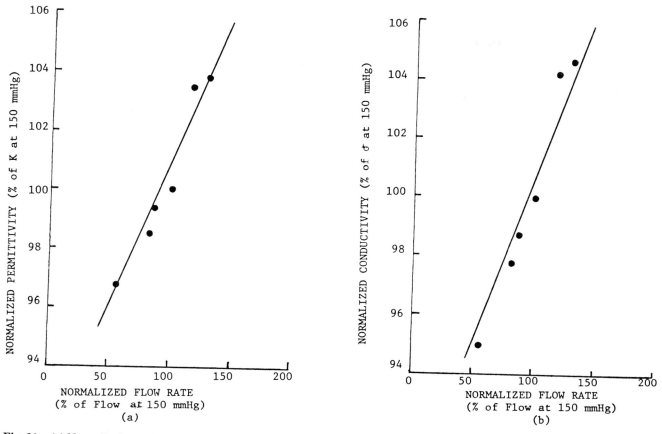

Fig. 24.   (a) Normalized permittivity, K and (b) normalized conductivity, σ, the normalized flow rate for the case shown in Figure 4.

± 9 mmHg are presented in Figure 25. Control values of all parameters prior to and following stimulation are also indicated. No significant changes in renal hemodynamics were observed at the 2 Hz stimulus frequency. Only minor perfusion circuit resistance changes were needed to maintain RPP at or near 150 mmHg. At stimulus frequencies of 6 and 12 Hz, significant responses in all parameters measured were observed and maintenance of RPP near control values required that the adjustable in-line resistance be increased. Using this approach, it was possible to hold the perfusion pressure relatively constant near 150 mm Hg. Increasing the stimulus frequency produced increased renal vasoconstriction with correspondingly larger increases in resistance being required to maintain constant RPP. At a stimulus frequency of 12 Hz, RPP was reduced to 52% of control values and changes in relative dielectric constant and conductivity were 3.7% and 4.1% respectively. These results are shown in Figure 26. The renal vascular response began with 1–2 seconds after onset of stimulation and was maximum 40–50 seconds after beginning stimulation. Stimulation was usually halted after 90 seconds and renal flow returned to control values within 60–90 seconds. GFR decreased in parallel to renal flow at the higher stimulus frequencies. At 12 Hz, GFR as reduced to approximately 20% of control values. Measured changes in relative dielectric constant and conductivity closely followed

the changes in RPF, returning to initial values shortly after cessation of stimulation. It is readily observed from the results shown in Figures 25 and 26 that the renal vascular response to graded renal sympathetic nerve stimulation and the renal dielectric response correspond closely with each other. Although the dielectric property changes are not as great as the renal vascular changes, they are great enough to be readily measured and to establish significance. An ability to use such information to follow/measure physiological changes is very important in establishing electromagnetic diagnostic methods.

Alpha adrenergic blockade with phentolamine in three kidneys produced renal vasodilation and a sudden decrease in RPP. Perfusion pressure was stabilized by decreasing the in-line variable resistance in the perfusion circuit and increasing pump speed. Figures 27 and 28 show the renal vascular antagonism with phentolamine. RPP was maintained at 150 ± 5 mmHg and renal flow, GFR, and dielectric properties were recorded. Renal flow increased 25% from baseline control values with the infusion of: 35 mg (0.05 mg/100 cc) of phentolamine into the renal perfusion circuit. GFR increased an average of 30% within 10 minutes thereafter. Increases in both relative dielectric constant and conductivity were measured which were proportional to the measured vascular changes. The relative dielectric constant

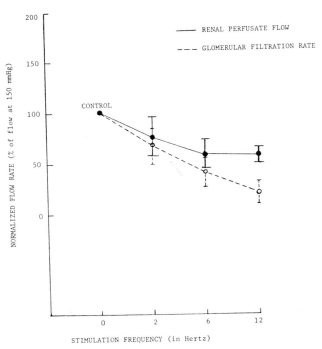

Fig. 25. Normalized flow rate as a function of stimulation frequency (5V stimulation intensity) for three kidneys.

increased approximately 4% and the conductivity increased approximately 6% from control values following stabilization of vascular parameters. These results again indicate that the effects of vasoactive drugs can be examined from measured dielectric property changes using the probe technique.

8.3 *In-Situ* Permittivity Measurements of Neoplastic Tissues

Additional work has been performed in our laboratory involving *in-situ* tissue permittivity measurements for dosimetry determination relating to the development of EM-induced hyperthermia as a method for treating cancer [47]. Tumor dielectric properties in seven different tumor tissues in mice and properties of surface lesions in humans have been studied with the intent of not only taking advantage of thermal regulatory differences in tumor and in normal tissue but also taking advantage of possible dielectric differences causing differential absorption of EM energy between normal and malignant tissues. If certain frequency ranges could be determined to be more useful for differential hyperthermia than others for specific tumor types and geometries, this would provide essential information for the selection of an appropriate frequency or frequencies for a particular tumor type which produces maximum differential

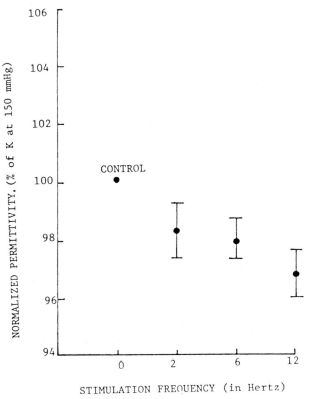

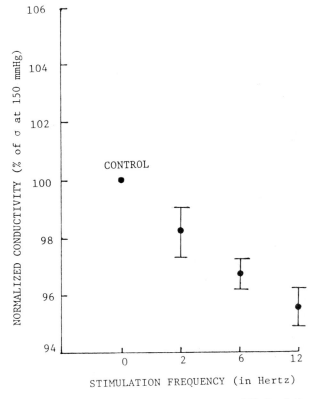

Fig. 26. (a) Normalized permittivity, K, and (b) normalized conductivity, σ, as a function of renal nerve stimulation frequency (5V stimulation intensity; 5 ms duration).

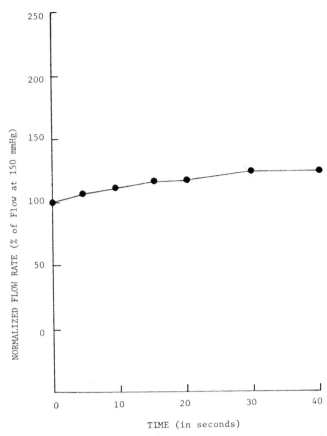

Fig. 27. Normalized flow rate vs. time following the injection of regitine (pressure is kept constant).

heating of the tumor with respect to normal tissues. *In-situ* dielectric properties of human and animal tumors and normal tissues have been measured over a frequency range of 10 MHz − 40 GHz. The results indicate that substantial differences exist between the dielectric properties of the measured tumors (C3HBA, Mendecki, and MA16/C mammary adenocarcinoma, B16 melanoma, Lewis lung carcinoma, glioblastoma, and ependymoblastoma) and the normal tissues which would typically surround the tumors. Dielectric differences between normal animal tissues and tumors of up to a factor of three are exhibited for some tumor types. Similar differences were also seen in measurements of human tumors. *In-situ* measurements of breast carcinoma and normal skin and breast tissue indicate that over the 0.5 to 4.0 GHz frequency range maximum differential power absorption (30% higher in tumor) occurs between 1.0 and 2.0 GHz. These dielectric measurements results for human breast carcinomas are similar to those obtained for mammary adenocarcinomas in experimental animals [47]. Significant differences (35%) in the dielectric characteristics of melanoma and normal skin were also measured, with melanoma having the lower dielectric constant and conductivity.

The above information coupled with considerations of

tumor location, size and geometry is necessary in the design of effective clinical, EM-induced differential hyperthermia procedures for the treatment of malignant tumors. In addition, tissue dielectric properties could be useful in investigations of normal and pathological tissue processes such as malignancy, fibrosis, and edema. Once changes in tissue dielectric properties are correlated to a specific pathological process, then this information could be used as a diagnostic aid.

9. SUMMARY AND DISCUSSION

A brief accounting of much of existing dielectric property data in the literature has been presented and *in-situ* permittivity measurements work has been described. Different methods investigated by researchers for performing *in-situ* permittivity measurements were discussed. These methods included efforts using a monopole antenna probe for the 10–100 MHz frequency range [23,24], modification of the same method for performing tumor dielectric measurements over the 3–100 MHz frequency range [25], development of a "four-electrode" method for tissue conductivity measurement at low frequencies [28], investigation of an "open-line resonator" technique for permittivity measurements in the 1–4 GHz frequency range [26], and an *in-situ* measurement probe technique with automated data acquisition/data processing capability for permittivity measurements over the 1 MHz to 10 GHz frequency range [30], [32], [40]. The most significant advantages of the "*in-vivo* probe" technique [30] over any heretofore available dielectric property measurement techniques are (1) the ability to perform measurements in-situ for a wide range of sample volumes, (2) the ability to obtain continuous electrical property data over a wide range of frequencies (1 MHz − 10 GHz), (3) the capability to perform measurements very rapidly, and (4) a very simple and flexible measurement procedure with respect to many other techniques. *In-vivo* measurements of muscle, kidney, brain, fat, and tumor data have been measured in laboratory animals [30]–[32], [47] and data for skin and several types of surface lesions were measured in humans [47]. Effects of changing physiological conditions on tissue dielectric properties, specifically dog brain and kidney [40], were examined. It is intended that this work [40,43] lead to the development of new diagnostic methods for use in medical applications and dosimetric analysis.

The research results demonstrate the existence of a direct relationship between tissue dielectric properties and blood flow (perfusion). Thus, measurement of tissue dielectric property changes using the probe technique has developed as a means for examining local/regional blood flow. The new method has potential advantages over conventional flow methods including radioactive microspheres, ultrasonic flow measurement, electromagnetic flowmeter, indicator-dilution methods, and direct collection. The principal advantages of the new method are (1) it permits characterization of blood flow on a dynamic basis and (2) it can be used to spatially

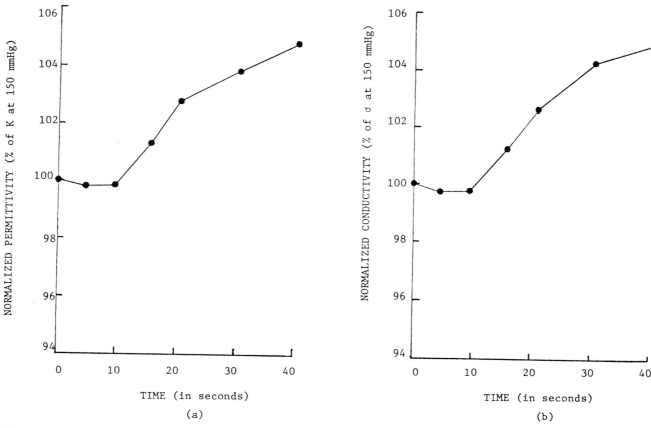

Fig. 28. (a) Normalized permittivity, K, and (b) normalized conductivity, σ, versus time following the injection of regitine (pressure kept constant).

"map" regional flow distribution. Finally, it should be pointed out that this new method for flow/perfusion measurement based on electromagnetic and physical principles has the potential for being developed into a totally non-invasive method for imaging regional flow differences within organs or relative flow differences between organs on a real-time basis.

Accurate dielectric property information that reflects *in-vivo* characteristics is necessary for successful application of electromagnetic diagnostic and treatment techniques in medicine and for accurately determining RF or microwave power absorption in tissues. The capability for accurate *in-situ* measurements of tissue permittivity provided by minimally-perturbing probe measurement systems could readily be used in dosimetry determinations for aiding in the establishment of a radiation level with respect to personnel safety and in treatment planning for cancer patients using hyperthermia induced by RF or microwave fields. Further, this *in-situ* measurement technique represents a potentially useful diagnostic tool for measuring changes in certain physiological processes, for differentiating between normal and diseased tissue, or for elucidating pharmacologically induced physiological effects.

Finally, it is important to note that tissue dielectric properties are of key importance to microwave imaging ap-

plications, both dosimetric and diagnostic in nature. The measurement of tissue permittivity *in-situ* (and an ability to relate physiological or pathophysiological alterations to changes in the dielectric properties of living tissue) is also effectively a measure of the ability of microwave imaging methods to discern the existence of pathophysiological conditions in the intact tissue, organ, or organism. A few applications of *in-situ* permittivity information would be in dosimetry determination (both intensity and distribution), diagnosis and management of lung disorders [48], measurement of cardiac performance, [49], [50] detection of arterial disease [51], [52], and diseased tissue diagnosis (such as cancer) [53]. Recently, significant strides in microwave imaging have been reported by Larsen and Jacobi [54]–[56] and Jacobi and Larsen [57]–[59]. In their investigations, reported elsewhere in this volume, good-quality images through whole kidneys were produced. These studies demonstrated the feasibility and future potential of electromagnetic imaging methods for medical applications. Microwave imaging is certainly worthy of adequate resource allocation for further development [60], and of crucial importance to such development is a better knowledge (and data base) of *in-situ* dielectric properties and their relationship to physiological and pathophysiological processes.

References

1. A. R. Von Hippel, *Dielectric Materials and Applications*, M.I.T. Press, 1954, pp. 47–122, 301–425.
2. H. P. Schwan, "Electrical properties of tissue and cell suspensions," *Adv. Biol. Med. Phys.*, Vol. 5, pp. 147–209, 1957.
3. P. Debye, *Polar Molecules*. New York: Chemical Catalog, 1929.
4. C. J. F. Boettcher, *Theory of Electric Polarization*. Houston, TX: Elsevier, 1952.
5. K. S. Cole and R. H. Cole, "Dispersion and absorption of dielectrics. I. Alternating current characteristics," *J. Chem. Phys.*, Vol. 9, pp. 341–351, 1941.
6. A. R. Von Hippel, *Dielectrics and Waves*, M.I.T. Press, 1954, pp. 93–104, 174–181.
7. B. Rajewsky, *Ultrakurzwellen, Ergebnisse der biophysikalischen Forschung, Bd. I*. Leipzig, Germany: Georg Thieme, 1938.
8. K. Osswald, *Hochfrequenz, Elektroakustik*, Vol. 49, pp. 40–50, 1937.
9. J. F. Herrick, D. G. Jelatis, and G. M. Lee, "Dielectric properties of tissues important in microwave diathermy," *Fed. Proc.*, Vol. 9, p. 60, 1950.
10. E. H. Grant, S. E. Keefe, and S. Takashima, "The dielectric behavior of aqueous solutions of bovine serum albumin from radiowave to microwave frequencies," *J. Phys. Chem.*, Vol. 72, pp. 4373–4380, 1968.
11. H. Fricke, "The Maxwell-Wagner dispersion in a suspension of elipsoids," *J. Chem. Phys.* Vol. 57, pp. 934–937, 1953.
12. H. F. Cook, "The dielectric behavior of some types of human tissues at microwave frequencies," *Br. J. Appl. Phys.*, Vol. 2, pp. 295–296, Oct. 1951.
13. J. E. Roberts and H. F. Cook, "Microwave in medical and biological research," *Br. J. Appl. Phys.*, Vol. 3, pp. 33–40, Feb. 1952.
14. K. R. Foster, J. L. Schepps, and H. P. Schwan, "Microwave dielectric relaxation in muscle: A second look," *Biophys. J.*, Vol. 29, pp. 271–281, 1980.
15. L. E. Larsen, J. H. Jacobi, and A. K. Krey, "Preliminary observations with an electromagnetic method for the noninvasive analysis of cell suspension physiology and induced pathophysiology," *IEEE Trans. Microwave Theory Tech.*, Vol. MTT-26, pp. 581–595, 1978.
16. W. R. Tinga and S. O. Nelson, "Dielectric properties of materials for microwave processing—tabulated," *J. Microwave Power*, 8(1), pp. 29–61, 1973.
17. C. C. Johnson and A. W. Guy, "Nonionizing electromagnetic wave effects in biological materials and systems," *Proc. IEEE*, Vol. 60, No. 6, pp. 694–695, June 1972.
18. L. A. Geddes and L. E. Baker, "The specific resistance of biological material—a compendium of data for the biomedical engineer and physiologist," *Med. Biol. Eng.*, Vol. 5, pp. 271–293, 1967.
19. H. P. Schwan and K. R. Foster, "RF-field interactions with biological systems: electrical properties and biophysical mechanisms," *Proc. IEEE*, Vol. 68, No. 1, pp. 104–113, 1980.
20. H. P. Schwan and K. Li, "Capacity and conductivity of body tissues at ultrahigh frequencies," *Proc. IRE*, Vol. 41, pp. 1735–1740, Dec. 1953.
21. W. J. V. Osterhout, *Injury, Recovery and Death in Relation to Conductivity and Permeability*. Philadelphia, PA: Lippincott, 1922.
22. E. C. Burdette, J. Seals, and S. P. Auda, "Dielectric and resistive characteristics of sarcoma and normal muscle, liver, and brain in the dog and the rat," submitted to IEEE Trans. Eng. Med. Biol.
23. R. L. Magin and C. P. Burns, "Determination of biological tissue dielectric constant and resistivity from *in-vivo* impedance measurements," *Region Three Conference of the IEEE*, April, 1972.
24. J. C. Toler and J. Seals, "RF dielectric properties measurement system: human and animal data," Final Technical Report, Project A-1862, National Institute for Occupational Safety and Health, Contract No. 210-76-0136, July, 1977.
25. G. M. Hahn, "Radiofrequency, microwaves and ultrasound in the treatment of cancer: some heat transfer problems," Abstracts of *NYAC Conference on Thermal Characteristics of Tumors: Applications in Detection and Treatment*, New York, March 14–16, 1979.
26. Eiji Tanabe and William T. Joines, "A nondestructive method for measuring the complex permittivity of dielectric materials at microwave frequencies using an open transmission line resonator," *IEEE Trans. Instrum., Meas.*, Vol. IM-25, No. 3, pp. 222–226, Sep., 1976.
27. W. T. Jones, Eiji Tanabe, and Raymond U., "Determining the electrical properties of normal and malignant biological tissue in-vivo," *Proc. 1976 IEEE Southeastern Conference and Exhibit*, Clemson, S.C., 1976.
28. A. W. Guy, "High frequency electromagnetic fields in phantom models of man and measured electrical properties of tissue materials," Final Technical Report, U.S. Air Force School of Aerospace Medicine, Contract No. F41609-73-C-0002, July, 1974.
29. J. Edrich and P. C. Hardee, "Complex permittivity and penetration depth of muscle and fat tissues between 40 and 90 GHz," *IEEE Trans. Microwave Theory Tech.*, Vol. MTT-24, pp. 273–275, May, 1976.
30. E. C. Burdette, F. L. Cain, and J. Seals, "*In-vivo* probe measurement technique for determining dielectric properties at VHF through microwave frequencies," *IEEE Trans. Microwave Theory Tech.*, Vol. MTT 28, No. 4, pp. 414–423, 1980.
31. F. L. Cain, E. C. Burdette, and J. Seals, "In-vivo determination of energy absorption in biological tissue," Annual Technical Report, Project A-1755, U.S. Army Research Office Grant No. DAAG29-75-G-0182, July, 1977.
32. E. C. Burdette, F. L. Cain, and J. Seals, "In-vivo determination of energy absorption in biological tissue," Final Technical Report, Project A-1755, U.S. Army Research Office Grant No. DAAG29-75-G-0182, Jan. 1979.
33. G. A. Deschamps, "Impedance of antenna in a conducting medium," *IRE Transactions on Antennas and Propagation*, pp. 648–650, Sep. 1962.
34. C. T. Tai, "Characteristics of linear antenna elements," *ANTENNA ENGINEERING HANDBOOK*, Chap. 3, H. Jasik, Ed., McGraw-Hill, 1961, p. 2.
35. *Semi-Automated Measurements Using the 8410B Microwave Network Analyzer and the 9825A Desk-Top Computer*, Hewlett-Packard Application Note 221, March, 1971.
36. F. Buckley and A. A. Maryott, "Tables of dielectric dispersion data for pure liquids and dilute solutions," *Nat. Bur. Stand. Cir. 589*, Nov. 1958.
37. H. F. Cook, "A comparison of the dielectric behavior of pure water and human blood at microwave frequencies," *Br. J. Appl. Phys.*, Vol. 3, pp. 249–255, Aug. 1952.
38. J. B. Hasted, *Water: A Comprehensive Treatise (The Physics of Physical Chemistry of Water)*, Ed., F. Franks, Plenum Press, New York—London, Vol. 2, pp. 255–305, 1972.
39. H. A. Ecker, E. C. Burdette, F. L. Cain, and J. Seals, "*In-vivo* determination of energy absorption in biological tissue," Annual Technical Report, Project A-1755, U.S. Army Research Office Grant No. DAAG29-75-G-0182, July, 1976.
40. E. C. Burdette, R. L. Seaman, J. Seals, and F. L. Cain, "*In-vivo* techniques for measuring electrical properties of tissues," Annual Technical Report No. 1, Project A-2171, U.S. Army Medical Research and Development Command, Contract No. DAMD17-78-C-8044, July, 1979.
41. T. W. Athey, "Automated dielectric measurements with a small monopole impedance probe," *1979 USNC/URSI-Bioelectromagnetic Society Symposium Digest*, pp. 375, Seattle, June, 1979.

42. K. R. Foster, J. L. Schepps, R. D. Stoy, and H. P. Schwan, "Dielectric properties of brain tissue between 0.01 and 10 GHz," *Phys. Med. Biol.*, Vol. 24, No. 6, pp. 1177–1187, 1979.

43. E. C. Burdette, P. G. Griedesick, and A. K. Moser, *"In-vivo* technique for measuring electrical properties of tissues," Annual Technical Report No. 2 U.S. Army Medical Research and Development Command Contract No. DAMD17-78-C-8044, Sept. 1980.

44. O. Sakurada, C. Kennedy, J. Jehle, J. D. Brown, G. L. Corbin, and L. Sokoloff, "Measurement of local cerebral blood flow with iodo ^{14}C antipyrine," *Am. J. Physiol.* 234, pp. H59–H66, 1978.

45. W. A. Pulsinelli and T. E. Duffy, "Local cerebral glucose metabolism during controlled hypoxemia in rats," *Science* 204, pp. 626–6269, 1979.

46. J. C. Lin, "Microwave properties of fresh mammalian brain tissues at body temperature," *IEEE Trans. Biomed. Eng.*, BME-22, pp. 74–76, 1975.

47. E. C. Burdette, J. Seals, and R. L. Magin, "Dielectric properties of normal and neoplastic animal and human tissues determined from *in-vivo* probe measurements," submitted to *Cancer Research.*

48. C. Susskind, "Possible use of microwaves in the management of lung disease," *Proc. IEEE* (Letters), Vol. 61, p. 673, 1973.

49. J. C. Lin, "Cardiopulmonary interrogation," presented at Electromagnetic Dosimetric Imagery Symposium, Washington, D.C., May, 1980.

50. I. Yamaura, "Measurement of heart dynamics using microwaves—microwave stethoscope," *Inst. Electron. Commun. Eng. Japan*, Vol. TG-EMC-J78-15, pp. 9–14, 1978.

51. M. I. Skolnik, "Radar measurements, resolution, and imaging of potential interest for the dosimetric imaging of biological targets," presented at Electromagnetic Dosimetric Imagery Symposium, Washington, D.C., May, 1980.

52. S. S. Stuchly, et al., "Monitoring of arterial wall movement by microwave doppler radar," presented at the Symp. EM Fields in Biol. Sys., Ottawa, Ont. Canada, June 27–30, 1978.

53. A. H. Barrett and P. Meyers, "Microwave Thermography," presented at Electromagnetic Dosimetric Imagery Symposium, Washington, D.C., May, 1980.

54. L. E. Larsen and J. H. Jacobi, "Microwave interrogation of dielectric targets. Part I: By scattering parameters," *Med. Phys.*, Vol. 5, No. 6, pp. 500–508, 1978.

55. L. E. Larsen and J. H. Jacobi, "Microwave scattering parameter imagery of an isolated canine kidney," *Med. Phys.*, Vol. 6, No. 5, pp. 394–403, 1979.

56. L. E. Larsen and J. H. Jacobi, "The use of polarization diversity in microwave transmission imaging of isolated canine kidney," presented at Electromagnetic Dosimetric Imagery Symposium, Washington, D.C., May, 1980.

57. J. H. Jacobi and L. E. Larsen, "Microwave interrogation of dielectric targets. Part II: By microwave time delay spectroscopy," *Med. Phys.*, Vol. 5, No. 6, pp. 509–513, 1978.

58. J. H. Jacobi and L. E. Larsen, "Microwave time delay spectroscopic imagery of isolated canine kidney," *Med. Phys.*, Vol. 7, No. 1, pp. 1–7, 1980.

59. J. H. Jacobi and L. E. Larsen, "Linear FM pulse compression radar techniques applied to biological imaging," presented at Electromagnetic Dosimetric Imagery Symposium, Washington, D.C., May, 1980.

Basic Principles and Applications of Microwave Thermography

Alan H. Barrett and Philip C. Myers

The basic principles of radiative transfer in the microwave domain, where $h\nu \ll kT$, are presented. The application of microwave thermography to the detection of breast cancer is discussed and the results of clinical tests at 6 GHz are presented. Other applications of microwave thermography, including the addition of microwave-induced hyperthermia, are briefly discussed although these other applications have not been subjected to clinical evaluations.

1. INTRODUCTION

This paper will present the basic principles of microwave thermography, review prior work, and present the recent results of a clinical program evaluating this technique as a means of detecting breast cancer. As the name implies, microwave thermography is the microwave analog of infrared thermography, i.e., the measurement of the human body's thermal emission at centimeter or millimeter wavelengths. However, since the wavelengths of infrared and microwave radiation differ by a factor of ca. 10^3 to 10^4, and since the electromagnetic properties of human tissue are a function of wavelength, microwave and infrared thermography provide measures of very different thermal properties of the body. For example, the penetration depth of an electromagnetic wave into a lossy dielectric varies with wavelength as $\lambda^{+1/2}$. Therefore, microwave radiation penetrates some 30–100 times deeper than infrared radiation. This implies that microwave radiation generated internally within the body can escape from a depth 30–100 times deeper than infrared radiation. In principle, the difference between infrared and microwave thermography is one of wavelength only but this difference implies different observing techniques, different instrumentation and different interpretations of the observations.

2. BASIC PRINCIPLES

Since in microwave and infrared thermography the thermal radiation emitted by the body is detected, the power received by the measuring instrument must be related to the temperature of the emitter. The intensity of radiation $I(\nu)$ emitted by a body at temperature T is given by the well-known Planck Radiation Law given by

$$I(\nu) = \frac{2h\nu^3}{c^2}[e^{h\nu/kT} - 1]^{-1} \qquad (1)$$

where ν is the frequency of the emitted radiation, c is the velocity of propagation, h is Planck's constant, and k is Boltzmann's constant. The units of intensity are watts per square meter per Hertz per steradian. In the infrared domain Eq. (1) must be used as it stands to relate the intensity of the detected radiation to the temperature. In the microwave domain, however, the Rayleigh-Jeans approximation can be made because $h\nu \ll kT$: This allows Eq. (1) to be written as

$$I(\nu) = \frac{2kT\nu^2}{c^2} \qquad (2)$$

which follows from Eq. (1) simply by expanding the exponent in the denominator. Thus we see that in the microwave range the radiation is directly proportional to the temperature of the emitter. This simplifies the calibration of microwave radiometers. Equations (1) and (2) are valid only for a perfect emitter, the so-called black-body radiator. In actuality both equations must be multiplied by a numerical factor known as the emissivity, $e(\nu)$, a pure number varying between 0 and 1. Typical values of the emissivity at microwave frequencies would be approximately 0.5 and at infrared frequencies would be 0.9 or higher.

The propagation of radiation through a non-magnetic absorbing, dielectric can be described by the Equation of Radiation Transfer [1], [2]. This is a differential equation which takes into account the radiation emitted as well as absorbed by each volume element of the medium. The equation can be written as

$$\frac{d}{dl}\left(\frac{I(\nu)}{n^2}\right) = \frac{\eta_\nu}{n^2} - \frac{\kappa_\nu}{n^2}I(\nu) \qquad (3)$$

where η_ν is the volume emissivity in watts/meter^3, κ_ν is the absorption coefficient representing the fractional loss in intensity per unit length, and n is the index of refraction. The left-hand side of Eq. (3) represents the change in intensity in the path length dl, the first term on the right-hand side

Department of Physics and Research Laboratory of Electronics, Massachusetts Institute of Technology, Cambridge, MA 02139

represents the intensity emitted in the volume element located at dl, and the second term on the right-hand side represents the attenuation of the intensity entering the volume element at dl. Equation (3) can be cast in the form

$$\frac{1}{\kappa_\nu}\frac{d}{dl}\left(\frac{I(\nu)}{n^2}\right) - \frac{I(\nu)}{n^2} = \frac{\eta_\nu}{n^2\kappa_\nu} = J(\nu) \qquad (4)$$

where $J(\nu)$, known as the source function, must be specified in order to proceed further. In a medium radiating thermal emission, the source function is given by Eq. (1) and may be approximated in the microwave domain by Eq. (2). Also, the intensities may be expressed in terms of temperature by using Eq. (2). When one uses Eq. (2) to convert intensities to temperatures, it is customary to call that temperature the brightness temperature, T_B. This may be taken as the definition of brightness temperature. One may now express the equation of radiative transfer entirely in terms of temperature as

$$\frac{1}{\kappa_\nu}\left(\frac{dT_B}{dl}\right) - T_B = T(l) \qquad (5)$$

This equation may now be integrated directly to give

$$T_B(O) = T_B(L)e^{-\kappa_\nu L} + \int_0^L T(l)e^{-\kappa_\nu l}\kappa_\nu dl \qquad (6)$$

In this result $T_B(O)$ is the temperature which characterizes the intensity at $l = 0$, $T_B(L)$ is the temperature which characterizes the intensity entering the dielectric at depth L and is subsequently attenuated between L and O, and the final term is the self-emission of the medium between O and L where $T(l)$ is the physical temperature of that medium. If it is assumed that L is sufficiently large that the entire intensity represented by $T_B(L)$ is completely attenuated and never reaches $l = 0$ than Eq. (6) simplifies to

$$T_B(O) = \int_0^\infty T(l)e^{-\kappa_\nu l}\kappa_\nu dl \qquad (7)$$

This is the fundamental equation relating the brightness temperature of the microwave emission to the physical temperature of a lossy dielectric medium and is applicable only to a medium characterized by a single dielectric constant. It should be emphasized that $T_B(O)$ is the brightness temperature of the radiation *in the medium* at $l = 0$. The brightness temperature of the radiation that emerges from the medium, $T_B(l < 0)$, is $T_B(O)$ multiplied by the emissivity, thus allowing for radiation reflected back into the dielectric at the boundary at $l = 0$.

Equation (7) can be integrated directly once the temperature as a function of l is known. As an example, if one has a linear variation of temperature versus l as given by

$$T(l) = T_o + \left(\frac{\partial T}{\partial l}\right)l \qquad (8)$$

then one can integrate Eq. (7) directly to get

$$T_B(O) = T\left(\frac{1}{\kappa_\nu}\right) = T(d); \quad T_B(l < 0) = e(\nu)T(d) \qquad (9)$$

This equation shows that, in this case, the brightness temperature of the emerging radiation, aside from the emissivity factor, is the temperature at the penetration depth, $d = \kappa_\nu^{-1}$. This may be taken as the justification for the statement that one "sees" into a depth typical of the penetration depth in the dielectric.

The microwave power actually delivered to the radiometer must depend on the properties of the antenna. The expression given above for the brightness temperature may be regarded as the brightness temperature of each individual ray emitted by the medium. If $G(\theta,\psi)$, the gain of the antenna, is a function of the polar angles θ,ψ measured from the axis of the antenna, then the antenna temperature T_A is given by

$$T_A = \frac{e(\nu)}{4\pi}\int_{4\pi} G(\theta,\psi)T_B(\theta,\psi)d\Omega \qquad (10)$$

where $d\Omega$ is the element of solid angle. The actual power delivered to the radiometer is $kT_A \Delta\nu$ where $\Delta\nu$ is the bandwidth of the radiometer.

Equation (10) is very difficult to evaluate in general and is especially difficult to evaluate in the case of microwave thermography because the radiating medium is located in the near-field of the antenna aperture. An additional complication is the fact that the body must be considered multi-layered because of the inhomogeneity due to the layer of skin, muscles, blood vessels, etc. Evaluation of this equation can only be done by assuming a model of the body and using computer techniques to numerically evaluate the integral in Eq. (10). An example of this approach has been given by Guy [3] in which he evaluated the radiation patterns, for diathermy purposes, from an open-ended waveguide in contact with biological materials.

A crude approximation of Eq. (10) can be made by assuming that the near-field radiation pattern occupies a volume whose dimensions are the aperture size of the antenna d and a depth given by $2 d^2/\lambda$ where λ is the wavelength in the dielectric medium, and the far-field may be represented by a diverging cone whose angular dimension is given by d/λ. One should be cautioned, however, that this technique represents a considerable oversimplification and will certainly give erroneous results when one considers the multi-layered aspects of the dielectric.

The penetration depth, defined as the depth at which the power will be reduced by e^{-1}, may be readily calculated from standard electromagnetic theory once the dielectric constant and the conductivity have been determined as a function of frequency. Taking the values given by Schwan [4] or Johnson and Guy [5] as being representative of typical biological tissues, the penetration depths may be evaluated. The results are given in Fig. 1. It can be seen from these results that microwave thermography offers the potential of sensing subcutaneous temperatures to a depth of several centimeters within the human body, whereas infrared radiation, a frequency so high as not to be shown in Fig. 1, will only give a measure of the skin or surface temperature. Once again, caution must be exercised in using the values of Fig. 1 because the body is anything but homogeneous and is also multi-layered.

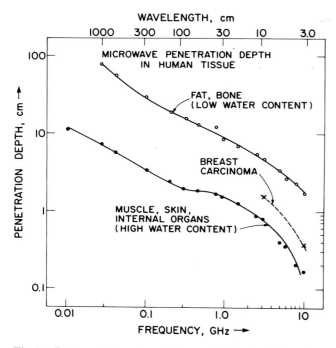

WAVELENGTH, cm

Fig. 1.   Representative values of the penetration depth in various types of human tissue as a function of frequency.

3. INSTRUMENTATION

Although the power radiated from the human body at 10 cm wavelength in a 1 MHz bandwidth is approximately 10^{-16} watts/cm^2, this power is readily detectable by using standard radioastronomy techniques. Many types of radiometers have been developed for this purpose and the interested reader is referred to the summary by Tiuri on receivers used in radioastronomy [6]. A typical receiver, used for microwave thermography, is shown in Fig. 2 [7]. The radiometer is a conventional Dicke-switched superheterodyne with a bandwidth of 100 MHz, an IF frequency of 60 MHz and operates at a frequency of 3.3 GHz. The radiometer input is alternately switched, at a rate of 10 Hz, between a load kept at a fixed temperature of 22°C and the antenna terminal. The resulting signal is synchronously demodulated and displayed on a strip chart recorder or supplied to a digital processor. Gain calibration is supplied by a solid state noise diode which provides a known signal at the radiometer input terminals.

As is apparent from the equations of Sec. 2, the detected microwave signal may be regarded as the product of the emissivity of the body and the temperature of interest. This implies that variations in the radiometer output as one moves the antenna to different positions on the body may

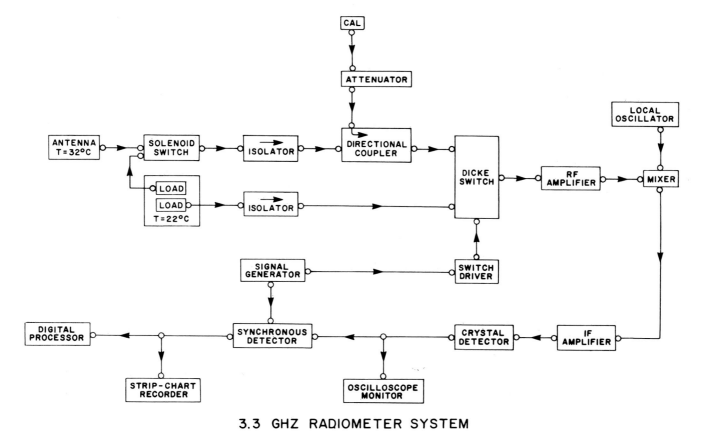

3.3 GHZ RADIOMETER SYSTEM

Fig. 2.   A block diagram of a typical Dicke-switched superheterodyne radiometer.

be due to variations in subcutaneous temperature or to variations in the emissivity. Radiometers designed to differentiate between these two effects have been developed by Ludeke et al. [8], [9] and by Mamouni et al. [10]. These devices, known as radiation balance microwave thermographs, feed a small amount of microwave power to the body and use the reflected power to correct the measurement of the body's emission for the emissivity. In this manner one is able to determine not only the microwave brightness temperature of the body but also the emissivity. There may be useful diagnostic information in the emissivity as well as the brightness temperature but this remains to be demonstrated.

One of the primary problems in microwave thermography is the coupling of the radiation from the body into the measuring instrument. Two techniques are in common usage:

(1) An open-ended waveguide may be placed in direct contact with the body and this serves as the antenna. The waveguide is generally filled with a dielectric in order that its physical size may be reduced from that required for propagation in an air-filled waveguide [7], [11], [12].

(2) An alternate technique, used by Edrich and co-workers, is to use a small parabolic reflector separated by some distance from the body and then to scan the body by moving the reflector [13], [14].

Both techniques have their own set of advantages and disadvantages over one another and both must contend with the impedance mismatch between the body and the radiometer input. Details are found in the original papers cited above.

4. APPLICATIONS OF MICROWAVE THERMOGRAPHY

Perhaps it is not surprising that the initial applications of microwave thermography lean heavily upon the results and applications of infrared thermography. One of the most widely tested applications of infrared thermography has been in the detection of breast cancer and it is in this area that microwave thermography also has been most widely tested. The MIT group has been taking data for five years at one or more frequencies at Faulkner Hospital in Boston [7], [11], [15]. The frequencies used have been 1.3, 3.3 and 6 GHz. In a typical microwave examination, a medical technician places the antenna at 9 points on each breast. The antenna is held in position for approximately 15 seconds at each position during which the radiometer integrates the received signal. At the end of the integration time the associated microprocessor converts the detected power to a temperature which is then displayed on a cathode ray tube. The antenna is then moved to the symmetrically opposite position on the other breast for another 15 second measurement. This process is continued until 18 data points are taken. The spacing between adjacent points is typically 3 cm and the antenna aperture dimensions are typically 1×2 cm.

A microwave thermographic examination requires approximately 10 minutes for each wavelength.

Since a microwave processor is a part of the instrumentation, the data is digitized so that the data analysis can be quantitative and not require the interpretation by a skilled reader. This also allows us to subject our data to statistical analyses in order to establish thresholds whereby a microwave examination would be regarded as positive or negative. Since there are 18 data points per patient, there are many ways in which the data can be analyzed. Many mathematical combinations of the measured temperatures have been tested in order to find the best discriminator between breast cancers, as determined by biopsy, and normal cases. Three quantities have been found to be important:

(1) The temperature difference between symmetrically opposite points on the right and left breast (9 such differences per patient).

(2) The average temperature of the right breast minus the average of the left breast.

(3) The temperature of the hottest position minus the average temperature of that same breast.

Items (1) and (2) reveal right-left asymmetries, whereas (3) may indicate a region of anomonously high temperature. The results of the computations (2) and (3) and the maximum difference in (1) are displayed on the CRT terminal at the conclusion of the examination, are printed out on a paper copy, and become a part of the patient's permanent record.

The 1.3 and 3.3 GHz results of the MIT group have been reported previously [7], [11], [15]. The results at 6 GHz are presented here for the first time. The data base consisted of 960 patients of which 35 had malignancies confirmed by biopsy. Each patient received microwave, infrared, and x-ray examinations. Unlike the microwave thermogram, the infrared and x-ray results must be interpreted by a trained radiologist. The infrared thermograms are graded 1 for normal or negative, 2 for suspicious, or 3 for abnormal or positive. The mammograms are graded 1 for negative, 2 for positive benign, 3 for suspicious, and 4 for positive malignant. These subjective grades leave little room for statistical analyses, but the microwave examinations with 18 digitized data points per patient may be subjected to many different types of statistical analyses.

Following the approach of Duda and Hart [16], Rosen [17] has performed a "three-dimensional" analysis of the microwave data. The three dimensions are the maximum temperature difference between symmetrically opposite points on the right and left breasts, designated m_1, the absolute value of the average temperature of the right breast minus the average temperature of the left breast, designated m_2, and the temperature of the hottest position minus the average temperature on the same breast, designated m_3. One may then write a linear equation relating these three measures as follows

$$a\,m_1 + b\,m_1 + c\,m_3 - t = M \qquad (11)$$

where the constants a, b, c, and t are to be determined to maximize the true negative detection rate for any given true

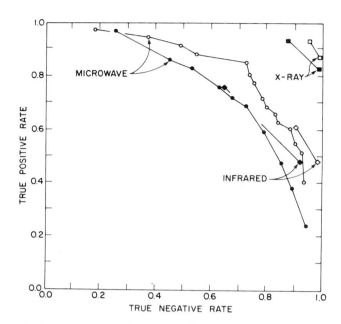

Fig. 3. Breast cancer detection rates for microwave and infrared thermography and mammography. The open symbols represent studies on a set of patients where the microwave frequency was 6 GHz. The filled symbols represent studies on a different set of patients where the microwave frequency was 1.3 GHz. Both sets of patients numbered approximately 1000 of which 30–35 had breast cancer as confirmed by biopsy.

position detection rate. If the value of M is positive, the result of the microwave examination is consisdered positive and if the value of M is negative the result of the microwave examination is considered negative.

The results of this analysis are shown in Fig. 3. The open symbols show the microwave, infrared, and x-ray results on the same group of patients, and the filled symbols show the similar results on an earlier series of patients where, in this case, the microwave results are at the frequency of 1.3 GHz. The two data points for the infrared detection statistics are those determined by using, in one case, only the infrared thermograms graded 3, and in the other case, using those graded 2 and 3. Likewise, the x-ray points are determined using those mammograms graded 4 in one case and 3 and 4 in the other case. As stated above, each point on the microwave curve is determined by choosing a true-positive detection rate and then choosing the coefficients in Eq. (11) to maximize the true negative detection rate.

The results shown in Fig. 3 seem to indicate that microwave thermography at 6 GHz is more effective in detecting breast cancers than thermography at 1.3 GHz. However, this conclusion may not be valid because the 6 GHz data were recorded on magnetic tape and the data could be analyzed on a dedicated microprocessor. Therefore, it is believed that the 6 GHz results shown in Fig. 3 are close to the optimum results which can be obtained, with the given data base of patients, at that frequency. The same is not necessarily true for the 1.3 GHz data.

Microwave thermography has been utilized in other studies at millimeter wavelengths but in no case has it been applied in a statistically significant number of cases to draw valid conclusions. In one study, carried out at 30 and 68 GHz on 14 women with known breast carcinomas, it was concluded that it was necessary to extend the study to lower frequencies [14]. This particular study was made with a parabolic antenna mechanically scanned across the patient. Millimeter wavelengths are particularly attractive for this mode of operation because one can obtain relatively high spatial resolution without having a physically large and cumbersome antenna. However, the penalty one must pay for using millimeter wavelengths in microwave thermography is apparent in Fig. 1. The penetration depth in tissue of high-water content, such as skin, is of the order of 1 mm, or less, at frequencies higher than 10 GHz, therefore it is not surprising that the desirability of using lower frequencies was apparent. A second result of this study was recognizing the need to examine a greater number of patients in order to establish statistically significant conclusions about the diagnostic efficacy of millimeter-wave thermography.

Since microwave thermography represents an entirely new method of determining localized body temperatures, in this case subcutaneous temperatures, it is not surprising that there are many other applications of microwave thermography. For example, it has been applied in a preliminary manner to studies of arthritic knee joints, abnormalities of the spine, and thyroidal and intracranial pathologies [13], [18].

Another promising application of microwave thermography is in conjunction with microwave-induced hyperthermia [12], [19]. In one study it is planned to do the microwave heating at a frequency of 1.6 GHz and do the microwave thermography at 4.7 GHz [12], and in the other experiments planned the heating will be done at 4.6 GHz and the thermography at 9.6 GHz [19]. These techniques have not been applied to human patients yet. An alternate way of conducting similar experiments would be to heat and receive at the same frequency but to time share the receiver and transmitter. This might be advantageous in that the antenna patterns in the two cases could be made essentially identical.

5. SUMMARY

Microwave thermography, a relatively new technique, has been applied over a frequency range from 1.6 GHz to 68 GHz. At the lower frequencies, the body's microwave radiation has been coupled to the receiver by using waveguide in direct contact with the patient; at the higher frequencies, it has been found convenient to use a lenses or parabolic reflector, not in contact with the body, and to scan this antenna across the body. The only definitive study of the diagnostic ability of microwave thermography has been in the detection of breast cancer at frequencies of 1.3, 3.3 and 6 GHz using the direct-contact type of antenna. These studies should be repeated at other frequencies, using other types of antennae, and extended to other diseases using a large number of patients so that meaningful statistics may be gathered. Multi-frequency observations should be encouraged for even though the depth resolution may be poor, they offer the possibility of determining subcutaneous temperature gradients, a quantity which may turn out to be of significant diagnostic value. Also, the diagnostic value of determining the emissivity should be established. Finally, one should consider various types of interferometric techniques to improve spatial resolution within the body.

6. ACKNOWLEDGMENT

The authors were supported in part by Grant 5 R01 GM20370-05 from the U.S. National Institute of General Medical Sciences.

References

1. S. Chandrasekhar, *Radiative Transfer*, Dover Publications, New York, 1963.
2. V. Kourganoff, *Basic Methods in Transfer Problems*, Dover Publications, New York, 1963.
3. A. W. Guy, "Electromagnetic Fields and Relative Heating Patterns Due to a Rectangular Aperture Source in Direct Contact with Bilayered Biological Tissue," *IEEE Trans. Microwave Theory Tech.*, Vol. MTT-19, pp. 214–223, Feb., 1977.
4. H. P. Schwan, "Interaction of Microwave and Radio Frequency Radiation with Biological Systems," *IEEE Trans. Microwave Theory Tech.*, Vol. MTT-19, pp. 146–152, Feb. 1971.
5. C. C. Johnson, and A. W. Guy, "Monionizing Electromagnetic Wave Effects in Biological Materials and Systems," *Proc. IEEE.* Vol. 60, pp. 692–718, June, 1972.
6. M. E. Tiuri, Chapter 7 of *Radio Astronomy* by J. D. Kraus, McGraw-Hill Book Co., New York, 1966.
7. A. H. Barrett, P. C. Myers, and N. L. Sadowsky, "Detection of Breast Cancer by Microwave Radiometry," *Radio Sci.* Vol. 12, No. 6(S), 167–171, Nov.–Dec., 1977.
8. K. M. Ludeke, B. Schiek, and J. Koehler, "Radiation Balance Microwave Thermograph for Industrial and Medical Applications," *Electron. Lett.*, Vol. 14, pp. 194–196, March 16, 1978.
9. K. M. Ludeke, J. Koehler and J. Kanzenbach, "A New Radiation Balance Microwave Thermograph for Simultaneous and Independent Temperature and Emissivity Measurements," *J. Microwave Power*, Vol. 14, plp. 117–122, June, 1979.
10. A. Mamouni, F. Bliot, Y. Leroy, and Y. Moschetto, "A Modified Radiometer for Temperature and Microwave Properties Measurements of Biological Substances," *Proc. 7th European Microwave Conference*, Microwave Exhib. & Publ. Ltd., Kent, pp. 703–708, 1977.
11. P. C. Myers, N. L. Sadowsky, and A. H. Barrett, "Microwave Thermography Principles, Methods, and Clinical Applications," *J. Microwave Power*, Vol. 14, plp. 105–115, June, 1979.
12. K. L. Carr, A. M. El-Mahdi, and J. Shaeffer, "Dual Mode Microwave System to Enhance Early Detection of Cancer," paper presented at IEEE Microwave Symposium, Washington, D.C., May 28–30, 1980.
13. J. Edrich, "Centimeter and Millimeter-Wave Thermography—A Survey on Tumor Detection," *J. Microwave Power*, Vol. 14, pp. 95–104, June, 1979.
14. M. Gautherie, J. Edrich, R. Zimmer, J. L. Guerguin-Kern, and J. Robert, "Millimeter-Wave Thermography Application to Breast Cancer—Preliminary Results," *J. Microwave Power*, Vol. 14, pp. 123–130, June, 1979.
15. A. H. Barrett, P. C. Myers, and N. L. Sadowsky, "Microwave Thermography in the Detection of Breast Cancer," *Am. J. Roentgenol.*, Vol. 134, pp. 365–368, Feb., 1980.
16. R. O. Duda, and P. E. Hart, *Pattern Classification and Scene Analysis*, John Wiley and Sons, New York, 1973.
17. B. R. Rosen, "Microwave Thermography for the Detection of Breast Cancer: A Discussion and Evaluation of a 6 GHz System," Master of Science Thesis, Physics Department, M.I.T., January, 1980.
18. J. Robert, J. Edrich, P. Thouvenot, M. Gautherie, and J. M. Escanye, "Millimeter-Wave Thermography: Preliminary Clinical Findings in Head and Neck Diseases," *J. Microwave Power*, Vol. 14, pp. 131–134, June, 1979.
19. D. D. N'Guyen, A. Mamouni, Y. Leroy, and E. Constant, "Simultaneous Microwave Local Heating and Microwave Thermography Possible Clinical Applications," *J. Microwave Power*, Vol. 14, pp. 135–138, June, 1979.

Microwave Propagation in Biological Dielectrics with Application to Cardiopulmonary Interrogation

James C. Lin

The use of microwave radiation in cardiopulmonary interrogation is described in this paper. The emphasis is on the basic tissue properties and fundamental propagation phenomena that govern microwave diagnostic procedures as well as the physical processes involved in the measurements. A number of experimental systems and representative results are included to highlight recent advances and to demonstrate the feasibility of using microwave radiation for interrogating the cardiovascular and respiratory systems.

1. INTRODUCTION

The need for noninvasive, transcutaneous methods of interrogating deep-lying body organs is becoming increasingly important. Current methods of detection and imaging range from plethysmography to computer assisted tomography. In fact, recent advances in the field of computerized x-ray and ultrasound tomography have opened a new dimension to the field of medical imaging. The success of these popular techniques for making tomographic images have prompted a number of researchers to explore other physical properties of biological materials for noninvasive tissue characterization. Electromagnetic techniques using radio and microwave radiation appear to possess some unique features that may allow them to become as useful as the current methods. Specifically, several diagnostic uses of microwaves have been developed in recent years for remote interrogation of cardiovascular and pulmonary functions [1]–[9].

Knowledge of the physiologic or pathophysiologic status of the heart as a pump for blood, and the lungs as the site of gas exchange are factors that can greatly assist physicians in the management of cardiopulmonary disease. In light of prevalence of mortality and morbidity associated with cardiopulmonary disease, considerable efforts have been devoted to the development of noninvasive diagnostic techniques which not only are safe but also offer the possibility of earlier detection as well as quantification of these disease states.

In the application of electromagnetic waves for cardiopulmonary interrogation, we are interested in the problems associated with energy transfer between a source and a receiver. This involves the analysis of coupling characteristics of electromagnetic waves into tissue media and the determination of the behaviors of waves that propagate from the transmitter to the receiver.

There are two basic electromagnetic methods which may be used for cardiopulmonary interrogation, depending on whether one is primarily interested in reflected or transmitted wave. In the case of reflection (back-scattering) measurement, the receiving antenna is placed by the transmitting antenna or the same antenna is employed for both transmitting and receiving functions. The backscattered waves provide information on the biological target and on the factors that govern the propagation to and from the biological target. In contrast, transmission measurement involves placing a separate receiving antenna beyond the biological target and usually along the axis of the transmitting antenna. In propagation through the biological target, the transmitted wave is affected by the same factors as in reflection measurement, and also the propagation phenomena from the target to the receiver. Thus, in order to extract diagnostic information from waves reflected by and transmitted through a biological target it is necessary to ascertain the characteristics of propagation mechanisms, the scattering properties of biological targets and the radiation behaviors of antennas or applicators.

This paper presents a discussion of these factors and gives a description of the characteristics of the principal biological targets of interest in cardiopulmonary interrogation. In addition, several experimental diagnostic procedures are reviewed to provide examples of the types of instrumentation and techniques that are useful in electromagnetic cardiopulmonary interrogation.

2. ORGANS IN THE THORAX

The biological targets of primary interest to cardiopulmonary interrogation are the heart and the lungs in the thoracic

Department of Electrical and Computer Engineering and Department of Physical Medicine and Rehabilitation, Wayne State University, Detroit, Michigan 48202. Presently with the Bioengineering Program, University of Illinois at Chicago Circle, Chicago, Illinois 60680.

cavity. Both of these organs exhibit characteristic dimensional and functional changes in health and in disease. They may, therefore be particularly amenable to electromagnetic interrogation. However, the cardiovascular and respiratory systems are extremely complex; a detailed description is beyond the scope of this paper. In this section we shall briefly describe some of the dynamics involved in cardiopulmonary functions.

2.1 The Heart

The heart is a muscular organ located in the chest cavity. In humans, it is divided longitudinally into left and right halves, each consisting of two chambers, an atrium and a ventricle. The chambers on each side of the heart communicate with each other, but the left chambers do not communicate directly with those on the right. Between the chambers in each half of the heart are valves which open to permit blood to flow from atrium to ventricle but not vice versa. There are also valves at the entrances of pulmonary artery and aorta which permit blood to flow into these arteries but close immediately preventing the reflux of blood in the opposite direction. In general, blood flows from superior or inferior venae cavae to right atrium, right ventricle, pulmonary arteries, left atrium, left ventricle and the aorta. This orderly flow of blood through the various parts of the heart is accomplished by the active contraction of the cardiac muscle, which is triggered by a coordinated process of depolarization of the muscle membrane. It is interesting to note that during contraction the heart shortens by approximately 75% and rotates by about 4° [10]. In addition, contraction of the heart muscle sets in motion the precordium overlying the apex. These movements correspond to physiological changes that occur during the cardiac cycle and may be correlated with abnormalities of cardiac contraction in patients with various types of heart disease.

2.2 The Lungs

There are two lungs, the left and right, each divided into several lobes. Together with the heart, great vessels and esophagus, the lungs completely fill the chest cavity. The lungs consist of air-containing tubes, blood vessels and connective tissues. The smallest tubes end in tiny sacs, the alveoli, which are the sites of gas exchange within the lungs. The lungs, however, lack muscle and are therefore passive elastic containers with no inherent ability to change their volume. Lung expansion is accomplished by action of the diaphragm (the muscle which separates the chest and abdominal cavities) and muscles which move the ribs.

The normal pulmonary capillary pressure is only 10–15 mm Hg. This in combination with osmotic gradients and alveolar surface tension allows the alveoli to remain dry, a feature essential for normal gas exchange. However, if the pulmonary capillary pressure increases greatly, as a result of pulmonary venous hypertension (e.g. with left ventricular failure), or as a result of direct increases in capillary permeability, fluid may accumulate in the interstitial spaces or in the alveoli. Pulmonary edema with intra-alveolar components prevents gas exchange across the alveolar wall.

3. ELECTROMAGNETIC WAVE PROPAGATION IN TISSUES

Electromagnetic waves propagate in a material medium including biological materials at the velocity

$$v = \frac{1}{\sqrt{\mu\epsilon}} \tag{1}$$

where μ and ϵ are permeability and permittivity of the medium, respectively. Their behaviors, however, depend on the frequency, polarization and configuration of the source, and on the electrical properties (dielectric constant and conductivity, in particular) and the geometrical parameters of the tissue structure.

3.1 Dielectric Constant and Conductivity

Biological materials appear as lossy dielectrics to electromagnetic radiation, and consequently, have magnetic permeability equal to that of free-space and independent of frequency. In contrast, electrical properties (permittivity)

TABLE I
Dielectric Constant and Conductivity of Biological Tissues at 37°C [11]–[16]

Frequency f (MHz)	Wavelength λ_0 (CM)	Saline ϵ_r	Saline σ	Saline λ_t	Blood ϵ_r	Blood σ	Blood λ_t	Muscle (Skin) ϵ_r	Muscle (Skin) σ	Muscle (Skin) λ_t	Lung ϵ_r	Lung σ	Lung λ_t	Fat (Bone) ϵ_r	Fat (Bone) σ	Fat (Bone) λ_t
433	69.3	70	1.72	7.5	62	1.2	8.2	53	1.43	8.5	36	0.72	10.8	5.6	.08	28.2
915	32.8	70	1.80	3.8	60	1.4	4.1	51	1.60	4.4	35	0.73	5.4	5.6	.10	13.7
2450	12.3	69	3.35	1.5	58	2.13	1.6	49	2.21	1.8	32*	1.32*	2.2	5.5	.16	5.2
5800	5.2	63	6.42	0.6	51	5.33	0.7	43	4.73	0.8	28*	4.07*	1.0	5.1	.26	2.3
10000	3.0	53	17.2	0.4	45	11.7	0.4	40	10.3	0.5	25*	9.08*	0.6	4.5	.44	1.4

λ_0 = Wavelength in air; ϵ_r = Relative dielectric constant; σ = Conductivity (S/M); λ_t = Wavelength in tissue (CM).
* extrapolated value.

of tissues are very frequency dependent in that, dielectric constants decrease and conductivities increase with increasing frequency. Furthermore, in most cases dielectric constants exhibit a small negative temperature coefficient and conductivity displays a slightly larger positive temperature coefficient. Typical values of measured dielectric constant and conductivity at 37°C are given in Table I for selected thoracic tissues along with the calculated wavelengths in air and in tissue [11]–[16]. The effect of physiologic influences upon dielectric properties of biosystems is presented elsewhere in this volume by Burdette *et al.*

Note that the dielectric constant and conductivity of muscle and other tissues with higher water content (blood, muscle, lung, etc.) are an order of magnitude higher than the corresponding values for fat or tissues with low water content. This difference yields a wavelength in tissue (λ_t) for higher water content materials about one-third of the wavelength in tissues with low water content. Furthermore, λ_t is nearly ten times smaller than λ_0 (wavelength in air) at a given frequency. These factors will help to improve the resolving power of electromagnetic waves in medical diagnosis. For example, the wavelength in air at 2450 MHz is 12.3 cm. Referring to Table I, we see that the wavelength is reduced to 1.8 cm in muscle. This will improve the spatial resolution of 2450 MHz radiation in muscle by a factor of seven.

3.2 Reflection and Transmission of Plane Waves at Planar Tissue Interfaces

The reflection and transmission of electromagnetic waves at boundaries separating different tissue media is an important element in most phenomena associated with cardiopulmonary interrogation. A basic understanding of the phenomenon can be obtained from a consideration of the reflection and transmission of plane waves at a planar surface.

The reflection and transmission of plane wave at a plane interface depend on the frequency, polarization, and angle of incidence of the wave, and on the dielectric constant and conductivity of the tissue. A wave of general polarization usually is decomposed into its orthogonal linearly polarized components whose electric or magnetic field is parallel to the interface. These components can be treated separately and combined afterward. The reflection coefficients for H and E polarizations are [17]

$$R_h = \frac{-\left(\frac{\epsilon_2}{\epsilon_1}\right)\cos\theta + \left(\frac{\epsilon_2}{\epsilon_1} - \sin^2\theta\right)^{1/2}}{\left(\frac{\epsilon_2}{\epsilon_1}\right)\cos\theta + \left(\frac{\epsilon_2}{\epsilon_1} - \sin^2\theta\right)^{1/2}} \qquad (2)$$

$$R_e = \frac{\cos\theta - \left(\frac{\epsilon_2}{\epsilon_1} - \sin^2\theta\right)^{1/2}}{\cos\theta + \left(\frac{\epsilon_2}{\epsilon_1} - \sin^2\theta\right)^{1/2}} \qquad (3)$$

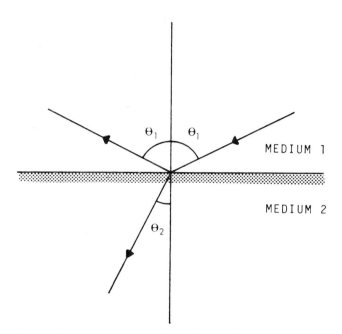

Fig. 1. A plane wave impinging at angle θ on a boundary separating two media.

and the transmission coefficients are given by

$$T_h = (1 + R_h)\frac{\cos\theta}{\left(1 - \frac{\epsilon_1}{\epsilon_2}\sin^2\theta\right)^{1/2}} \qquad (4)$$

$$T_e = 1 + R_e \qquad (5)$$

where θ is the angle of incidence (Fig. 1) and ϵ_1 and ϵ_2 are the complex permittivity of medium in front and the medium behind the interface, respectively. In particular, $\epsilon = \epsilon_0 (\epsilon_r - j\sigma/\omega\epsilon_0)$ with free-space permittivity ϵ_0 and radian frequency $\omega = 2\pi f$.

Figures 2 and 3 illustrate the magnitude and phase of the reflection coefficients of representative tissue interfaces at a temperature of 37°C for 2450 MHz. These figures clearly show the difference between E and H polarization. For E polarization, there is only a slight variation in magnitude and phase of the reflection coefficient with incidence angle. For H polarization, however, there is a pronounced dependence on incidence angle. The reflection coefficient reaches a minimum magnitude and has a phase angle of 90° at the Brewster angle. Thus, the H polarized wave is totally transmitted into the muscle medium at the Brewster angle.

For a plane wave impinging normally ($\theta = 0$), from a medium of complex permittivity ϵ_1, on a medium of complex permittivity ϵ_2, both Eqs. (2) and (3) reduce to

$$R = \frac{\sqrt{\epsilon_1} - \sqrt{\epsilon_2}}{\sqrt{\epsilon_1} + \sqrt{\epsilon_2}} \qquad (6)$$

Table II summarizes the magnitude of reflection coefficient for waves normally incident on the plane boundary separating various tissues in the thorax at five frequencies of most

REFLECTION COEFFICIENT

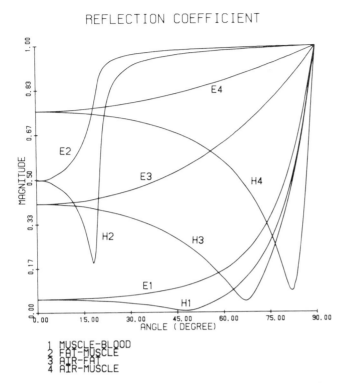

1 MUSCLE-BLOOD
2 FAT-MUSCLE
3 AIR-FAT
4 AIR-MUSCLE

Fig. 2. The magnitude of reflection coefficients of representative tissue interfaces in the thoracic cavity at a temperature of 37°C for 2450-MHz plane wave.

REFLECTION COEFFICIENT

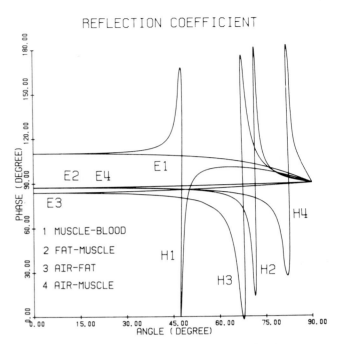

Fig. 3. The phase of reflection coefficients of representative tissue interfaces in the chest cavity at a temperature of 37°C and frequency of 2450-MHz.

interest to cardiopulmonary diagnosis. The fraction of normally incident power reflected by the discontinuity is obtained from R^2 and the transmitted fraction is related to $(1 - R^2)$. Clearly, the transmitted power at air-tissue interfaces is quite substantial at all frequencies and about one-half of the incident power is reflected at these boundaries. The reflection coefficients for tissue-tissue interfaces generally are smaller than air-tissue interfaces. The values range from a low of five for muscle-blood to a high of 60 for bone-biological fluid interfaces. This suggests that the closer the electrical properties across the interface, the higher the power transmission.

As the transmitted wave propagates in the tissue medium, energy is extracted from the wave and absorbed by the medium. This absorption will result in a progressive reduction of the wave's power density as it advances in the tissue. This reduction is quantified by the depth of penetration δ, which is the distance in which the power density decreases by a factor of e^{-2} [16]. Table III presents the calculated depth of penetration in selected thoracic tissues using the dielectric constants and conductivities provided in Table I. It is seen that δ is frequency dependent and takes on different values for different tissues. In particular, the penetration depth for fat and bone is nearly five times greater than for tissues with higher water content.

When there are several layers of different tissues, the reflection and transmission characteristics become more complicated. Multiple reflections can occur between the skin and subcutaneous tissue boundaries, with a resulting modification of the reflection and transmission coefficients [1], [13]. In general, the transmitted wave will combine with the reflected wave to form standing-waves in each layer. Figure 4 shows the distribution of electric field strength in a layer of muscle (heart) beneath layers of fat, muscle and bone for two frequencies. It is seen that in addition to frequency de-

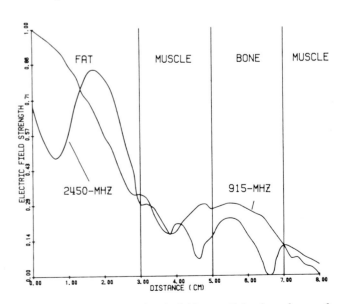

Fig. 4. Distributions of electric field strength in planar layers of fat, muscle, bone and muscle (heart) exposed to 915 and 2450-MHz plane waves.

TABLE II

Reflection Coefficient (Magnitude in Percentages) between Biological Tissue at 37°C.

	Frequency (MHz)	Air	Fat (Bone)	Lung	Muscle (Skin)	Blood	Saline
Air	433	0	46	76	82	81	83
	915	0	43	73	78	79	80
	2450	0	41	71	76	77	79
	5800	0	39	70	75	76	78
	10000	0	37	70	74	76	78
Fat (bone)	433		0	46	56	56	60
	915		0	43	52	54	57
	2450		0	42	50	53	57
	5800		0	42	50	53	56
	10000		0	45	52	54	58
Lung	433			0	14	13	19
	915			0	12	14	18
	2450			0	10	15	19
	5800			0	10	14	19
	10000			0	10	13	18
Muscle (skin)	433				0	4	6
	915				0	4	7
	2450				0	5	10
	5800				0	4	9
	10000				0	3	9
Blood	433					0	6
	915					0	4
	2450					0	5
	5800					0	5
	10000					0	6
Saline	433						0
	915						0
	2450						0
	5800						0
	10000						0

pendence, the electric fields exhibit considerable fluctuation within each tissue layer. While the standing-wave oscillations become bigger at 2450 MHz than 915 MHz, microwave energy at both frequencies can penetrate into more deeply situated tissues. This implies that at these frequencies sufficient energy may be transmitted and reflected to allow interrogation of the cardiopulmonary organs within the chest cavity.

3.3 Effect of Body Size and Curvature

Although depth of penetration, reflection and transmission characteristics in planar tissue structures provide considerable physical insight into coupling and distribution of microwave radiation, biological bodies generally are more complex in form and exhibit substantial curvature that can

TABLE III

Depth of Electromagnetic Wave's Penetration in Biological Tissues as a Function of Frequency

Frequency (MHz)	Tissue				
	Saline	Blood	Muscle (Skin)	Lung	Fat (Bone)
	Depth of penetration (cm)				
433	2.8	3.7	3.0	4.7	16.3
915	2.5	3.0	2.5	4.5	12.8
2450	1.3	1.9	1.7	2.3	7.9
5800	0.7	0.7	0.8	0.7	4.7
10000	0.2	0.3	0.3	0.3	2.5

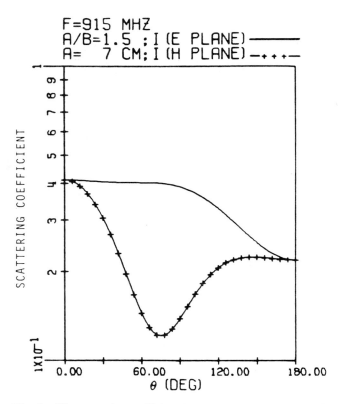

F=915 MHZ
A/B=1.5 ; I (E PLANE) ————
A= 7 CM; I (H PLANE) —+·+—

Fig. 5. The scattering coefficient for a 14cm long homogeneous prolate spheroidal lung model with a major-to-minor axis ratio of 1.5 at 915-MHz. The plane wave impinges along the major axis of the prolate spheroid.

The scattered energy varies widely with angle of observation, especially in the H-plane. The reason that the scattering coefficient for 915 MHz is smoother and fluctuates less than for 2450-MHz is because the 14-cm spheroid represents a smaller fraction of a wavelength at 915-MHz than at 2450-MHz. In general, the smaller the ratio of body size to wavelength the more uniform the distribution of scattering coefficient as a function of observation angle. It is significant to observe that the back-scattered microwave energy at 2450-MHz is less than one-tenth of the forward-scattered component, and the back-scattered microwaves at 915-MHz is on the same order of magnitude as the forward-scattered component. This suggests that while transmission measurement may be preferable at 2450 MHz, both transmission and reflection measurements should be equally applicable to 915-MHz. Furthermore, measurement in the E-plane usually would be more advantageous.

3.4 Effect of Target Motion

When electromagnetic wave is scattered from a biological target moving relative to a receiver, the received energy undergoes an apaprent frequency change, generally referred to as the Doppler shift. Using the scheme illustrated in Fig.

modify microwave transmission and reflection. For bodies with complex shape, the propagation characteristics depend critically on the polarization and orientation of the incident wave with respect to the body, and the ratio of body size to wavelength. These complications place severe limitations on transmission and reflection calculations for bodies of arbitrary shape and complex permittivity. We shall briefly describe some results that have been obtained from prolate spheroidal models of the lungs. This is chosen partly because the conclusion may serve as a pattern that can be used to estimate transmitted and reflected energy for other cases.

The scattering coefficient, defined as the ratio of scattered energy to incident energy, for a 14-cm long prolate homogeneous spheroid with a major-to-minor axis ratio of 1.5 is shown in Figs. 5 and 6 for 915- and 2450-MHz radiation, respectively [18]. The dielectric constant and conductivity for lung tissue given in Table I are used. In both cases the medium external to the lungs is assumed to be air and the plane wave impinges along the axial direction. It is seen that the scattering coefficients in the E-plane (plane parallel to electric field vector) and in the H-plane (plane parallel to magnetic field vector) usually differ from each other except for the forward- (0°) and back-scattered (180°) components.

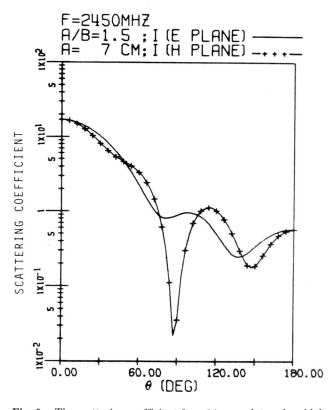

F=2450MHZ
A/B=1.5 ; I (E PLANE) ————
A= 7 CM; I (H PLANE) —+·+—

Fig. 6. The scattering coefficient for a 14cm prolate spheroidal model of the lungs at 2450-MHz. (See Fig. 5 for other details.)

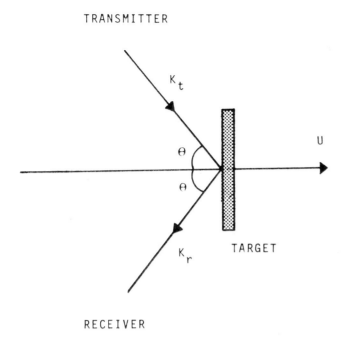

Fig. 7. Scattering of a plane wave by a moving biological target.

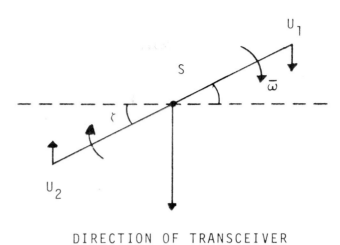

Fig. 8. Effect of target rotation on reflected microwave energy.

7 a relation can be obtained [19], [20] between the Doppler frequency change f_d and the target velocity u as

$$f_d = \frac{1}{2\pi}(k_r - k_t) \cdot u \qquad (7)$$

where k_r and k_t are propagation vectors associated with the receiver and transmitter, respectively. If the receiving and transmitting antennas are located in close proximity of each other or are the same, i.e., $k_r = -k_t$, Eq. (7) then reduces to

$$f_d = -\frac{1}{2\pi}k_r \cdot u = -2f\left(\frac{u}{v}\right)\cos\theta \qquad (8)$$

where f is the source frequency, v is the velocity of electromagnetic wave and θ is the angle between the target velocity vector and the direction of wave propagation. It is seen from Eq. (8) that f_d is directly proportional to the target velocity and takes on the largest value when $\theta = 0$ or 180° such that

$$f_d = \pm 2fu/v \qquad (9)$$

where the plus and minus signs account for movements toward and away from the transmitter, respectively.

In cardiopulmonary interrogation, various parts of the target fill all or an appreciable portion of the incident beam, the target velocity varies over the beam so that the Doppler component has a spectrum of frequencies. For example, during contraction the heart rotates anteriorly by about 4°. If we consider the situation depicted in Fig. 8 where a rotation imparts an angular velocity $\overline{\omega}$ of the target about its center of gravity, two fixed points on the target a distance

s apart will have a relative radial velocity toward the transmitter of

$$\Delta u = u_1 - u_2 = \overline{\omega}s \cos\zeta \qquad (10)$$

Hence from Eq. (9) the difference between the Doppler frequencies between these two points is

$$\Delta f_d = \frac{2f\Delta u}{v} = \frac{2\overline{\omega}s \cos\zeta}{v} \qquad (11)$$

Thus, the Doppler spectrum will be proportional to the angular velocity of the target and the gross aspect of the target. In a Doppler system the spectrum will be detected as frequency shifts relative to the transmitter frequency. This is usually accomplished through mixing the back-scattered wave with the transmitted wave and then measure the difference frequency by using a digital counter or by passing the demodulated signal through a set of bandpass filters.

If we multiply Eq. (9) by 2π and integrate over time, while neglecting the constant term, we obtain

$$\phi(t) = 2\pi \int f_d(t)dt = 4\pi fx(t)/v \qquad (12)$$

where $\phi(t)$ is the instantaneous phase variation corresponding to the distance x(t) traveled by the target. Thus, the Doppler phase shift is directly proportional to target displacement, while Doppler frequency is directly proportional to target velocity. If the demodulated waveform is fed through a low-pass filter, the output g(t) will be a signal that varies with time and target motion such that

$$g(t) = A \sin\phi(t) = A \sin[4\pi fx(t)/v] \qquad (13)$$

where A is the amplitude. A recording of g(t) may be made on a strip chart recorder or photographed from the screen of a video display unit. Since the instantaneous displacement of the target involved in cardiopulmonary interrogation is small compared with the wavelength in tissue (see Table I) for microwave frequencies of most interest, we obtain an approximate relation for g(t) as

$$g(t) = 4\pi Af x(t)/v = 4\pi Ax(t)/\lambda_t \qquad (14)$$

where we have used $f\lambda = v$. The displacement of the target is therefore directly proportional to the output of the low-pass filter as displayed on an oscilloscope or recorded on a strip chart. Doppler processing may also be considered from the perspective of the inverse synthetic aperture radar as discussed by Skolnik elsewhere in this volume.

Thus far the propagation phenomenon has been described in terms of plane waves impinging on parallel layers of tissues and simple surfaces isolated in free space. In many practical situations however interaction can occur in the near-field rather than the far-field where plane waves predominate. Furthermore, they involve applicators or antennas that are comparable in dimensions to the wavelength. This combination, along with the fact that most biological targets consist of complex surfaces of irregular shape makes the propagation characteristics too complicated to permit accurate description.

Because of the fundamental nature of plane wave interaction the understanding obtained serves a useful purpose when used with proper precautions, although to be used effectively they must be supplemented by further detailed information on the nature of the problem. For example, the data given in Table III shows that electromagnetic energy at 433-MHz can penetrate three times as deep into the tissues as energy at 5800-MHz. This implied advantage of lower frequencies is somewhat misleading in that as the frequency is decreased, the wavelength and the applicator size become increasingly large until it is no longer possible to direct the wave to a desired biological target with reasonable applicator dimensions. If the applicator is not increased in size as frequency is lowered, the radiated energy will rapidly diverge and be scattered by the target; only a small fraction of the available energy will penetrate the tissues.

4. EXPERIMENTAL TECHNIQUES

Several investigators have demonstrated successful detection of cardiovascular and respiratory signatures with microwave radiation [1]–[7]. These developments generally make use of amplitude and phase information contained in the transmitted and reflected waves and operate under continuous wave (CW) conditions. A simplified functional block diagram is shown in Fig. 9. The system consists of a microwave signal generator, a directional coupler, a pair of transmitting and receiving antennas, an amplitude or phase detector, a set of filters, a display unit and/or a loud speaker.

The microwave signal generator in the desired frequency range provides signal strength required for detection. The selection of microwave frequency depends on several factors. At higher frequencies where spatial resolution would be best, the penetration depth is very short (see Tables I–III) that detection of energy transmitted through and reflected from deep-lying tissue interfaces becomes impractical. As the frequency is lowered, the penetration depth increases. However, the antenna aperture needed to efficiently deliver the electromagnetic energy also increases. This along with

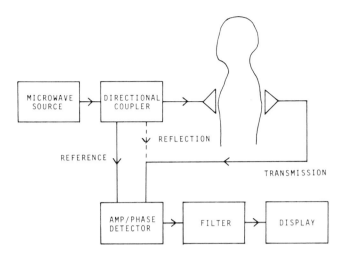

Fig. 9. A simplified block diagram of microwave instrumentation for measuring amplitude and phase variations in reflected and transmitted signals.

the increase in wavelength would degrade the spatial resolution of the system to the point where it becomes useless as a tissue interrogating device. The choice of operating frequency thus depends on a compromise between signal intensity that decreases and spatial resolution that increases with frequency [21]. Fortunately, wavelength contraction that naturally occurs in biological tissues, and by use of antennas loaded with high dielectric constant materials permit, microwave interrogation of deep-lying organs at frequencies between 1 and 6 GHz with manageable attenuation loss and practical spatial resolution in tissue media [8]. Similarly, experiences have shown that 10-GHz is a good frequency for sensing subcutaneous arterial wall motions [7].

The directional coupler diverts a small portion (30–50 db) of the forward power to serve as a continuous reference for the detector. It also samples the back-scattered signal for use in reflection measuring systems. The antennas couple the microwave energy into and from the biological targets. Microwave antennas developed for diagnostic applications utilize two basic schemes for coupling microwave energy into tissue media: non-contact and direct-contact methods. Non-contact antennas are normally spaced at distances of 1.0 to 5.0 cm from the subject and make use of the scattered field. Energy reaching a receiving antenna under these circumstances may follow multitudinous pathways in and around the body. This multipath propagation imposes a severe limitation on experiments involving transmission measurements [22]. Direct-contact methods that minimize scattered radiation are preferred over non-contact schemes. In contrast, back-scattered radiation is efficiently employed in some diagnostic applications associated with reflection measurement since multipath propagation contributes less significantly to back-scattered radiation. Hence both non-contact and direct-contact methods are useful.

The radiation pattern of an antenna determines the distribution of radiated energy. This pattern is the same for transmitting and receiving operations. Antennas developed

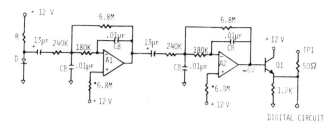

Fig. 10. The circuit of a practical amplifier and buffer for a CW microwave Doppler motion detector [20].

for biomedical applications usually have radiation patterns that give maximal intensity in the forward direction and the radiation is essentially confined to within a cylindrical column in the near-field. The finite beam-width permits electromagnetic energy to be directed to the biological target of interest and provides better spatial resolution. In the near-field, the electric and magnetic fields are out of time phase and the ratio of electric to magnetic field strengths varies from point to point, giving rise to a widely divergent wave impedance. Also, power density is not uniquely defined. However, it may be estimated from dividing the total power delivered by the antenna aperture. Using this relationship, the average radiated electromagnetic power density from present systems ranges from approximately 10^{-3} to one mW/cm^2. The antenna aperture ranges from 0.25 to 200 cm^2.

The reflected and transmitted signals from moving interfaces are amplitude as well as phase modulated. The Doppler shift of the received signal is detected by sensing the instantaneous phase difference between the received signal and the reference signal from the microwave generator. The output from the detector after passing through a low-pass filter or a band-pass filter having a low center frequency is a signal whose amplitude indicates the instantaneous displacement of the moving target and whose envelope variation corresponds to the Doppler frequency. In addition to visual display, most Doppler systems use an audio amplifier to raise the Doppler signal to suitable levels for listening.

The propagation characteristics through biological media with time-varying complex permittivity may also be investigated by the system shown in Fig. 9. This system performs amplitudes and phase measurements of microwave transmission and reflection coefficients which depict the relative attenuation loss and phase variation of microwave signals as it propagates through biological materials.

A practical CW microwave Doppler motion detecting device [20] is shown in Fig. 10. The circuit consists of a low power (3 mW) microwave circuit module (MCM) Doppler transceiver (GE C-2070M) which combines transmitting and receiving functions. The operating frequency is at 10.5 GHz. The reflected energy is mixed with the transmitted signal inside the bulk-effect diode of the MCM. The resultant Doppler signal is amplified by a pair of cascaded bandpass amplifiers A_1 and A_2, each having a voltage gain of 25 from 0.08 and 14 Hz. It is then buffered by the transistor Q_1 to

provide an analog output for display on an oscilloscope and recording on a strip chart. The output at transistor Q_1 may also be used for digital detection and counting. It should be noted that the upper frequency limit of the bandpass amplifiers was selected for testing the device with laboratory animals. For additional improvement in artifact rejection on human subjects, capacitors C_B may be increased to 0.02 μf, which enables the upper frequency limit to be lowered to 7 Hz.

5.   REPRESENTATIVE MEASUREMENTS

The ability to measure the displacement of a moving biological target and to detect its velocity, as well as to sense time-dependent permittivity change from outside the body opens up many fruitful areas of potential biomedical applications. Several research groups have used microwave energy for cardiovascular and respiratory measurements. Their results will be briefly discussed to indicate present efforts.

5.1   Ventricular Volume Change

The use of microwave reflection and transmission coefficients to measure volume changes in biological objects was first suggested nearly 20 years ago [23]. The method was adapted to assess the ventricular volume change in humans using 915MHz radiation [1]. The technique consisted of measuring transmission losses between a pair of 13 cm × 13 cm direct-contact antennas placed over the ventricles and on the back. Figure 11 shows the variation in transmission loss as the microwave beam propagates through the chest during several cardiac cycles. The comparison with a textbook trace of left ventricular volume change clearly demonstrates the feasibility of the technique. It is particularly interesting to note the correlation of details including the

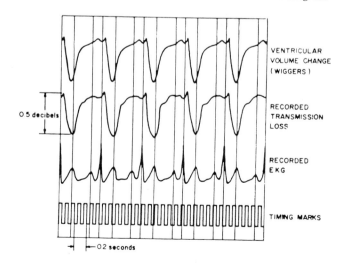

Fig. 11. Changes in ventricular volume and amplitude of transmitted 915-MHz microwave through the chest [1].

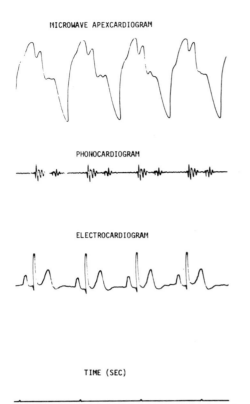

Fig. 12. Ventricular movement in a healthy young male as detected using microwave techniques [6].

small increase in the end-diastolic volume due to atrial contraction.

5.2   Ventricular Movement

Low-frequency displacements of the precordium overlying the apex of the heart is related to movements in the left ventricle and echoes the hemodynamic events within the left ventricle. A microwave method has been developed for recording these movements [6]. It involves a detecting phase variations in the signal reflected using a 10cm diameter antenna operating at 2450-MHz located over the apex of the heart. An example of microwave sensed ventricular movement in a healthy young male who held his breath throughout the measurement is shown in Fig. 12 along with simultaneously recorded electrocardiographic and phonocardiographic tracings. It is seen that toward the end of systole, a rapid rising wave occurs due to ventricular filling, which is completed by atrial contraction occurring between the P-wave and QRS-complex in the electrocardiogram. A rapid downward deflection represents maximal ventricular ejection following a period of isometric contraction, just after the QRS-complex. The ventricular movement reaches a plateau at the level of midsystole, and it is then followed by another downward deflection which coincides with the aortic

valve opening and completes the cardiac cycle. The method may therefore be used to delineate fine structures in ventricular movements.

5.3   Arterial Wall Motion

Microwave instruments have also been used to detect arterial wall movements [7]. Various information may be obtained, including frequency and regularity of the pulse, state and patency of the artery, and the characteristics of the arterial pressure pulse wave. The arterial wall motion was detected by measuring the Doppler shift between the 10.5 GHz microwave energy reflected by a peripheral artery and that transmitted through an open-ended dielectric-loaded rectangular waveguide antenna. Measurements were done by placing the antenna (10.7 × 4.3mm aperture) directly in contact with the skin over the peripheral artery of interest. Some typical experimental records are given in Fig. 13 for upper and lower extremities. Note that the amplitude of these signals is directly proportional to the arterial wall movement. Moreover, the characteristics closely resemble waveforms of arterial pulse pressure waves. Thus, microwave techniques are capable of detecting arterial wall motion and

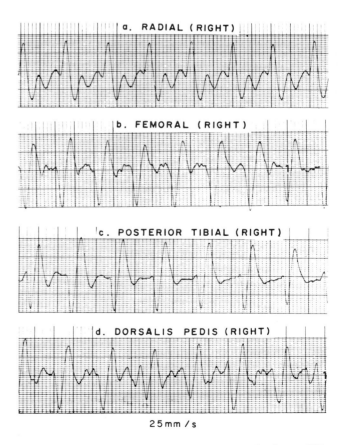

Fig. 13. Records of arterial wall motion measured using 10-GHz microwaves: radial artery in the arm; femoral, posterior tibial and dorsalis pedis arteries in the leg of a human subject [7].

could provide valuable information on the physical conditions of peripheral arteries.

5.4   Pulmonary Edema

The use of microwave techniques in the elderly detection and management of pulmonary abnormalities has been suggested by several investigators [3], [21], [23]. Many pulmonary abnormalities are associated with alterations in lung water content and/or water distribution.

Since electrical properties of biological tissues at microwave frequencies are directly related to their water content, measurement of transmission and reflection coefficients should provide direct information on the quantity of water present. One technique uses a pair of 915 MHz stripline antennas placed against the skin across the thoracic cavity and measures the amplitude and phase of the transmitted signal with respect to the source as a function of time [24]. The result obtained from artificially induced pulmonary edema in a dog is illustrated in Fig. 14 where the phase of the transmitted microwave and the pulmonary arterial pressure are plotted against time. Clearly, phase changes in the microwave signal paralleled the physiological indicator of pulmonary edema. It is therefore feasible to use microwave methods for noninvasive measures of changes in lung water.

5.5   Respiratory Movement

A microwave technique to remotely monitor the respiratory movement of humans and animals using continuous-wave system also has been reported in the literature [2], [20]. Low-power microwave energy at 10-GHz was directed by a standard gain horn toward the upper torso of the subject. The energy reflected from the chest is compared to the transmitted signal to give a measure of the respiratory movements. A record of microwave monitored respiratory activity of an anesthetized cat is shown in Fig. 15. It can be seen that the technique is capable of registering instanta-

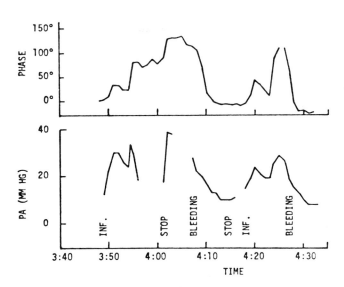

Fig. 14.   Experimental results showing correlation between the phase of transmitted 915-MHz signal and the pulmonary arterial pressure in a dog during induction of pulmonary edema [24].

neous changes in respiration including artifically induced apnea and hyperventilation.

6.   CONCLUSION

The use of microwave radiation in cardiopulmonary interrogation has been described in this paper. The emphases have been on the basic tissue properties and fundamental propagation phenomena that govern microwave diagnostic procedures as well as the physical processes involved in the measurements. A number of experimental systems and representative results have been included to high-light recent advances and to demonstrate the feasibility of using microwave radiation for interrogating the cardiovascular and respiratory systems.

It should be apparent from the microwave diagnostic

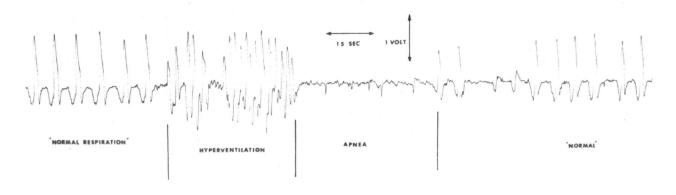

Fig. 15.   Microwave recorded respiratory activity of an anesthetized cat [20].

systems which have been discussed that these techniques are still experimental and the measurements were either qualitative or relative even though the signals were processed and recorded as analog waveforms. Thus, while microwave techniques are quite promising, a large amount of work is yet to be done to develop and to fully explore all the possibilities. As with all new techniques, many problems and questions must be resolved before the role of microwave cardiopulmonary interrogation in research and in clinical diagnosis can be defined. Nevertheless, the benefits that accrue to this approach including noninvasive examination and early detection of cardiopulmonary abnormalities justify the time and effort required for this development.

Some of the most important problems that remain include quantitative displacement and velocity, increased spatial resolution and correlation with physiological state and

function of the cardiopulmonary organs. The first of these might be accomplished through calibration procedures involving combinations of theoretical computation and laboratory evaluation. The last question can only be addressed by extensive laboratory testing.

A severe limitation of the CW microwave technique is its sensitivity to all motions in the microwave field. It is therefore impossible to determine the range to the moving biological target in a simple CW system. The range resolution problem could be overcome if the source is pulsed. Another technique which might provide the desired result involves multiple-frequency operation [25]. Also, lateral resolution is governed by the ratio of wavelength in tissue to the size of antenna aperture. The requirement for improved lateral resolution can therefore be satisfied by appropriate antenna design.

References

1. C. C. Johnson and A. W. Guy, "Nonionizing electromagnetic wave effects in biological materials and systems," *Proc. IEEE*, Vol. 60, pp. 692–718, 1972.
2. J. C. Lin, "Noninvasive microwave measurement of respiration," *Proc. IEEE*, Vol. 63, p. 1530, 1975.
3. P. C. Pedersen, C. C. Johnson, C. H. Durney, and D. G. Bragg, "An investigation of the use of microwave radiation for pulmonary diagnosis," *IEEE Trans. Biomed. Eng.*, Vol. 23, pp. 410–412.
4. D. G. Bragg, C. H. Durney, C. C. Johnson, and P. C. Pedersen, "Monitoring and diagnosis of pulmonary edema by microwaves: A preliminary report," *Invest. Radiol.*, Vol. 12, pp. 289–291, 1977.
5. D. W. Griffin, "Microwave interferometers for biological studies," *Microwave J.*, Vol. 21, pp. 69–72, 1978.
6. J. C. Lin, J. Kiernicki, M. Kiernicki, and P. B. Wollschlaeger, "Microwave Apexcardiography," *IEEE Trans. Microwave Theory Tech.*, Vol. 27, pp. 618–620, 1979.
7. S. S. Stuchly, M. Goldberg, A. Thansandote, and B. Carraro, "Monitoring of arterial wall movement by microwave Doppler radar," *Proc. 1978 Symp. Electromagnetic Fields in Biological Systems*, Ottawa, Canada, pp. 229–242, IMPI Ltd., Edmonton, 1979.
8. L. E. Larsen and J. H. Jacobi, "Microwave scattering parameter imagery of an isolated canine kidney," *Med. Phys.*, Vol. 6, pp. 394–403, 1979.
9. M. F. Iskander and C. H. Durney, "Electromagnetic Techniques for medical diagnosis: A review," *Proc. IEEE*, Vol. 68, pp. 126–132, 1980.
10. R. W. Brower and G. T. Meester, "Quantification of left ventricular function in patients with coronary artery disease," in *Quantitative Cardiovascular Studies*, N. H. C. Hwang, D. R. Gross and D. J. Patel, Eds., pp. 639–688, Univ. Park Pr., Baltimore, 1979.
11. A. R. VonHippel, *Dielectric Materials and Application*, New York, John Wiley, p. 361, 1954.
12. H. P. Schwan and K. Li, "Capacity and Conductivity of body tissues at ultrahigh frequencies," *Proc. IRE*, Vol. 41, pp. 1735–1740, 1953.
13. H. P. Schwan and G. M. Pierson, "The absorption of electromagnetic energy in body tissues," *Am. J. Phys. Med.*, Vol. 33, pp. 371–404, 1954.
14. H. P. Schwan, "Electrical Properties of tissues and cell syspensions," *Adv. Biol. Med. Phys.*, Vol. 4, pp. 147–209, 1957.
15. H. P. Schwan and K. R. Foster, "RF-field interaction with biological systems: electrical properties and biophysical mechanisms," *Proc. IEEE*, Vol. 68, pp. 104–113, 1980.
16. J. C. Lin, "Microwave Biophysics," in *Microwave Bioeffects and Radiation Safety*, M. Stuchly, Ed., International Microwave Power Institute, Alberta, Canada, pp. 15–54, 1978.
17. J. D. Kraus and K. R. Carver, *Electromagnetics*, New York, McGraw-Hill, pp. 445–465, 1973.
18. C. L. Wu and J. C. Lin, "Absorption and scattering of electromagnetic waves by prolate spheroidal models of biological structure," *International Antennas and Prop. Symp. Digest*, Stanford, pp. 142–145, 1977.
19. T. P. Gill, *The Doppler Effect*, London, Logos Pr., 1965.
20. J. C. Lin, E. Dawe, and J. Majcherek, "A noninvasive microwave apnea detector," in *Proc. San Diego Biomed. Symp.*, Vol. 30, pp. 441–444, 1977.
21. C. Susskind, "Possible uses of microwaves in the management of lung diseases," *Proc. IEEE*, Vol. 61, pp. 673–674, 1973.
22. I. Yamaura, "Measurement of 1.8–2.7-GHz microwave attenuation in the human torso," *IEEE Trans. Microwave Theory Tech.*, Vol. 25, pp. 707–710, 1977.
23. Y. E. Moskalenko, "Application of centimeter radio waves for electrodeless recording of volume changes of biological materials," *Biophysics (USSR)*, Vol. 5, pp. 259–264, 1960.
24. C. H. Durney, M. F. Iskander, and D. G. Bragg, "Noninvasive microwave methods for measuring changes in lung water content," *IEEE Electro/78 Proc. 30/6*, Boston, pp. 1–7, 1978.
25. J. H. Jacobi and J. E. Larsen, "Microwave interrogation of dielectric targets, Part II: by time delay spectroscopy," *Med. Phys.*, Vol. 5, pp. 509–513, 1978.

Radar Measurements, Resolution, and Imaging of Potential Interest for the Dosimetric Imaging of Biological Targets

Merrill I. Skolnik

A review is given of radar techniques that might have applicability for the dosimetric imaging of biological targets. The basic nature of radar echoes is described as scattering from discontinuities in the dielectric constant that are comparable to the radar wavelength. In biological targets the scattering is complicated by loss in the medium, by inhomogeneities, and by the change in wavelength within the medium. The basic radar point-target measurements of range, range rate, and angle are mentioned along with the less usual measurements more appropriate for finite targets such as size, shape, change of shape, symmetry, surface roughness, and dielectric constant. The ability of a radar to resolve two closely spaced objects is discussed. It is stated that accurate measurements of nearby targets can be obtained if resolution is achieved in any one coordinate. Methods for obtaining "super resolution," such as those based on maximum entropy, are considered as not providing significant improvement with the coherent echo sources characteristic of radar. Radar is able to obtain good images in range and cross-range by taking advantage of Doppler-frequency resolution to isolate individual parts of the target when there is relative motion between target and radar, as is done in synthetic aperture radar. It is suggested that areas where radar might have some potential for the imaging of biological targets are inverse synthetic aperture radar, operation in the millimeter wave region, and in the proper interpretation of image speckle.

1. INTRODUCTION

Radar has proven to be a unique sensor providing information regarding distant targets (such as aircraft, ships, spacecraft, birds and insects) as well as the natural environment (land, sea, and atmosphere). It is sometimes called a "remote sensor" since it can obtain this information from great or small distances without contact or disturbance of the target. The purpose of this paper is to discuss radar techniques that might be of potential interest for the observation of biological targets as considered under the topic of *electromagnetic dosimetric imaging*. (Dosimetric imaging implies the accurate measurement of the spatial distribution of energy dissipation in a biological dielectric.) This paper attempts to give a tutorial review of the measurement capabilities of radar, the ability of radar to resolve closely spaced objects, and its ability to image a scene. Although the prime emphasis of this paper is on reviewing the current state of radar measurements as applied by radar engineers, consideration also will be given to the possible application to biological targets.

Radar Division, Naval Research Laboratory, Washington, D.C.

2. RADAR SCATTERING

A radar obtains information about a remote object by transmitting a known signal and observing the nature of the signal scattered by the object. In almost all cases, it is the back-scattered signal which is of interest to the radar engineer (the echo that returns to the radar); although there are special cases when bistatic or forward-scatter signals are employed.

The strength of the backscattering from an object is defined by the radar cross section of the target

$$\sigma = \lim_{R \to \infty} 4\pi R^2 \frac{|E_s|^2}{|E_i|^2} \tag{1}$$

where R = distance to the object, E_s = scattered field strength received at the radar, and E_i = field strength incident on the object. Equation (1), which defines the target cross section, is equivalent to the usual radar range equation

$$P_r = \frac{P_t G^2 \lambda^2 \sigma}{(4\pi)^3 R^4} \tag{2}$$

where P_r = received signal, P_t = transmitted power, G = antenna gain, and λ = wavelength.

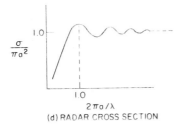

(a) SPECULAR SCATTER   (b) DIFFUSE SCATTER   (c) RESONANT SCATTER

(d) RADAR CROSS SECTION

Fig. 1. Radar cross section of a sphere.

Scattering of electromagnetic energy results from discontinuities in the dielectric constant of the scattering object that are comparable to the radar wavelength. For example, the echo from a smooth, metallic sphere whose radius is large compared to the wavelength occurs from a small spot (the first Fresnel zone) on the tip [Fig. 1(a)]. For scattering to occur from the entire sphere the surface would have to be rough, or diffuse [Fig. 1(b)], such as is a white billiard ball or the full moon when viewed at optical frequencies. When the radius of the sphere is comparable to the radar wavelength, there is a contribution to the backscatter due to a *creeping wave* that circles around behind the sphere as shown in Fig. 1(c). This creeping wave causes constructive and destructive interferences with the direct reflection so that the backscatter varies periodically as a function of a/λ. This is sometimes called Mie scatter. The classical description of the scattering from the sphere as a function of wavelength is sketched in Fig. 1(d).

Some of the energy incident on a non-metallic (dielectric) sphere is reflected similar to that of a metallic sphere, but reduced by the reflection coefficient of the dielectric [1]. Some energy penetrates the body of the spheres where it is partially reflected from the opposite side [Fig. 2(a)]. This partially reflected component interferes with the direct backscatter from the front surface to produce a back-scatter cross section that depends on the size of the sphere, the dielectric constant, and the loss in propagating through the dielectric material. An example of the variation of the scattering from a dielectric sphere is shown in Fig. 2(b).

The dielectric sphere is perhaps the simplest object that might represent scattering from biological targets, yet its scattering behavior is not simple. Inferring the nature of a biological target from the backscatter signal will be complicated by the shape of the target which causes the radar cross section to be a function of aspect, and by the inhomogeneity of both the dielectric constant and the loss properties of the material. (A high loss material, for example, will attenuate the internal signals and the scattering will be due mainly to the properties of the object at and near the sur-

face.) Also, the change in wavelength of the radar energy propagating in a dielectric medium changes the scale of the discontinuities responsible for scattering. It is far more difficult to infer the nature of a non-metallic target from its scattered signal than it is a metallic target. Fortunately, the size of most biological targets is relatively small so that in many cases the loss in propagating through them will not be prohibitively large.

There is another scattering model that is of interest when describing the scattering from a collection of randomly oriented scatters or from a rough surface such as the sea, land, or from atmospheric turbulence. This is called *Bragg scatter*, by analogy to the X-ray scattering from crystals. A rough scatterer can be decomposed by Fourier analysis into a spectrum consisting of a set of sinusoidal surfaces of different amplitudes and frequencies. The scattering from the rough surface occurs only from that Fourier component which is "resonant" with the radar wavelength; i.e., when the spectral component has a wavelength equal to half the radar wavelength. When this occurs all the scattering contributions add coherently (in phase) to produce a large echo.

When the object is small compared to the radar wavelength, the scattering follows the Rayleigh law [left hand

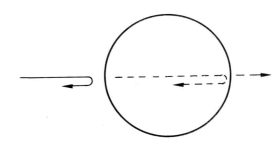

(a) RAYS IN DIELECTRIC SPHERE

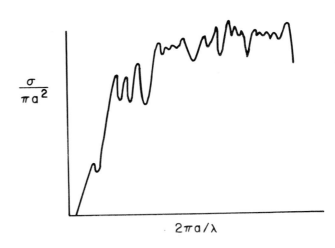

(b) EXAMPLE RADAR CROSS SECTION

Fig. 2. Radar cross section of a dielectric sphere.

region of Fig. 1(d)]. The scattering is small and decreases as f^4, where f = frequency. The scattered signal in the Rayleigh region is affected more by the volume of the object rather than by its shape. A comparison of forward and backscatter radar cross sections for canonical targets is presented elsewhere in this volume by Larsen et al.

3. BASIC RADAR MEASUREMENTS

The following measurements can be made by radar [2]:

Range—The accuracy of the range measurement is determined by the signal bandwidth. Accuracies of a few centimeters at a distance of several tens of miles are possible, with the basic limitation being the accuracy with which the velocity of propagation is known.

Range Rate—The Doppler-frequency shift provides a direct measure of the relative velocity. The longer the duration of the signal and the higher the frequency, the more accurate is the measurement. In radar, the Doppler-frequency shift is often used to separate desired moving targets from undesired fixed targets (clutter). The Doppler-frequency shift is also important for some types of radar imaging.

Angle of Arrival—The angular location of the target is determined by use of a directive (narrow beamwidth) antenna. The larger the size of the antenna aperture (measured in wavelengths) the more accurate the measurement.

The above three measurements are what are normally made by a radar. They assume the target is a point scatter, i.e., of infinitesimal size. When the target is distributed, of finite size, the following measurements apply:

Size—This is generally determined by use of a short pulse, or its equivalent (pulse compression). A measure of size can also be obtained by examining the variation of radar cross section with frequency.

Shape—The observation of the radar cross section from different aspect angles provides shape. Imaging of the target in range and angle also provides shape.

Change of Shape—This is provided by the time variation of the radar cross section. A classical example is the modulation of the echo by aircraft propellers or jet engines.

Symmetry—The symmetry of a scattering object can be determined from observation of the target as a function of the polarization (direction of the electric field) of the radar signal. Symmetrical targets are insensitive to changes in polarization, asymmetrical targets are not. For example, a sphere can be readily recognized from a rod by noting the change of echo amplitude as the plane of rotation is rotated.

In addition, the following two measurements can be made that describe something about the nature of the scattering material:

Surface Roughness—The scale of surface "roughness" is measured in terms of the wavelength. When the surface variations are small compared to the wavelength, the surface is considered to be smooth.

When the variations are large compared to the wavelength, it is rough. Measurements of the scattered signal as a function of wavelength can indicate the scale of surface roughness.

Dielectric Constant—The reflection coefficient of a scattering surface depends on the dielectric constant. The measurement of the dielectric constant of a radar scatter is difficult since it is necessary to know the surface roughness, shape, and the loss tangent. This severely restricts the utility of the measurement so it is not often made by a radar. However, the dielectric constant of the moon's surface was estimated many years ago from a series of earth-based radar measurements. The radar estimate was consistent with that measured with the moon rock samples brought back by Apollo. The measurement of dielectric constant is also of interest to NASA for estimating the soil moisture from a satellite-borne radar.

The above measurements are what a radar is capable of providing. Of all the measurements that can be made by a radar, those which provide target imaging and resolution are probably of most interest for biological targets.

4. RESOLUTION

The resolution of two equal noncoherent sources is a classical problem that has been attacked by several methods, of which the Rayleigh resolution criterion is one of the simplest. The resolution of coherent sources, such as radar echoes, is more difficult to treat analytically, especially when the radar echoes are of unequal strength. Nevertheless, there are some things that can be said.

By resolution is meant the ability of a radar to recognize as separate two closely spaced targets. Resolution may be achieved in range, Doppler velocity, and/or angle. Range resolution depends on the signal bandwidth, Doppler-velocity resolution depends on the signal duration, and angle resolution depends on the size of the antenna aperture. The subject of resolution is also important when multipath signals are present (multiple echoes from the same target that arrive at the radar via different paths). Resolution is required in order to separate the direct echo from the multipath echoes. (The presence of multipath echoes along with the direct echo can cause distortion of the received signal and an incorrect interpretation of the target's characteristics.)

The ability to resolve two targets depends, in part, on the method of processing. It has been found that two targets can be resolved in angle by relatively conventional means if they are separated by at least 0.8 beamwidth [3]. (A separation of 0.8 beamwidth means that the two targets can overlap and still be resolved.) In some cases of high resolution imaging, the resolution has been said to be as good as 0.2 beamwidth [4]; but this is rare. Similarly, two targets can be resolved in range if they are separated by at least 0.8 pulse-width.

Large signal-to-noise ratios are required if good resolution is to be obtained. Most high resolution methods do not yield satisfactory results if the signal-to-noise ratios are insufficient. A "minimum" signal-to-noise ratio is difficult to describe, but lacking any other information one should not expect the best results when the signal-to-noise ratio is considerably less than about 20 dB.

Good resolution implies precise location in the resolved coordinate, but the opposite need not be true. That is, a method for achieving precise location does not always provide good resolution. Good angular resolution, for example, requires large continuous antenna apertures. Precise angular location of a single target, however, can be obtained with widely spaced discrete apertures, as in an interferometer. But an interferometer does not normally provide resolution of multiple targets. Resolution is a prerequisite if accurate measurements are to be made in the presence of multiple targets. It is not necessary to resolve a target in all the radar coordinates in order to obtain accurate measurements in all coordinates. Good resolution is necessary in only one coordinate in order to adequately separate multiple targets. Once the targets are resolved in a single coordinate, accurate measurements can be made in the other coordinates even though resolution is not provided in these coordinates. For example, in a radar with poor angle resolution it is possible to isolate multiple targets by the use of either high range resolution or narrow-band Doppler filtering and then make precise measurements in angle without the degradation imposed by multiple targets within the same resolution cell [5].

In most radar applications, the angle resolution is usually poor compared to the range resolution. However, when there is relative motion between the target and radar it is possible to achieve the equivalent of resolution in the angle domain by achieving resolution in doppler frequency. That is, the various parts of a target can be resolved because they have different Doppler-frequency shifts that are sufficient for resolution. This is the basis for synthetic aperture radar described later.

Over the years there have been a number of techniques proposed to achieve "super resolution." They have been mainly applied to angular resolution, but the basic procedures can be applied to range and frequency resolution as well. They include such techniques as supergain, multiplicative arrays, data restoration, and image enhancement. Most employ some form of nonlinear processing and have been applied mainly to noncoherent sources. The current "super resolution" method enjoying favor is called Maximum Entropy Method (MEM) or Maximum Entropy Spectral Analysis (MESA). It is similar in its implementation to a least-mean-squares smoothing (filtering) and prediction, but the key characteristic of this technique is the assumption that the received angular spectrum can be represented by a finite number of poles in the complex plane just as an all-pole filter (a series of resonators) is used to represent a time waveform. The processing is nonlinear. The maximum entropy method appears to achieve "spectacular improvement" in resolution compared to that based on classical diffraction theory. The results, however, do not violate fundamental principles and do not offer new resolution capabilities [6, 7]. With noncoherent sources (a case with little interest in radar) they do provide significant resolution compared with classical diffraction theory, but the results are not much different from previous nonlinear resolution techniques. With coherent echoes, as in radar, they offer little or no improvement over classical methods. In general, the nonlinear processing associated with these "super resolution" techniques can cause spurious and erroneous information to be generated when there are multiple coherent echoes (those illuminated by the same radar signal), as do other nonlinear techniques.

5. RADAR IMAGING

A radar can be made to have excellent range resolution, but its resolution in the angular, or cross-range, coordinate is generally poor. For example, a radar with a one degree beamwidth has a resolution of about 350 m at a range of 20 km. In the range coordinate it could have a resolution of a meter or so, even as small as a few centimeters if desired. A one degree beamwidth is not the smallest that can be achieved. Radars have been built with 0.05° beamwidth, but these are special cases. A more practical minimum beamwidth is about 0.2°. Even so, the resolution in angle is much poorer than can be achieved in range. In spite of the limitation on resolution imposed by the antenna beamwidth, there have been important applications of imaging radar using a large antenna (narrow beamwidth) and high range resolution. One example is the use of classical imaging radars for commercial mineral prospecting.

The good range resolution possible with a radar can be used to recognize one type of target from another if different parts of the target can be resolved. Ships and aircrafts can be observed with radar whose range resolution cell is small enough to detect the major scattering centers. The target's projected size is also available as a by-product. It is also possible to measure the angular offset of each scattering center from the radar's boresight axis by obtaining a monopulse measurement of the angular displacement of each resolvable scatterer [5].

The equivalent of good angular resolution can be obtained by utilizing the fact that each part of a distributed scatter has a different Doppler-frequency shift when there is relative motion between radar and target. When the target is fixed and the radar is in motion, this technique is called *synthetic aperture radar*, or SAR [8]. The angular, or cross-range, resolution that can be obtained with SAR is theoretically equal to D/2, where D is the dimension of the real antenna used by the radar. Thus the angular resolution can be of the order of a few meters, which is comparable to the range resolution. The resolution in both angle and range in SAR is independent of the range, which is unlike optical or IR photography where the resolution worsens with disease. SAR is being considered by NASA for use in satellites to provide information on sea conditions and for crop census, soil moisture measurement, ice cover, and other earth resources factors.

If the radar is stationary but the target is in motion, the resolution in the doppler domain of individual parts of the target can also obtain an image. This is described in Fig. 3.

If the target rotates through an angle $\Delta\theta$, the resolution in the dimension perpendicular to the radar line of sight can be shown to be [8]

$$\Delta x = \frac{\lambda}{2\Delta\theta} \qquad (3)$$

where λ = radar wavelength, and $\Delta\theta$ is in radians. Thus the amount of rotation necessary to provide good resolution is small. The resolution using Doppler frequency in this manner is in one dimension only. Range resolution provides the orthogonal coordinate to give a two-dimensional image.

The maximum resolution is not always required (or desired) for the recognition of a target. There must be a minimum number of resolvable cells on the object to be imaged in order to properly recognize what is being looked for. Significant (order of magnitude) improvements in resolution do not always provide that much better recognition or information about the object. (Unfortunately precise guidelines as to the minimum resolution have not been formulated for general application, although the effect has been recognized experimentally.)

When a coherent radar (or laser) illuminates a distributed target the resultant image is likely to contain "speckle." Speckle is due to the constructive and destructive interference associated with scattering from multiple scatterers located within the resolution cell. The result is speckle, or bright spots, that might not be related to the actual scattering objects themselves and which can provide misleading information. If the viewing aspect, radar frequency, or relative position of the scatterers change, the speckle pattern will change. In fact, frequency diversity, can be used not only to suppress speckle, but also to provide a basis for reflection tomography as presented elsewhere in this volume by Farhat.

There are several responses the designer can take with regard to speckle, in addition to simply tolerating its presence. These are:

1. Use a resolution sufficient to isolate the major scattering centers. This will avoid "speckle" but since the major scattering centers have aspect sensitivity the nature of the image will still vary with changes in aspect. Thus the image will change in appearance with aspect angle as the various scattering centers fade in and out, but the apparent locations of the scattering centers will not change as when the resolution cell is large enough to include more than one scattering center.

2. The speckle can be smoothed, or averaged, to produce a more realistic looking image by incoherent superposition of several (many) images with different (independent) speckle patterns. These separate images are obtained by viewing the object at slightly different aspects or with different frequencies. In some radar applications it is claimed that better target recognition can be obtained by trading resolution for independent speckle patterns that can be superimposed noncoherently [9], [10]. For example,

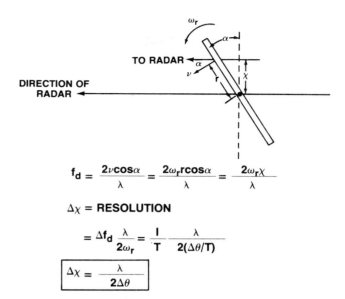

$$f_d = \frac{2\nu\cos\alpha}{\lambda} = \frac{2\omega_r r\cos\alpha}{\lambda} = \frac{2\omega_r \chi}{\lambda}$$

$$\Delta\chi = \text{RESOLUTION}$$

$$= \Delta f_d \frac{\lambda}{2\omega_r} = \frac{1}{T} \frac{\lambda}{2(\Delta\theta/T)}$$

$$\boxed{\Delta\chi = \frac{\lambda}{2\Delta\theta}}$$

Fig. 3.   Inverse SAR.

instead of a coherent radar image with 3 meters resolution, it might be better to use the available observation time in a SAR to noncoherently superimpose four independent images of 12 meters resolution, each with a different speckle pattern. (Noncoherent superposition means superposition of intensities and not phase.)

3. The speckle pattern can obtain useful information about the nature of the scattering object [11]. This has been taken advantage of at optical wavelengths, but not yet at microwaves.

The shape of the resolution cell, or pixel as it is called in optics, is generally made to be the same in both the range and azimuth coordinate. A more realistic image results with equal resolution. However, the resolution need not be equal for purposes of target resolution. Experiments show that the area of the resolution cell is more important than the linear dimensions as far as image interpretability is concerned, up to an aspect ratio of as much as 10 to 1 [9]. This is an important consideration when resolution in one coordinate is easier to achieve than in the other.

Although there has been considerable success in recent years in imaging with radar, it has been at distances and with resolutions that are hardly of interest for biological targets. Advantage has been taken in some biological imaging experiments [12] of the fact that in a medium of dielectric constant ϵ the wavelength of electromagnetic radiation in the medium is λ/ϵ, where λ is the wavelength in air. Thus greater angular resolution can be achieved in the dielectric with a given physical size aperture than when in air. The disadvantage of propagating in a dielectric medium are that the propagation losses are generally greater than in air, and it is difficult to transfer energy from air or vacuum to the dielectric medium without "mismatch loss" due to the difference in dielectric constants of two media. The mismatch loss can be minimized, however, by immersing the entire antenna in a matching liquid. With the distances involved in imaging biological targets, the propagation loss apparently has not been a serious problem. In fact it can be used to advantage since echoes from unwanted distant objects are attenuated to such a low level that they do not interfere with the desired signal [12].

Although the direct application of current microwave radar techniques to biological targets may be limited by the significant differences in resolution in the two cases, the use of the millimeter and submillimeter wave regions might be of interest. The wavelengths can range from a few millimeters to a fraction of a millimeter, thus permitting good resolution with a reasonably sized antenna aperture. Since operation is likely to be in the near field of the antenna, focussing of the aperture can be employed to achieve the theoretical resolution. The wide bandwidths offered by the millimeter-wave region will permit good range resolution either by the generation of very short pulses or by the use of pulse compression. The resonances of oxygen, water and other molecules found in this region of the spectrum might be of use for identification of the various constituents of a biological target. In recent years there have been considerable advances made in the development of the components and technology needed for operation in the millimeter-wave part of the EM spectrum. Components are still expensive, however, and they are more of an experimental nature than the components found in a microwave catalog.

6. OTHER FACTORS

In addition to the above, there are several other aspects of microwave radar that might have application to the dosimetric imaging of biological targets. These are pulse compression, forward scatter, and multipath.

Pulse compression is a radar technique for achieving the resolution of a short pulse but with the energy of a long pulse [13]. In radar it is used when the peak power is limited by breakdown, and a short pulse of sufficient energy cannot be obtained. The short ranges of biological target imaging might not make it necessary to use pulse compression. In the millimeter region, however, it might be necessary with some transmitter techniques if exceptionally good resolution is required. Because of the short distances, it might be necessary to operate in a CW manner with isolation between separate transmitting and receiving antennas. The transmitted waveform can be a linearly frequency modulated pulse, modulated over a band Δf [17]. After passing through a matched filter, or its equivalent, the width of the compressed output pulse will be $1/\Delta f$.

Radar as described in this paper, and as almost always employed in practice, operates with backscatter echoes from the target. That is, the transmitter and receiver are co-located, usually sharing the same antenna. It is also possible to operate the radar bistatically, with transmitter and receiver widely separated so that the scattering angle θ_s is not 180° as in backscatter, but can be any other angle [14]. Bistatic radar provides no information about a conventional target that cannot be obtained using monostatic (single site) radar. In fact, the difficulty involved in extracting mea-

surements with a bistatic radar means that it might produce less information than can a monostatic radar. It can be shown that when physical optics assumptions apply, the bistatic cross section is equivalent to the radar cross section seen by a monostatic radar viewing the target at an angle that bisects the angle formed by the incident and scattered rays [15]. This does not apply, however, in the vicinity of forward scatter where the scattering angle θ_s approaches zero. Therefore, forward scatter, or direct transmission measurements, can possibly provide information not readily available from a monostatic radar. In the case of a turbulent scatterer (atmospheric turbulence, for example), the above rule regarding the equivalence of bistatic and monostatic cross sections does not apply. The bistatic cross section of turbulent media is considerably larger than the monostatic and might prove useful in some cases [18].

Multipath echoes are those received from the same object that arrive back at the radar via different paths. The superposition of the multipath echoes can cause erroneous measurements. It was noted earlier in this paper that one solution to this problem is to provide sufficient resolution to separate each multipath echo. Unfortunately, this cannot always be done. In some cases, however, multipath can be of use rather than hindrance. It is actually used in some radar applications to good purpose. Multipath cannot always be turned around for good in the majority of cases, but its potential usefulness cannot be dismissed out of hand.

7. DISCUSSION

Radar has proven to be quite useful as a remote sensor capable of extracting information from a great distance. The sensing from great distance is an attribute not of interest to the imaging of biological objects. The biological imaging that has already been performed using microwave sensors evidences considerable innovation and sensitivity to the unique aspects associated with the geometry and configuration of biological targets [16], [17]. It is doubtful that there exists in radar technology some major, important technique with which those engaged in biological imaging are not already aware.

Of the several potential radar technologies mentioned in this paper, those which seem to offer possible merit in further investigation are:

— Inverse synthetic aperture, in which rotation of the object provides imaging in one coordinate better than can be achieved with angle resolution.
— Operation in the millimeter wave region to take advantage of the higher resolution possible in both range and angle.
— Use of the information contained in the speckle pattern found in images obtained with coherent radar to ascertain something regarding the nature of the target.

It appears that a basic area needing further study for the better understanding of biological imaging is the nature of propagation and scattering of electromagnetic energy in biological materials.

References

1. D. E. Barrick, Dielectric Spheres—Lossy and Lossless, Sec. 3.3 of *Radar Cross Section Handbook*, Vol. 1, G. T. Ruck, Ed., Plenum Press, New York, 1970.

2. M. I. Skolnik, An Introduction to Radar, Sec. 1.3, *Radar Handbook*, McGraw-Hill Book Co., New York, 1970.

3. M. I. Skolnik, "Comment on the Angular Resolution of Radar," *Proc. IEEE*, Vol. 63, No. 9, pp. 1354–1355, Sep., 1975.

4. F. Richey, "Radar Contrast Patterns of Airport Runways," *J. Opt. Soc. Am.*, Vol. 52, pp. 51–57, Jan., 1962.

5. D. D. Howard, "High Range Resolution Monopulse Tracking Radar," *IEEE Trans.*, Vol. AES-11, pp. 749–755, Sep., 1975.

6. W. F. Gabriel, "Superresolution of Coherent Sources by Adaptive Array Techniques," *IEEE International Radar Conference*, April 29–30, 1980, Arlington, VA.

7. W. D. White, "Angular Spectra in Radar Applications," *IEEE Trans.*, Vol. AES-15, pp. 895–899, Nov., 1979.

8. M. I. Skolnik, *Introduction Radar Systems*, Second Ed., Sec. 14.1, McGraw-Hill Book Co., New York, 1980.

9. R. K. Moore, "Tradeoff Between Picture Element Dimensions and Noncoherent Averaging in Side-Looking Airborne Radar," *IEEE Trans.*, Vol. AES-15, pp. 697–708, Sep., 1979.

10. L. J. Porcello, N. G. Massey, R. B. Innes, and J. M. Marks, "Speckle Reduction in Synthetic-Aperture Radars," *J. Opt. Soc. Am.*, Vol. 66, pp. 1305–1311, Nov., 1976.

11. R. K. Erf, Ed., *Speckle Metrology*, Academic Press, New York, 1978.

12. J. H. Jacobi, L. E. Larsen, and C. T. Hast, "Water-Immersed Microwave Antennas and Their Application to Microwave Interrogation of Biological Targets," *IEEE Trans.*, Vol. MTT-27, pp. 70–78, Jan., 1979.

13. M. I. Skolnik, *Introduction to Radar Systems*, Second Ed., Sec. 11.5, McGraw-Hill Book Co., New York, 1980.

14. ———, Sec. 14.6.

15. J. W. Crospin, Jr. and K. M. Siegel, *Methods of Radar Cross-Section Analysis*, Chapter 5, Academic Press, New York, 1968.

16. L. E. Larsen and J. H. Jacobi, "Microwave Scattering Parameter Imagery of an Isolated Canine Kidney," *Med. Phys.*, Vol. 6, No. 5, pp. 395–403, Sep./Oct., 1979.

17. J. H. Jacobi and L. E. Larsen, "Microwave Time Delay Spectroscopic Imagery of Isolated Canine Kidney," *Med. Phys.*, Vol. 7, No. 1, pp. 1–7, Jan./Feb., 1980.

18. V. I. Tatarski, *Wave Propagation in a Turbulent Medium*, McGraw-Hill Book Co., New York, 1961.

Microwave Holography and Coherent Tomography

N. H. Farhat

A review of the theory of optical holography is presented in a form suitable for further comparative discussion of microwave holography. The advantages and motivations for extending holography to the microwave range of the spectrum are given. Practical constraints arising from this extension are identified by referring to key equations. These include: (a) usually prohibitive cost, and often cumbersome size, of high-resolution microwave hologram apertures capable of real-time data acquisition, (b) inability to view a true 3-D image as in optical holography because of a wavelength scaling problem, (c) degradation of image quality because of coherent noise and speckle caused by the small numerical apertures normally realizable and the bipolar nature of the impulse response characteristic of coherent imaging systems. Methods for overcoming these constraints are discussed with attention given to the use of wavelength diversity to enhance resolution, reduce coherent noise, and furnish a tomographic 3-D imaging and display capability. Finally a brief qualitative examination of the potential of microwave holographic imaging techniques in noninvasive medical imaging is given.

1. INTRODUCTION

The extension of Gabor's holographic imaging principle [1] to the microwave range of the spectrum has been considered in several review papers [2]–[4]. The motivation for microwave holography stems from the following considerations:

(a) *Favorable Propagation Properties*—By virtue of their longer wavelengths, centimeter and millimeter microwaves can penetrate a variety of optically opaque substances offering a "seeing through" capability unattainable at optical wavelengths. This capability can be useful in a variety of applications such as nondestructive evaluation, noninvasive medical imaging, all weather imaging radar (cf. Skolnik elsewhere in this volume) and the imaging of concealed and buried objects [2]–[4].

(b) *Availability of a Highly Sophisticated Microwave Measurement Technology*—Modern microwave measurement techniques provide efficient, sensitive and accurate means for the acquisition of microwave holographic (amplitude and phase) data over a broad range of frequencies in the (0.1–100) GHz range. An increasing number of commercially available computerized microwave network analyzers permit rapid automated hologram data acquisition in the (0.1–40) GHz range specially when they are used in conjunction with automated spatial field mapping scanners. Data acquisition with unprecedented accuracy becomes now feasible because of the availability of error correction routines that can be carried out rapidly by the controlling computers to remove system errors. In addition, because of

the high spectral purity of available microwave phase-locked sources and synthesizers, electromagnetic wave-fields with high or any desired degree of coherence that is required in the holographic process can be produced. This leads to large coherence volumes permitting holographic imaging and analysis of large bodies even when the optical path differences of the interfering wavefronts is quite large.

(c) *Signal Processing Flexibility and Advantages*—Because of the two step nature of the holographic process (recording and reconstruction steps), the raw hologram data, representing in some situations either the Fresnel or Fourier transform of the object, can be subjected to a variety of preprocessing and filtering operations directly in the *transform domain* that can enhance reconstructed image quality. For example in the scanned mode of microwave hologram data acquisition where the complex field amplitude (CFA) over the hologram recording aperture is mapped by mechanical scanning of a suitable sensor, the hologram data is available in the form of an electronic signal that lends itself to a variety of signal processing operations. For example logarithmic compression or hard-clipping can be applied as a means of accommodating the wide dynamic range of the signal into the finite dynamic range of the measurement system which could be limited especially when optical computing (optical storage and data processing) is utilized rather than digital storage and processing. Alternately the signal can be differentiated to cause an enhancement of edges and fine details in the ultimately reconstructed image [5]. In yet another example of transform plane signal processing, electronic wavefront subtraction can be employed to discern the difference between two scenes [6], [7].

(d) *Prospects for Real-Time Operation*—The density of

University of Pennsylvania, The Moore School Graduate Research Center, Philadelphia, Pa. 19174. April 6, 1980. Revised June 10, 1980.

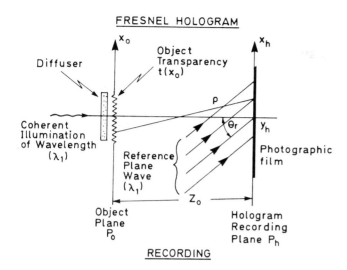

FRESNEL HOLOGRAM

RECORDING

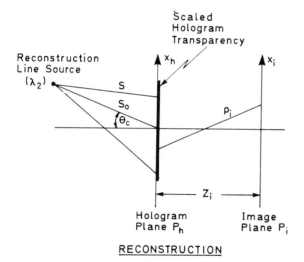

RECONSTRUCTION

Fig. 1. Two dimensional model for the recording and reconstruction of a Fresnel hologram.

fringes and therefore the information content in microwave holograms are typically much less than in optical holograms because of the longer wavelengths used. This indicates that real-time image retrieval should be feasible either digitally or in a hybrid (digital/optical) data processing mode which combines the flexibility of digital computations with the parallel nearly instantaneous processing capabilities of an optical computer employing a rapid recyclable spatial light modulator (SLM) or noncoherent to coherent image converter [2].

In the following sections of this paper we will first review the principles of optical holography leading to the derivation of basic equations. The results will then be examined in the context of microwave holography to identify those constraints that must be overcome before full practical potential of high resolution microwave holographic imaging can be realized. Ways for overcoming these constraints are then

discussed including a wavelength diversity holographic imaging method which can also be viewed as a coherent tomography imaging method. As such the method could be of particular interest in noninvasive medical imaging.

2. FRESNEL HOLOGRAMS AND THEIR PROPERTIES

In this section a physical optics treatment of the holographic process is given to provide the theoretical basis for further discussion. The approach is similar to that first given by Armstrong [8] in determining the imaging properties of optical Fresnel holograms. Although the discussion is in terms of optical holograms, the results of the analysis are equally applicable to microwave holography. The term Fresnel hologram stems from the fact that the dimensions and distances of significance in the recording and reconstruction arrangements are such that the CFAs (complex field amplitudes) in the various relevant planes are arrived at by means of the Fresnel diffraction integral. For the sake of mathematical simplicity, a two-dimensional model of the problem is analyzed without sacrificing the generality of the results obtained. An off-axis reference plane wave will be employed in the recording step while an off-axis cylindrical wave produced by a line source will be employed in the reconstruction step. The general case when the recording is made at one wavelength λ_1, and the reconstruction at a second wavelength λ_2, after changing the size of the hologram transparency, is considered to make the results useful for the discussion of microwave holography and to demonstrate certain properties of the holographic process pertaining to image distortion namely the *wavelength scaling problem*.

2.1 Recording Step

Referring to the recording geometry shown in Fig. 1 we start by characterizing the diffuse illumination of the object transparency disposed in the object plane P_o. The diffuser,* a plate of ground or fogged glass, opal glass or simply scotch tape scatters light transmitted through it in all directions nearly equally as depicted in Fig. 2.

By placing the transparency close to the right hand surface of the diffuser, we insure that any point on the transparency will be subjected to light rays that pass through it in a variety of directions. Thus, information about the transmittance at a particular point of the object is conveyed to many regions of the hologram recording plate and gets stored over a wide area of the plate. This leads to redundant storage of information in the hologram and eventually to retrieved images that are clearer and can be observed from a wider angular range with greater ease than images retrieved from holograms made without the diffuser. This later characteristic is similar to the more familiar practice of observing ordinary

---

* In the microwave region the diffuser would consist of a corrugated or perforated plexiglas plate [9].

photographic color slides by placing them against a back illuminated fogged glass panel for better viewing.

The action of the diffuser can be expressed mathematically by involking the useful concept of the angular spectrum of plane waves (see appendix I or for example, reference [10] for more detail). The roughened surface of the diffuser produces nearly random fluctuations in its thickness. A plane wave transmitted through the diffuser will emerge, therefore, with nearly random spatial phase modulation. If we designate by $f(x_o)$ the CFA in the object plane produced by the diffuse illumination, then due to its white noise nature, $f(x_o)$ will possess a very wide angular spectrum of plane waves, in which each component wave gets spatially modulated by the object transmittance function $t(x_o)$. Radiation emerging from any portion of the transparency will be diffracted over a wide range of angles that subtend the hologram recording film lying in the hologram recording plane. In interfering with the reference beam some of these planes wave components produce net intensity distributions with spatial frequencies that are within the resolution limit of the film and thus get recorded. Loss of some information because of the finite resolution capability of the recording medium is thus also minimized when a diffuser is used.

We express now the CFA $f(x_o)$ produced by the diffuse illumination at the object transparency in terms of its angular spectrum $A(k_{1x},x)$ through.

$$f(x_o) = \frac{1}{2\pi} \int_{-\infty}^{\infty} A(k_{1x})e^{-jk_{1x}x_o} \, dk_{ax}$$

where

$$A(k_{1x}) = \int_{-\infty}^{\infty} f(x_o)e^{jk_{1x}x_o} \, dx_o \tag{2}$$

and where $k_1 = 2\pi/\lambda_1$ and k_{1x} is the component of the wave vector of the particular plane wave component $\bar{k}_1$ along x_o.

The CFA in the wavefield emerging from the object transparency will be,

$$D(x_o) = f(x_o)t(x_o) \tag{3}$$

The CFA $O(x_h)$ produces by the object wave in the hologram recording plane P_h can now be obtained making use of the Fresnel-Kirchhoff diffraction integral,

$$O(x_h) = \left(\frac{j}{\lambda Z_o}\right)^{1/2} \overbrace{\int_{-\infty}^{\infty} \underbrace{\left\{\int_{-\infty}^{\infty} A(k_{1x})e^{-jk_{1x}x_o} \, dk_{1x}\right\}}_{\text{diffuse illumination}} \Big| \underbrace{t(x_o)e^{-jk_{1\rho}} \, dx^o}_{\substack{\text{object} \\ \text{trans-} \\ \text{mittance}}}}^{D(x_o)} \tag{4}$$

where $D(x_o)$ will be carried for the sake of clarity in its explicit form through the analysis and where,

$$\rho \simeq Z_o + \frac{1}{2Z_o}(x_h - x_o)^2 \tag{5}$$

The CFA $R(x_h)$ produced by the off-axis plane reference beam in the P_h plane will be

$$R(x_h) = R_o e^{-jk\sin\theta_r x_h} \tag{6}$$

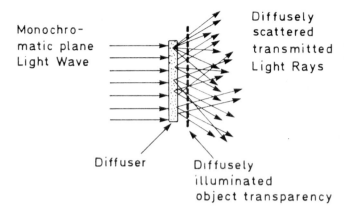

Fig. 2. Illustrating the action of a diffuser.

Thus the total CFA in P_h is,

$$H(x_h) = O(x_h) + R(x_h) \tag{7}$$

$$= \left(\frac{j}{\lambda Z_o}\right)^{1/2} e^{-jk_1 z_o} \int_{-\infty}^{\infty} \{\int A(k_{1x})e^{-jk_{1x}x_o} \, dk_{1x}\} t(x_o)_x$$
$$\times e^{-j(k_1/2Z_o)(x_h-x_o)^2} \, dx_o + R_o e^{-jk_1\sin\theta_x x_h} \tag{8}$$

Being a square-law or intensity detector, the photographic plate in the hologram recording plane will record the intensity distribution,

$$I(x_h) = HH^* = |O|^2 + |R|^2 + OR^* + O^*R \tag{9}$$

resulting after correct development in a hologram transparency with amplitude transmittance,

$$t(x_h) = \beta I(x_h) = \underbrace{\beta(|O|^2 + |R|^2)}_{\text{zero order}} + \underbrace{\beta OR^*}_{\substack{\text{primary} \\ \text{image}}} + \underbrace{\beta O^*R}_{\substack{\text{conjugate} \\ \text{image}}}$$

$$= t_o(x_h) + t_p(x_h) + t_c(x_h) \tag{10}$$

where β is a real constant dependent on film characteristics and,

$$t_o = \beta(|O|^2 + |R|^2), \quad t_p = \beta OR^*, \quad t_c = \beta O^*R \tag{10a}$$

give rise, respectively, in the reconstruction step to zero order undiffracted light, to a primary image, and to a conjugate image.

2.2 Reconstruction or Image Retrieval Step

The developed hologram transparency is scaled next in size by a factor $1/\gamma$ such that a newly scaled replica $t(\gamma x_h)$ results.

If $\gamma > 1$, the scaled transparency has been demagnified or reduced in size, while for $\gamma < 1$ the transparency has been magnified or enlarged. This latter statement can easily be remembered by noting that given, for example, $t(x_h) = \sin(x_h)$, then $\sin(\gamma x_h)$ is of higher frequency when $\gamma > 1$ and as such represents a compressed or demagnified version of $t(x_h)$ while if $\gamma < 1$, $\sin(\gamma x_h)$ is of lower frequency representing a magnified, stretched, or enlarged version of $t(x_h)$. The scaled hologram transparency is then illuminated as shown in the reconstruction geometry of Fig. 1 by a cylindrical wave of wavelength λ_2 produced by an off-axis reference line source. The CFA $C(x_h)$ in P_h produced by the reconstruction line source will be,

$$C(x_h) \simeq \left(\frac{j}{\lambda S_0}\right)^{1/2} e^{-jk_s S_o} e^{-j(k_2/2S_o)x_h^2} e^{jk_2 \sin\theta_c x_h} \quad (11)$$

The last term in Eq. (11) accounts for the off-axis location of the reconstruction line source.

We concentrate next on the reconstruction of the conjugate image term in Eq. (10) which would give rise as we shall see to a real image of the diffusely illuminated object transparency $D(x_o)$. A similar treatment could be applied to the primary image term resulting in similar properties and conclusions.

The CFA $\Psi_c(x_i)$ in the image plane P_i due to the conjugate term $t_c(x_h)$ of Eq. (10) will, therefore, be

$$\Psi_c(x_i) = \left(\frac{j}{\lambda_2 Z_i}\right)^{1/2} \int_{-\infty}^{\infty} C(x_h) t_c(\gamma x_h) e^{-jk_2\rho_i} dx_h \quad (12)$$

where

$$\rho_i = Z_i + \frac{1}{2Z_i}(x_i - x_h)^2 \quad (12a)$$

and where $t_c(\gamma x_h)$ represents the scaled version of $t_c(x_h)$ and is given by

$$t_c(\gamma x_h) = \beta O^*(\gamma x_h) R(\gamma x_h) \quad (13)$$

Substituting Eq. (13) in Eq. (12) and making use of Eq. (4) for 0 and Eq. (11) for R we obtain,

$$\Psi_c(x_i) = C_1 \int_{-\infty}^{\infty} dx_h \underbrace{\int_{-\infty}^{\infty} dx_o \left\{ \int_{-\infty}^{\infty} A^*(k_{1x}) e^{jk_{1x}x_o} dk_{1x} \right\} t^*(x_o) e^{j(k_1/2Z_o)(\gamma x_h - x_o)^2}}_{\text{term due to } t_c(\gamma x_h)}$$

$$\times \underbrace{e^{-jk_1 \sin\theta_r \gamma x_h}}_{\substack{\text{term due to} \\ R(x_h)}} \underbrace{e^{-j(k_2/2S_o)x_h^2} e^{jk_2 \sin\theta_c x_h}}_{\text{term due to } C(x_h)} \underbrace{e^{-j(k_2/2Z_i)(x_i - x_h)^2}}_{\text{diffraction}} \quad (14)$$

where

$$C_1 = \beta C_o R_o \left(\frac{j}{\lambda Z_o}\right)^{1/2} \left(\frac{j}{\lambda_2 Z_i}\right)^{1/2} \left(\frac{j}{\lambda S_o}\right)^{1/2} e^{-jk_1 Z_o} e^{-jk_2 Z_i} e^{-jk_2 S_o}$$

$$(14a)$$

The terms containing x_h^2 in the exponentials in Eq. (14) collect to $(k_1\gamma^2/2Z_o - k_2/2S_o - k_2/2Z_i)x_h^2$. This expression can be made to vanish by choosing

$$\frac{k_1\gamma^2}{2Z_o} - \frac{k_2}{2S_o} - \frac{k_2}{2Z_i} = 0 \quad (15)$$

which simplifies greatly the integration of Eq. (14) with respect to x_h. If the remaining integrals in Eq. (14) after making use of Eq. (15) lead to image retrieval (as will indeed be shown), then expression 15 can be regarded as a *focusing condition* that will give the image plane distance Z_i in terms of the other parameters ($k_1, k_2, Z_o; S_o$, and γ). By substituting for k_1 and k_2, we can rewrite Eq. (15) in the form

$$\frac{\gamma^2}{\gamma_1 Z_o} - \frac{1}{\lambda_2 S_o} - \frac{1}{\lambda_2 Z_i} = 0 \quad (15a)$$

or

$$\frac{1}{Z_i} + \frac{1}{S_o} = \left(\gamma^2 \frac{\lambda_2}{\lambda_1}\right) \frac{1}{Z_o} \quad (15b)$$

The last relation is similar to the focusing condition of the thin lens [11]. Thus the quantity

$$\left(\gamma^2 \frac{\lambda_2}{\lambda_1} \frac{1}{Z_o}\right)^{-1} \triangleq F_{eq} \quad (16)$$

may be regarded as an equivalent focal length F_{eq} in the holographic process.

Making use of the focusing condition, Eq. (15), in Eq. (14) we obtain,

$$\Psi_c(x_i) = C_1 \int_{-\infty}^{\infty} dx_h \int_{-\infty}^{\infty} dx_o \left\{ \int_{-\infty}^{\infty} A^*(k_{1k}) e^{jk_{1x}x_o} dk_{1x} \right\}$$
$$\times t^*(x_o) e^{j(k_1/2Z_o)(x_o^2 - 2\gamma x_o x_h)}$$
$$\times e^{j(k_1 \sin\theta_r - k_2 \sin\theta_c)x_h} e^{-j(k_2/2Z_i)(x_i^2 - 2x_i x_h)} \quad (17)$$

Carrying out the integration with respect to x_h first we have,

$$\int_{-\infty}^{\infty} e^{j[-(k_1\gamma/Z_o)x_o + (k_2/Z_i)x_i z - k_i\gamma \sin\theta_r + k_2 \sin\theta_c]x_h} dx_h$$

$$= \delta\left[-\frac{k_1\gamma}{Z_o}x_o + \frac{k_2}{Z_i}x_i - k_1\gamma \sin\theta_r + k_2 \sin\theta_c\right] \quad (18)$$

where $\delta[\cdot]$ is the Dirac delta "function." The length of subsequent expressions can be shortened somewhat without altering the generality of the results if we choose,

$$k_1\gamma \sin\theta_r = k_2 \sin\theta_c \quad (19)$$

i.e., by positioning the reconstruction line source at an angle

$$\theta_c = \sin^{-1}\left[\gamma \frac{k_1}{k_2} \sin\theta_r\right] \quad (20)$$

Then Eq. (18) becomes,

$$-\frac{Z_o}{\gamma k_1} \delta\left[x_o - \frac{Z_o}{Z_i} \frac{k_2}{k_1} \frac{1}{\gamma} x_i\right] = -\frac{Z_o}{\gamma k_1} \delta[x_o - M^{-1} x_i] \quad (21)$$

where

$$M = \gamma \frac{Z_i}{Z_o} \frac{k_1}{k_2} \quad (22)$$

In writing down Eq. (21) use was made of the relationship

$$\delta(a\xi + b) = \frac{1}{a}\left(\xi + \frac{b}{a}\right) \tag{23}$$

Substituting next Eq. (21), which is the result of the integration w · r to x_h, back into Eq. (17) we obtain

$$\Psi_c(x_i) = -\frac{Z_o}{\gamma k_1} C_1 \int_{-\infty}^{\infty} dx_o \left\{ \int_{-\infty}^{\infty} A^*(k_{1x}) e^{jk_{1x}x_o} dx_o \right\}$$
$$\times\, t^*(x_o) e^{j(k_1/2Z_o)x_o^2} e^{-j(k_2/2Z_i)x_i^2} \delta[x_o - M^{-1} x_i] \tag{24}$$

which yields by the "sifting" property of the delta "function,"

$$\psi_c(x_i) = -\frac{Z_o}{\gamma k_1} C_1 e^{j(k_1/2Z_o)M^{-2}x_i^2} e^{-j(k_2/2Z_i)x_i^2}$$
$$\times \left\{ \int_{-\infty}^{\infty} A^*(k_{1x}) e^{jk_{1x}M^{-1}x_i} dk_{1x} \right\} t^*(M^{-1} x_i) \tag{25}$$

making use of Eq. (22) we note now that in Eq. (25),

$$e^{j(k_1/2Z_o)M^{-2}x_i^2} = e^{j(k_1/2Z_o)(1/\gamma^2)(Z_o/Z_i)^2(k_2/k_1)^2 x_i^2}$$
$$= e^{j(k_2/2Z_1)(M^{-1}/\gamma)x_i^2} \tag{26}$$

This further simplifies Eq. (25) to

$$\Psi_c(x_i) = \frac{Z_o}{\gamma k_1} C_1 e^{-j(k_2/2Z_i)[1-1/\gamma M]x_i^2} \left[\left\{ \int_{-\infty}^{\infty} \right.\right.$$
$$\left.\left. A^*(k_{1x}) e^{jk_{1x}M^{-1}x_i} dk_{1x} \right\} \cdot t^*(M^{-1} x_i) \right] \tag{27}$$

The term within the square brackets is seen to be,

$$D^*(M^{-1} x_i) = f^*(M^{-1} x_i) t^*(M^{-1} x_i) \tag{27a}$$

Since $D(x_o) = f(x_o)t(x_o)$ was the original diffusely illuminated object, we see in view of this result that a scaled replica $D^*(M^{-1} x_i)$ of $D^*(x_o)$ has been reconstructed in an image plane a distance Z_i from the hologram where Z_i satisfies the focusing condition of Eq. (15).

Equation (27) shows that the image scale obtained by the correspondence of the variable x_o and $M^{-1} x_i$ is,

$$\frac{\Delta x_i}{\Delta x_o} = M \tag{27b}$$

Thus M represents the lateral scale of the image (i.e., its lateral magnification or demagnification) and is seen according to Eq. (22) to be a function of Z_i. Substituting for Z_i from Eq. (15b) in Eq. (22) for M we find,

$$M = \frac{1/\gamma}{1 - \frac{1}{\gamma^2}\frac{\gamma_1}{\gamma_2}\frac{Z_o}{S_o}} \tag{28}$$

As before, $M > 1$ would indicate a laterally demagnified image, while $M < 1$ a laterally magnified image. A negative M would indicate the image is inverted.

The conjugate image observed in the image plane by either projecting it on a screen or by recording it on a photographic plate, is actually the intensity $|\Psi_c(x_i)|^2$. Thus by referring to Eq. (27) and noting that ordinary amplitude modulating

object transparencies have a real transmittance function for which $t^* = t$, we arrive at the conclusion that the image reconstructed from the conjugate term is a diffusely illuminated scaled replica of the original diffusely illuminated object transparency.

2.3 Extension of Results to Three Dimensional Objects

The extension of the previous treatment to three dimensional objects introduces the concept of longitudinal scaling of the image and the subject of image distortion. To obtain an expression for the longitudinal scaling, we consider two planar objects consisting of two longitudinally displaced slits illuminated from the left as shown in Fig. 3 and make use of the results obtained previously.

In accordance to focusing condition Eq. (15b) the retrieved real images of the two slits will be situated at distances Z_i and Z_i' from the hologram plane given by,

$$\frac{1}{Z_i} + \frac{1}{S_o} = \gamma^2 \frac{\lambda_2}{\lambda_1}\frac{1}{Z_o} \tag{29}$$

and

$$\frac{1}{Z_i'} + \frac{1}{S_o} = \gamma^2 \frac{\lambda_2}{\gamma_1}\frac{1}{Z_o'} \tag{30}$$

Subtracting Eq. (30) from Eq. (29) we obtain,

$$\frac{1}{Z_i} - \frac{1}{Z_i'} = \gamma^2 \frac{\lambda_2}{\lambda_1}\left(\frac{1}{Z_o} - \frac{1}{Z_o'}\right) \tag{31}$$

or in the limit when $Z_o' \to Z_o$ and $Z_i' \to Z_i$,

$$\Delta\left(\frac{1}{Z_i}\right) = \gamma^2 \frac{\lambda_2}{\lambda_1}\Delta\left(\frac{1}{Z_o}\right)$$

or

$$\frac{\Delta Z_i}{Z_i^2} = \gamma^2 \frac{\lambda_2}{\lambda_1}\frac{\Delta Z_o}{Z_o^2} \tag{32}$$

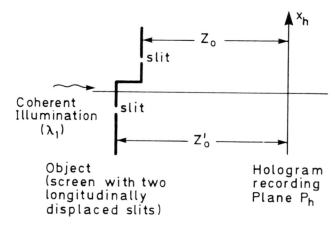

Fig. 3. Geometry used in analyzing longitudinal image scaling.

from which we obtain the longitudinal scaling of the image of a three dimensional object,

$$M_{long} \triangleq \frac{\Delta Z_i}{\Delta Z_o} = \gamma^2 \frac{\lambda_2}{\lambda_1} \frac{Z_i^2}{Z_o^2} \qquad (33)$$

Comparing the longitudinal scaling M_{long} with the lateral scaling M of Eq. (22) we see that

$$M_{long} = \gamma \frac{Z_i}{Z_o} M \qquad (34)$$

which can be further simplified through elimination of Z_i making use of the focusing condition Eq. (15b),

$$M_{long} = \frac{\dfrac{\lambda_1}{\lambda_2} \dfrac{1}{\gamma}}{1 - \dfrac{1}{\gamma^2} \dfrac{\lambda_1}{\lambda_2} \dfrac{Z_o}{S_o}} M \qquad (35)$$

In general, therefore, the longitudinal and lateral scaling (magnification or demagnification) of a three dimensional object are not equal leading to image distortion. This would mean that the image of a sphere would be a spheroid. However, when $S_o = \infty$ as would happen when the reconstruction line source is situated at infinity or equivalently and more practically, when a plane wave is used in the reconstruction we see that $M_{long} = M$ if $\gamma = \lambda_1/\lambda_2$ i.e., if the hologram transparency is demagnified by the ratio of the recording to the reconstruction wavelengths 3-D image distortion does not occur.

2.4   Resolution in Holography

Factors influencing the resolution capability in holography (or the ability to discern fine object detail) include:

  a) Size of the hologram recording aperture (e.g., size of photographic plate in optical holography)—diffraction limited resolution.
  b) In some practical cases the resolution capability of the recording device (e.g., resolution of film in lines/mm) can limit the maximum useful recording aperture and therefore ultimately the image resolution.
  c) Aberrations of the holographic process [8].

In practice these factors combine in a complicated manner in determining ultimate resolution. The basic limiting factor in many instances however is aperture size. This is particularly true in microwave imaging situations. In the following we shall consider the influence on resolution by the recording aperture alone which so far in the analysis has been assumed to be infinite.

To determine quantitatively the diffraction limited resolution capability of the holographic imaging process we will compute the line spread function or impulse response of the system. We retrace our steps to Eq. (18), which we rewrite using Eq. (19) as

$$\int_{x_h=-\infty}^{x_h=\infty} e^{j[-(k_1\gamma/Z_o)x_o+(k_2/Z_i)x_i]x_h} \, dx_h = \delta\left[-\frac{k_1\gamma}{Z_o}x_o + \frac{k_2}{Z_i}x_i\right] \qquad (36)$$

If we assume now a finite recording aperture of extent $2L$ such that $|x_h| < L$ and reevaluate the integral in Eq. (36), we obtain instead,

$$\int_{-L}^{L} e^{j[-(k_1\gamma/Z_o)x_o+(k_2/Z_i)x_i]x_h} \, dx_h = \left[\frac{e^{j[\ldots]x_h}}{j[\ldots]}\right]_{-L/\gamma}^{L/\gamma}$$

$$= \frac{2L}{\gamma} \frac{\sin[\ldots]L/\gamma}{[\ldots]L/\gamma} = \frac{2L}{\gamma} \operatorname{sinc}\left\{\left(\frac{k_2}{Z_i}x_i - \frac{k_1\gamma}{Z_o}x_o\right)\frac{L}{\gamma}\right\} \qquad (37)$$

We assume now $t(x_o) = \delta(x_o) + \delta(x_o - \Delta x_o)$ which represents an opaque object transparency with two transmitting slits located respectively at $x_o = 0$ and $x_o = \Delta x_o$ or two coherent object line sources located at these points such that the separation between the two slits is Δx_o. Then by making use of Eqs. (37) and (17) we find the conjugate image retrieved from a scaled replica of the finite sized hologram transparency of this object to be

$$\Psi_c(x_i) = C_1 \int_{-\infty}^{\infty} dx_o \left\{\int_{-\infty}^{\infty} A^*(k_{1x}) e^{jk_{1x}x_o} \, dk_{1x}\right\}$$

$$\cdot \underbrace{[\delta(x_o) + \delta(x_o - \Delta x_o)]}_{t(x_o)}$$

$$\times e^{j(k_1/2Z_o)x_o^2}e^{-j(k_2/2Z_i)x_i^2}\frac{2L}{\gamma}\operatorname{sinc}\left\{\left(\frac{k_2}{Z_i}x_i - \frac{k_1\gamma}{Z_o}x_o\right)\frac{L}{\gamma}\right\} \qquad (38)$$

which yields through the "sifting" property of the delta "function,"

$$\Psi_c(x_i) = 2L\, C_1\, e^{-j(k_2/2Z_i)x_i^2}\left\{C_2 \operatorname{sinc}\left[\frac{k_2}{Z_i}\frac{Lx_i}{\gamma}\right]\right.$$

$$\left. + C_3 \operatorname{sinc}\left[\frac{k_1}{Z_i}\frac{Lx_i}{\gamma} - \frac{k_1L}{Z_o}\Delta x_o\right]\right\} \qquad (39)$$

where

$$C_2 = \int_{-\infty}^{\infty} A^*(k_{1x}) \, dk_{1x} \qquad (39a)$$

and

$$C_3 = \int_{-\infty}^{\infty} A^*(k_{1x})e^{j\Delta x_o k_{1x}} \, dk_{1x} \qquad (39b)$$

Thus, the CFA of the retrieved conjugate image of two slits or line sources consists of two $\operatorname{sinc}[\cdot]$ functions separated laterally in the spatial frequency coordinate $q = (k_2L/Z_i\gamma)x_i$ by a distance

$$\Delta q = \frac{k_i}{Z_o}L\,\Delta x_o \qquad (40)$$

as shown in Fig. 4 assuming for simplicity $C_2 = C_3$. Note that since $\gamma = \lambda_1/\lambda_2 = k_2/k_1$, the variable q can also be expressed as $q = (k_1L/Z_i)x_i$. If we assume that the images of the two slits remain distinguishable or resolved when their minimum separation is such that the peak in the $\operatorname{sinc}[\cdot]$ function of the one falls on the first zero of the $\operatorname{sinc}[\cdot]$ function of the other and vice versa as drawn in Fig. 4, then $\Delta q = \pi$ since the first zero of $\operatorname{sinc}[q]$ occurs for $q = \pi$. Thus, we have under this condition which is commonly known as the Rayleigh resolution criterion,

$$\Delta q = \frac{k_1 L}{Z_o} \Delta x_o = \pi \qquad (41)$$

or

$$\Delta x_o = \frac{\lambda_1 Z_o}{2L} \qquad (42)$$

Thus, the larger the extent 2L of the hologram recording aperture the smaller is the separation Δx_o of the two slits in the object plane for which distinguishable images are obtained. It is important to note that changing the hologram in size (through the parameter γ) does not affect the object resolution expressed by Eq. (42).

It is worthwhile to note that in practice the image is observed or recorded with a square law detector such as the eye or a photographic plate. This means, as pointed out earlier, that the observable is not the image CFA $\Psi_c(x_i)$ but its intensity $I_c(x_i) = \Psi_c \Psi_c^*$. When this fact is applied to Eq. (39) we note that cross products of the two sinc[·] functions arise which, unless negligible, can alter the resolution limit of Eq. (42) as it was derived here assuming that such cross products are zero or negligible, i.e. only when radiation from the two slits is uncorrelated. Generally, therefore, the use of coherent radiation leads as shown in Fig. 4 to a bipolar* impulse response and resolution capability which is either slightly better or worse than that predicted by the "incoherent" resolution limit of Eq. (42) depending on the relative phase between radiation emanating from the two slits [11].

The corresponding resolution capability in the image plane is obtained by recalling the lateral scaling M and the correspondence

$$D^*(x_o) \to D^*(M^{-1} x_i) \qquad (43)$$
$$x_o = M^{-1} x_i \qquad (43a)$$

or

$$\Delta x_i = M \Delta x_o \qquad (43b)$$

which yields making use of Eqs. (22) and (42),

$$\Delta x_i = M\Delta x_o = \left(\gamma \frac{Z_i}{Z_o} \frac{\lambda_2}{\lambda_1}\right) \frac{\lambda_1 Z_o}{2L} = \gamma \frac{\lambda_2 Z_i}{2L} \qquad (44)$$

Therefore, unlike the resolution capability in the object space, changing the hologram in size before reconstruction will affect the separation Δx_i in the image plane through γ. This is not unexpected since by changing γ the image scale factor M is also changed and the image is magnified or demagnified depending on whether M is smaller or greater than unity.

3. MICROWAVE CONSIDERATIONS

Extension of the principles of optical holography presented in the preceding section to microwaves gives rise to the following new set of considerations:

---

* The term bipolar refers to the presence of both positive and negative values in the sinc q function.

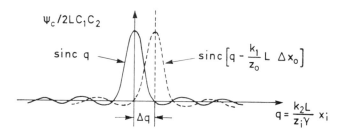

Fig. 4. Image of two line sources used in determining the diffraction limited resolution capability of the holographic process.

(a) Sensitive area detectors, analogous to the photographic plate, that can be useful in the mapping of power density or intensity distributions of microwave fields are not available. Several attempts have been made to realize such a detector, for example in the form of liquid crystal panels [12], [13]. The sensitivity of such devices is, however, so low that the high irradiance levels needed for their activation are in many applications unacceptable. The recording of microwave holograms over usefully sized apertures remains best performed with discrete sensors such as the point-contact crystal diode or mixer, thermistors, or bolometers. Of these, the point-contact diode is the most widely used because of its high sensitivity and short response time. Two modes of wavefront mapping can be employed: (a) by electronic interrogation of a planar array of sensors forming the hologram recording aperture or (b) by mechanical scanning of a single sensor over the recording aperture in a raster or other suitable format to form in time a synthetic aperture. For economical reasons the majority of nonreal-time microwave holography work is conducted using the latter mode of data acquisition. Real-time operation on the other hand would require the use of holographic arrays with electronic interrogation of sensor elements. Although the use of microwave microstrip circuits and antennas has the potential of reducing the cost of such arrays, their cost at the present time remains quite high because of the large number of elements needed. Waveguide arrays are discussed elsewhere in this volume by Fofi *et al* and Guo *et al*. Lensless Fourier transform hologram recording arrangements [14] in which the reference wave is furnished by a coherent point source located in the vicinity of the object, can be used to reduce the fringe density of the interference pattern in the hologram plane. Then the number of elements required to map the fringe pattern without causing undersampling and aliasing* is drastically reduced [15]. It has been shown by Mackovski [15] that the minimum number of elements in a Lensless Fourier-transform hologram recording array arrangement equals roughly the number of resolvable resolution cells in object space. Thus, even in the lensless Fourier transform arrangement, the number of elements in the hologram recording array is not expected to be less than a few hundreds or thousands with the exact figure being dependent on object size and desired resolution.

---

* Undersampling and aliasing lead to the retrieval of multiple overlapping images. An extensive treatment of this aspect is found in Refs. [19] and [34].

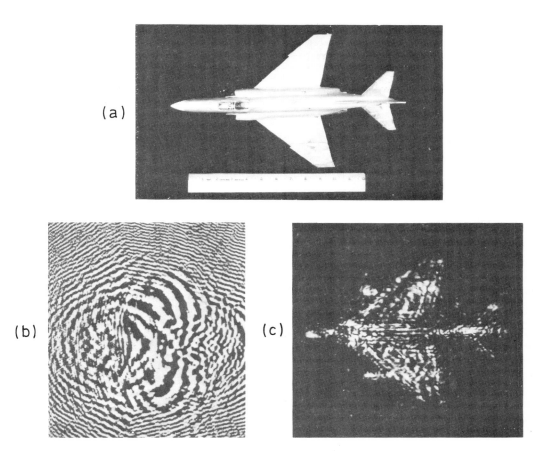

Fig. 5. Example of the imaging capabilities of close-range microwave holography: (a) test object, (b) hologram, (c) retrieved image showing bothersome speckle.

(b) The relation between longitudinal and lateral image scale (magnification or demagnification) derived in the preceding section [Eq. (35)] indicates that distortionless 3-D image reconstruction can be obtained when a collimated laser beam is used in the reconstruction process by scaling the hologram transparency by a factor $(1/\gamma)$ where $\gamma = \lambda_1/\lambda_2$. In microwave holography λ_1 is the wavelength of microwaves used to record the hologram and λ_2 is the wavelength of the laser beam employed in the optical image retrieval from the scaled microwave hologram transparency. The ratio λ_1/λ_2 can therefore be of the order of several thousands implying that the scaled optical transparency equivalent of a microwave hologram recorded for example over a few meters aperture would be less than a millimeter in size. It is certainly difficult (if not impossible) to view, as in optical holography, a 3-D virtual image though such a minute hologram even with the aid of optical instruments (e.g. a microscope) since these introduce their own longitudinal distortion. Furthermore, for such a high value of γ the image is extremely small because both the lateral and longitudinal scales M and M_{long} are extremely small as indicated by Eqs. (28) and (35). To bypass this scaling problem, microwave holographers have learned to settle for viewing 2-D imagery obtained by projecting the reconstructed real image on a screen. This allows

relaxing the microwave hologram reduction requirement so that the size of the resulting optical transparency equivalent can be somewhat increased. This permits the use of photographic film with lower resolution such as highly convenient Polaroid films.

(c) Because of the small size (measured in wavelengths) of microwave apertures that can be realized at present in practice, and because of the bipolar* nature of their coherent impulse response which was referred to previously, images formed by such apertures are normally degraded by coherent noise or speckle. Shown in Fig. 5 is an example of microwave holography of a test object which illustrates coherent noise or speckle effect in the image. The test object was a 38cm-long metalized scaled aircraft model [Fig. 5(a)]. The hologram shown in Fig. 5(b) was recorded using 4mm microwaves and a scanned transmitter-receiver (T-R) mode of aperture synthesis [16]. The scanned T-R mode of recording with the object fixed is exactly equivalent to scanning the object in front of a stationary T-R. For practical reasons the scanned object mode is usually preferred. Both of these modes of scanned holography usually employ a raster scan format but other scan formats have been used [17]–[19]. Both have the advantages of (a) furnishing multiaspect illumination and (b) enhancing the effective resolution of the aperture by a factor of two without decreasing the wavelength [16]. This property of scanned object or scanned T-R holography is very useful in some microwave imaging situations where a compromise between attenuation due to propagation and resolution is to be found as is true in microwave imaging of the human body. Indeed it was used by Larsen and Jacobi for microwave imagery of isolated organs as presented elsewhere in this volume. In the present case, the scanned aperture size was 38 cm × 42 cm and the object was centered in front of the synthesized recording aperture at a distance of 50 cm. The effective size of a resolution cell on the object according to Eq. (42) should thus be of the order of a few millimeters. The image of Fig. 5(c) was obtained by optical computing using the instantaneous Fourier transform properties of the convergent lens in coherent (laser) light [11]. Further detail on both the recording and reconstruction arrangements briefly described here can be found in [16].

(d) Unlike optical holography, where in many instances the vector nature of the wavefield can be overlooked, microwave holography must, strictly speaking, account for the vector nature of the illuminating and object scattered wavefields. This calls for polarization sensitive measurement techniques that can convert polarization diversity data into useful information. A scheme for the inclusion of polarization diversity data to enhance image quality is given briefly at the end of the next section and in the chapter by Larsen *et al.*

One can see at this point that the development of microwave holographic imaging systems that possess resolution and image quality comparable to those of optical systems is hampered by several limitations that can be summarized as follows: (a) conventional approaches can not produce cost-effective real-time microwave apertures, (b) it is not possible to view a virtual 3-D image as in optical holography, (c) because of the low numerical apertures attainable in practice image quality is degraded by the presence of speckle and coherent noise and (d) polarization effects must be accounted for. It is clear that a new approach is needed to overcome these limitations.

3.1 Frequency Diversity Imaging

In an ongoing program initiated in 1974, at the Electro-Optics and Microwave Optics Laboratory of the University of Pennsylvania we have been studying wavelength or frequency diversity holography as a means of overcoming or bypassing the limitations cited above. Frequency diversity was studied as a means of increasing the amount of information collected by a recording aperture in order to enhance resolution. We commenced by inquiring into the conditions under which the data from N holograms of the same nondispersive object recorded over the same aperture, each at a different wavelength, can be combined to yield a single image superior in resolution and quality to the image retrieved from any of the individual holograms. Since such a process converts spectral degrees of freedom of the wavefield scattered from the object into spatial image detail one can expect an enhancement in image resolution. Furthermore, because of

wavelength diversity, the effect of object resonances is minimized and behavior similar to that encountered in noncoherent (white light) imaging is expected to emerge. Particularly, suppression of coherent noise or speckle will lead to further enhancement of image quality. Thus we would be combining the best of two worlds: the superior properties of coherent detection and the noise combating properties of noncoherent imaging systems.

One way of approaching the question posed above would be to determine the conditions under which the known formulas for the focusing condition, magnification, and impulse response in holography can be rendered independent of wavelength. This quickly leads to the conclusion that wavelength independence can be met if a reference point source centered on the object is used. This is, of course, the condition for recording a lensless Fourier transform hologram [14] where the presence of the reference point source in the object plane leads to the recording of a Fraunhofer diffraction pattern of the object rather than its Fresnel diffraction pattern because of the elimination of a quadratic phase term in the object wavefield in the recorded hologram. This is known to result, as pointed out earlier, in a highly desirable reduction in the resolution required from the hologram recording medium and is therefore of practical importance especially in nonoptical holography. More detail of the processing involved in combining the data in multi wavelength holography can be found elsewhere [20].

Additional insight into the process of attaining superresolution by wavelength diversity can be obtained by considering the concept of wavelength or frequency synthesized aperture [21]–[25]. The synthesis of a one-dimensional aperture by wavelength diversity is based on the simple fact that the Fraunhofer (or far field) diffraction pattern of a non-dispersive planar object changes its scale, i.e. it "breathes," but does not change its shape (functional dependence), as the wavelength is changed. A stationary array of coherent broadband sensors distributed in this breathing diffraction pattern at suitably chosen locations would "sense" different parts of the diffraction pattern as the wavelength is altered collecting thereby more information on the detail of the diffraction pattern and therefore on the object that gave rise to it than if the wavelength was fixed and the diffraction pattern was stationary. Each stationary sensor in the array is thus able to collect CFA data as the wavelength is changed, and the breathing diffraction pattern sweeps over it. This is the same set of data or information that can be collected by a movable sensor mechanically scanned over the appropriate part of the diffraction pattern if this was being kept stationary by fixing the wavelength. Hence the term wavelength or frequency synthesized aperture.

The orientation and location of the wavelength synthesized aperture for any planar distribution of sensors deployed in the Fraunhofer diffraction pattern of a planar object and for the retrieval of an image has been determined [22], [22]. It was clear, however, that extension of the wavelength diversity concept to the case of 3-D objects is necessary before its generality and practical importance could be established.

For this purpose we considered [26] an isolated planar object of finite extent with reflectivity $D(\bar{\rho}_o)$, where $\bar{\rho}_o$ is a two dimensional position vector in the object plane (x_o, y_o). The object is illuminated as shown in Fig. 6(a) by a coherent plane wave of unit-amplitude and of wave vector $\bar{k}_i = k\bar{1}_{k_i}$ produced for example by a distant source located at $\bar{R}_T$. The wavefield scattered by the object is monitored at a receiving point at $\bar{R}_R$ belonging to a recording aperture lying in the far field region of the object. The receiving point will henceforth be referred to as the receiver and the source point as the transmitter. The position vectors $\bar{\rho}_o$, $\bar{R}_T$ and $\bar{R}_R$ are measured from the origin of a cartesean coordinate system (x_o, y_o, z_o) centered in the object. The object is assumed to be nondispersive i.e., D is independent of k. However, when the object is dispersive such that $D(\bar{\rho}_o, k) = D_1(\bar{\rho}_1)D_2(k)$ and $D_2(k)$ is known, the analysis presented here can easily be modified to account for such object dispersion by correcting the data collected for $D_2(k)$ as k is changed.

Referring to Fig. 6(a) and ignoring polarization effects, the field amplitude at $\bar{R}_R$ caused by the object scattered wavefield may be expressed as,

$$\psi(k, \bar{R}_R) = \frac{jk}{2\pi} \int D(\bar{\rho}_o)e^{-j\bar{k}_i \cdot \bar{r}T} \frac{e^{-jkr_R}}{r_R} \, d\bar{\rho}_o \quad (45)$$

where $d\bar{\rho}_o$ is an abbreviation for $dx_o dy_o$ and the integration is carried out over the extent of the object. Noting that $\bar{r}_T = \bar{\rho}_o - \bar{R}_T$, $\bar{R}_T = -R_T \bar{1}_{k_i}$ and using the usual approximations valid here: $r_R \simeq R_R + \rho_o^2/2R_R - \bar{1}_R \cdot \bar{\rho}_o$ for the exponential and $r_R \simeq R_R$ for the denominator in Eq. (45) where $\bar{1}_R = \bar{R}_R/R_R$ and $\bar{1}_{k_i} = \bar{k}_i/k$ are unit vectors in the $\bar{R}_R$ and $\bar{k}_i$ directions respectively, one can write Eq. (45) as,

$$\psi(k, \bar{R}_R) = \frac{jk}{2\pi R_R}e^{-jk(R_T+R_R)} \int D(\bar{\rho}_o)e^{-j\bar{p}\cdot\bar{\rho}_o} \, d\bar{\rho}_o, \quad (46)$$

where we have used the fact that the observation point is in the far field of the object so that $\exp(-jk\rho_o^2/2R_R)$ under the integral sign can be replaced by unity. In Eq. (46), $\bar{p} = k(\bar{1}_{k_i} - \bar{1}_R) = p_x \bar{1}_x + P_y\bar{1}_y + p_z \bar{1}_z$ is a three dimensional vector whose length and orientation depend on the wavenumber k and the angular positions of the transmitter and the receiver. For each receiver and/or transmitter present, p indicates the position vector for data storage. An array of receivers for example would yield, therefore, as k is changed (frequency diversity) or as $\bar{k} (= k\bar{1}_{k_i})$ is changed (wave-vector diversity) a 3-D data manifold. The projection of this 3-D data manifold on the object plane yields $\psi(k, \bar{R}_T)$ because $\bar{p} \cdot \bar{\rho}_o = \bar{p}_t \cdot \bar{\rho}_o = p_x x_o + p_y y_o$ where $p_x = k(\bar{1}_{k_i} - \bar{1}_R)_x$ and $p_y = k(\bar{1}_{k_i} - \bar{1}_R)_y$ are the cartesian components of the projection $\bar{p}_t$ of $\bar{p}$ on the object plane. Accordingly Eq. (46) can be expressed as,

$$\psi(k, R_R) = \frac{jk}{2\pi R_R}e^{-jk(R_T+R_R)}$$

$$\times \int D(x_o, y_o) \, e^{-j(p_x x_o + p_y y_o)} \, dx_o \, dy_o \quad (47)$$

Because of the finite extent of the object, infinite limits can be assumed and the integral in Eq. (47) is recognized as the two dimensional Fourier transform $\tilde{D}(p_x, p_y)$ of $D(x_o, y_o)$. It is seen to be dependent on the object reflectivity function, the angular positions of the transmitter and the receiver and on the values assumed by the wavenumber k; but it is entirely independent of range. Information about D can thus be collected by varying these parameters. Note that the range information is contained solely in the factor $F = jk \exp[-jk(R_T + R_R)]/2\pi R_R$ preceding the integral. The field observed at $\bar{R}_R$ has thus been separated into two terms one of which, $\tilde{D}$, contains the lateral object information and the other F, contains the range information. The presence of F in Eq. (47) hinders the imaging process since it complicates data acquisition and if not removed, gives rise to image distortion because R_R is generally not the same for all receivers. To retrieve an image of the object via a 2-D Fourier transform of Eq. (47), which can be carried out optically, the factor F must first be eliminated. Holographic recording of the complex field amplitude given in Eq. (47) using a reference point source located at the center of the object will result in the elimination of the factor F and the recording of a Fourier transform hologram. This operation yields $\tilde{D}$ over a two dimensional region in the p_x, p_y plane. The size of this region, which determines the resolution of the retrieved image, depends on the angular positions of the transmitter and the receiver and on the values assumed by k, i.e., the extent of the spectral window used. The later dependence on k implies super-resolution imaging capability because of the frequency synthesized dimension of the 3-D data manifold generated. Because of the dependence of resolution on the relative position of the object, the transmitter, and the receiving aperture, the impulse response is clearly spatially variant. In fact, a receiver point situated at $\bar{R}_R$ for which $\bar{p}$ is normal to the object plane can not collect any lateral object information because for this condition $(\bar{p} \cdot \bar{\rho}_o = 0)$ the integrals in Eqs. (46) and (47) yield a constant. Such a receiving point is located in the direction of specular reflection from the object where the diffraction pattern is stationary, i.e., does not change with k. In this case, the observed field is solely proportional to F containing thus range information only. Obviously this case can easily be avoided through the use of more than one receiver which is required anyway when 2-D or 3-D object resolution is sought [25], [27].

The analysis presented here can be extended to three dimensional objects by viewing a 3-D object as a collection of thin meridional slices [see Fig. 6(b)] each of which representing a two-dimensional object of the type analyzed above. With the n-th slice we associate a Cartesian coordinate system x_o, y_{o_n}, z_{o_n} that differ from other slices by rotation about the common x_o axis. Since the vectors $\bar{p}, \bar{R}_T$ and $\bar{R}_R$ are the same in all n-coordinate systems, Eq. (46) holds. $\psi_n(k, P_R)$ is then obtained from projection of the three dimensional data manifold collected for the 3-D object on the x_o, y_{on} plane associated with the n-th slice. An image for each slice can then be obtained as described before. An inherent assumption in this argument is that all slices are illuminated by the same plane wave. This is a reasonable approximation when the 3-D object is weakly scattering and the Born approximation is applicable or when the 3-D object is perfectly reflecting and does not give rise to multiple reflections between its parts. Additional discussion on the Born approximation may be found in the chapter by Slaney *et al* elsewhere in this volume. For metallic targets, the two dimensional meridional slices $D_n(\overline{\rho}_{0_n})$ deteriorate into contours, such as C in Fig. 6(b) defined by the intersection of the meridional planes with the illuminated portion of the surface of the object. Accordingly we can write for the n-th meridional slice or contour.

$$\psi_n(k,R_R) = F \int D_n(\overline{\rho}_{0_n})e^{-j\overline{p}\cdot\overline{\rho}_{0_n}} \quad (48)$$

We can regard $D_n(\overline{\rho}_{0_n})$ as the n-th meridional slice or contour of a three-dimensional object of reflectivity $U(\overline{r})$ where $\overline{r}$ is a three-dimensional position vector in object space. This means that $D_n(\overline{\rho}_{0_n}) = U(\overline{r})\delta(z_{0_n})$ where δ is the Dirac delta "function." Consequently Eq. (48) becomes

$$\psi_n(k,R_R) = F \int U(\overline{r})\delta(z_{0_n})e^{-j\overline{p}\cdot\overline{\rho}_{0n}} \, d\overline{\rho}_{0_n}$$

$$= F \int (\overline{r})\delta(z_{0_n})e^{-j\overline{p}\cdot\overline{r}} \, d\overline{r} \quad (49)$$

where $d\overline{r}$ designates an element of volume in object space and, where the last equation is obtained by virtue of the sifting property of the delta function.

Summing up, the data from all slices or contours of the object we obtain,

$$\sum_n \psi_n = F \int U(\overline{r})e^{-j\overline{p}\cdot\overline{r}} \, dr = \psi(\overline{p}) \quad (50)$$

because

$$\sum_n U(\overline{r})\delta(z_{0_n}) = U(\overline{r}).$$

Assuming that the Factor F in Eq. (50) is eliminated as before, this equation reduces to

$$\psi(\overline{p}) = \int U(\overline{r})e^{-j\overline{p}\cdot\overline{r}} \, d\overline{r} \quad (51)$$

which is the 3-D Fourier transform of the object reflectivity $U(\overline{r})$.

Wavelength diversity permits, therefore, access to the 3-D Fourier space of a nondispersive object providing thereby the basis for 3-D Lensless Fourier transform holography. An alternate formulation to that given above (i.e. super-resolved wave vector diversity imaging of 3-D perfectly conducting objects) is possible [28] by extending the formulation of the inverse scattering problem [29], [30] to the multistatic case along lines that are somewhat different than those given by Raz [31]. The resulting scalarized formulas are identical to Eq. (51) establishing thus the connection between the holographic and the inverse scattering approaches to the imaging problem for reflecting targets.

The above considerations of multiwavelength holography have lead us to determine a way by which the 3-D Fourier space of a perfectly reflecting objects or a weakly scattering object with known dispersion law can be assessed employing coherent broadband techniques. It is clear that once the 3-D Fourier space data is available, 3-D image detail can be retrieved by means of an inverse 3-D Fourier transform which can be carried out digitally. Alternately holographic concepts can be invoked again. Fourier doman *projection theorems*

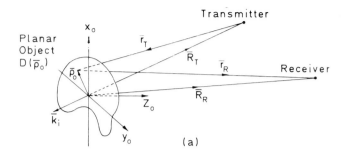

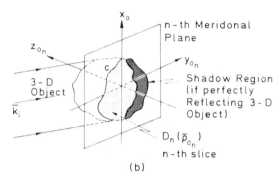

Fig. 6. Geometries for wavelength diversity imaging: (a) two dimensional object, (b) three dimensional object with the n-th meridional slice and cross sectional outline c shown.

(cf Slaney *et al* elsewhere in this volume) can be applied to the Fourier space data to produce a series of *projection holograms* from which 2-D images of meridional or parallel slices of the object can be retrieved on the optical bench [26]. This procedure does not involve specific scaling of the size of the optical hologram transparency relative to the size of the original recording aperture thus bypassing the *wavelength scaling problem* of conventional longwave holography discussed earlier which prevented the viewing of a 3-D image free of longitudinal distortion. The lateral and longitudinal resolutions in the retrieved image depend now on the dimensions of the volume in Fourier space accessed by wavelength diversity. This volume depends on the wavelength range, and on the recording geometry. Thus the longitudinal resolution does not deteriorate as rapidly with range as in conventional monochromatic imaging systems.

An example of a computer simulation of frequency diversity holographic imaging of a 3-D object consisting of eight point scatterers distributed as shown in Fig. 7 is shown in Fig. 8. Shown in Fig. 8 are three weighted Fourier domain projection holograms and the corresponding optically retrieved images for three equally spaced parallel slices of the object containing distinguishable 2-D distributions of scatterers. The simulated recording arrangement shown in Fig. 9 consisted of an array of 16 receivers equally distributed on an arc extending from $\phi = 40°$ to $\phi = 77.5°$ surrounding a central transmitter capable of providing plane wave illumination of the object. The results shown were obtained assuming a frequency sweep of (2–4) GHz. They clearly indicate a lateral and longitudinal resolution capability of the

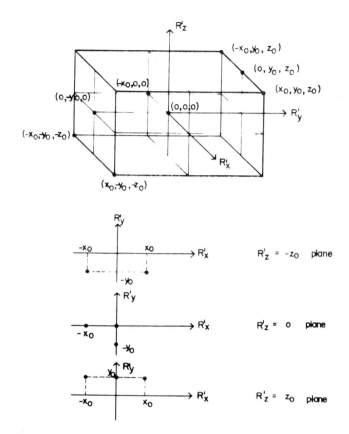

Fig. 7. 3-D object consisting of a set of eight point scatterers shown in isometric and Rx′ − Ry′ plane views at $R_z' = -z_o$, O, z_o, $x_o = y_o = z_o = 100$ cm.

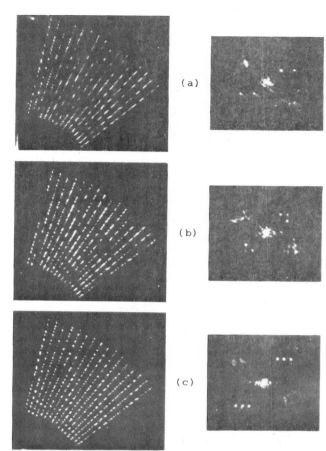

Fig. 8. Weighted projection holograms and their optical reconstructions for the set of point scatterers in Fig. 7 at different R_z' planes. (a) Hologram and reconstructed image of scatterers at $R_z' = -z_o$ plane. (b) Hologram and image at $R_z' = 0$ plane. (c) Hologram and image at $R_z' = z_o$ plane, $x_o = y_o = z_o = 100$ cm.

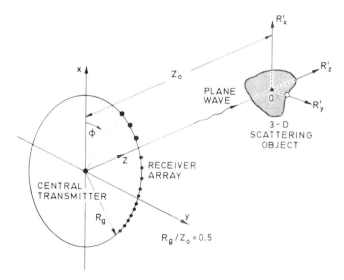

Fig. 9. Arrangement assumed in the computer simulation of frequency diversity 3-D imaging.

order of 25 cm. Wider sweep width yield better the resolution. For example a (1–18) GHz sweep would yield a 3-D resolution of the order of 1.5 cm. Note that these values apply to imaging in air. The ability to produce 3-D images in slices from coherently detected fields enable us to view the method also as *coherent tomography*. The Fourier space accessed in the above fashion by wavelength diversity can be considered as a *generalized 3-D hologram* in which one dimension has been synthesized by wavelength diversity. Such a hologram not only contains spatial amplitude and phase data as in conventional holography but also spectral information and hence can yield better resolution than the classical Rayleigh limit of the available aperture operating at the shortest wavelength of the spectral window used. This *super-resolving* property is further enhanced by suppression of the effects of object resonances and coherent noise in the retrieved image, the latter being so because frequency diversity tends to make the impulse response of the system unipolar resembling that of a non-coherent imaging system that is free of speckle and coherent noise artifacts [20]. Further enhancement of information content and resolution can be achieved by *polarization diversity* where the $\bar{p}$ space can be multiply accessed for different nonredundant polarizations of the illumination and the receivers and the resulting polarization diversity images added either coherently

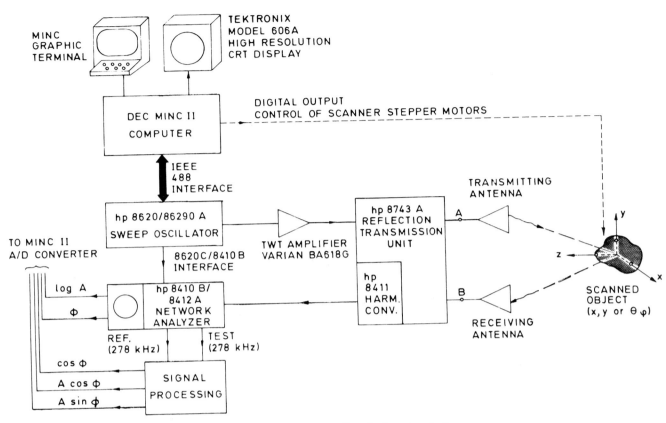

Fig. 10.   Automated microwave network and analyzer for frequency diversity imaging studies.

or non-coherently in order to obtain noise averaging as in [20]. This scheme is presently under study and results are to be published elsewhere.

There are intriguing parallels between wavelength diversity imaging and some known and postulated features of the echo-location in the dolphin and the bat. The most interesting of these is that both of these mammals employ "chirp" signals to enrich the echo with information enhancing thereby their ability to resolve objects in their environment beyond the classical Rayleigh limit of any possible physical receiving aperture that they may possess. In fact our study of wavelength diversity techniques as a means of enhancing resolution in conventional holography without increasing available aperture size was motivated to a large extent by noting these extraordinary features that have long been perfected by nature in these mammals.

4. POTENTIAL FOR MEDICAL IMAGING

The preceding discussion of microwave imaging assumed that the medium between the object and the measurement system is lossless and homogeneous. This assumption is naturally not true when we consider microwave imaging in biological systems where one is attempting to image internal structures. At microwave frequencies these systems are generally highly lossy, inhomogeneous in nature and dispersive. These characteristics make the imaging problem much more difficult as described elsewhere in this volume by Slaney *et al.*

Although the attenuation of microwave radiation in biological systems is quite high [32] the sensitivity of current sophisticated coherent microwave detection techniques permits useful data acquisition and information extraction [32], [33]. Because of the water dominated environment of biological systems, the wavelengths at microwave frequencies are about 9 times shorter than in air. This *wavelength contraction* implies a potential for higher resolution. In the context of frequency diversity imaging, this means that a frequency sweep of (0.1–3) GHz in a water dominated environment can yield resolutions equivalent to those attainable in air with a (0.9–27) GHz sweep. Thus the potential for achieving useful resolutions is quite good provided that means can be found to account for nonhomogeneous propagation and to correct for it in the reconstructed image. An example of a versatile automated measurement system suitable for the study of frequency diversity imaging in air or in a water dominated environment is shown in Fig. 10. The system which has been configured and assembled at our Electro-Optics and Microwave Optics laboratory consists of a computer controlled microwave network analyzer capable of making vector (amplitude and phase) measurements in the (0.1–18) GHz range. The computer/controller, a Digital Equipment DEC 11, provides automatic frequency stepping, object positioning, data acquisition and storage. It is also used in the computation of projection holograms and their display on a high resolution Tektronix CRT. The recognized power of such automated data acquisition systems stems from their speed and ability to analyze and correct for system errors yielding thereby unprecedented accuracies. This system is presently used in the study of various aspects of frequency and polarization diversity imaging with particular attention to imaging through aberrating inhomogeneous layers of known dispersion.

5. CONCLUSIONS

The removal of several longstanding constraints on conventional microwave holography through the use of wavelength diversity described here leads to a new class of imaging systems. These are capable of converting spectral degrees of freedom in the object wavefield recorded over an aperture into a 3-D spatial image with detail furnished by true super-resolution. Wavelength diversity is applicable to the imaging of two classes of objects: perfectly reflecting objects of the type encountered in radar and weakly scattering objects of low or known dispersion of the type encountered in biology and medicine. The potential application of wavelength diversity imaging to biosystems depends on the development of methods to correct image distortions caused by inhomogeneities of the medium. Image slicing and iterative corrections in what might be called an *image pealing* process are being considered along with means of synthesizing a movable point source reference within the object to realize the conditions for 3-D lensless Fourier transform holography.

6. ACKNOWLEDGMENT

This study has been sponsored by the U.S. Air Force Office of Scientific Research, Air Force Systems Command, USAF under grant No. AFOSR-77-3256A and the U.S. Army Research Office, Durham under grant No. DAAG-76-G-0230.

References

1. D. Gabor, "A new Microscopic Principle," *Nature*, Vol. 161, No. 4098, pp. 777–778, 1948.
2. G. Tricoles and Nabil H. Farhat, "Microwave Holography: Applications and Techniques," *Proc. IEEE*, Vol. 65, pp. 108–121, 1977.
3. A. P. Anderson, "Microwave Holography," *Proc. IEE*, Vol. 124, pp. 946–962, 1977.
4. E. N. Leith, "Quasi-holographic Techniques in the Microwave Region," *Proc. IEEE*, Vol. 59, pp. 1305–1318, 1969.
5. J. D. Blackwell and N. H. Farhat, "Image Enhancement in Longwave Holography by Electronic Differentiation," *Opt. Commun.*, Vol. 20, pp. 76–80, 1977.
6. M. A. Kujoory and N. H. Farhat, "Electronic Wavefront Subtraction in Longwave Holography," *Proc. IEEE* (Letters), Vol. 63, pp. 1258–1260, 1975.
7. M. A. Kujoory and N. H. Farhat, "Microwave Holographic Subtraction for Imaging Burried Objects," *Proc. IEEE* (Letters), Vol. 66, pp. 94–96, 1978.
8. J. A. Armstrong, "Fresnel Holograms: Their Imaging Properties and Aberrations," *IBM J. Res. Dev.*, Vol. 9, pp. 171–178, 1965.
9. E. L. Rope and G. Tricoles, "Microwave Holography: Difussers: Binary Detour-Phase Holography," Applications of Holography, J-Ch. Vienot, Ed., University of Besancon, 1970.
10. P. C. Clemmow, *The Plane Wave Spectrum Representation of Electromagnetic Fields*, Pergamon Press, New York, (1966).
11. J. W. Goodman, Introduction to Fourier Optics, New York: McGraw-Hill, 1968, Ch. 8.
12. C. F. Augustine, "Field Detector Works in Real Time," *Electronics*, Vol. 41, pp. 118–121, 1968.
13. C. F. Augustine, C. Deutsch, D. Fritzler, and E. Marom, "Microwave Holography Using Liquid Crystal Area Detectors," *Proc. IEEE*, Vol. 57, pp. 1333, 1969.
14. G. W. Stroke, *Introduction to Coherent Optics and Holography.* 2nd Ed. Academic Press, New York, p. 114, 1969.
15. A. Makovski, "Hologram Information Capacity," *J. Opt. Soc. Am.*, Vol. 60, pp. 21–29, 1970.
16. N. H. Farhat, "Microwave Holographic Imaging—Prospects for Real-Time Camera," SPIE Vol. 150, *Real-Time Signal Processing II*, pp. 10–14, 1979.
17. N. H. Farhat and A. H. Farhat, "Double Circular Scanning in Microwave Holography," *Proc. IEEE*, Vol. 61, pp. 509–510, 1973.
18. N. H. Farhat, "Quasi Real-Time Microwave Holography," *J. Franklin Inst.*, Vol. 296, pp. 393–402, 1973.
19. N. H. Farhat, W. Guard and A. H. Farhat, "Spiral Scanning in Longwave Holography," in *Accustical Holography*, Vol. 4, G. Wade (Ed.), Plenum Press, pp. 267–276, 1972.
20. N. H. Farhat and C. K. Chan, "Super-resolution by Wavelength Diversity in Longwave Imaging Systems," in *Optica Hoy Y Manāna*, J. Bescos et al. (Eds), Sociedad Espanola De. Optica, Madrid, 1978.
21. N. H. Farhat, "High Resolution Longwave Frequency Swept Holographic Imaging," Ultrasonics Symposium Proceedings, IEEE Cat. No. 75 CHO 944-4SU, 1975.
22. N. H. Farhat, "Frequency Synthesized Imaging Apertures," Proc. International Optical Computing Conference, IEEE Cat. No. 76 CH 1100-7C, 1976.
23. N. H. Farhat, "New Imaging Principle," *Proc. IEEE* (Letters), Vol. 64, pp. 379–380, 1976.
24. N.H. Farhat, "Principles of Broadband Coherent Imaging," *J. Opt. Soc. Am.*, Vol. 67, pp. 1015–1020, 1977.
25. N. H. Farhat, "Comment on a New Imaging Principle," *Proc. IEEE* (Letters), Vol. 66, pp. 609–700, 1978.
26. N. H. Farhat and C. K. Chan, "Three Dimensional Imaging by Wave-vector Diversity," in *Acoustical Imaging*, A. Metherell (Ed.), Plenum, New York, 1980.
27. W. M. Waters, "Comment on a New Imaging Principle," *Proc. IEEE* (Letters), Vol. 66, pp. 609–700, 1978.
28. C. K. Chan, "Analytical and Numerical Studies of Frequency Swept Imagery," Ph.D. Dissertation, Univ. of Pennsylvania, Philadelphia, 1978.
29. N. H. Bojarski, "Inverse Scattering," Final Report, Contract B000-19-73-C-0316, Naval Air Syst. Command, 1974.
30. R. M. Lewis, "Physical Optics Inverse Diffraction," *IEEE Trans. Antennas Propag.*, Vol. AP-24, pp. 66–70, 1969.
31. S. R. Raz, "On Scatterer Reconstruction from Far Field Data,"

IEEE Trans. Antennas Propag., Vol. AP-24, pp. 66–70, 1976.

32. L. E. Larsen and J. H. Jacobi, "Microwave Interrogation of Dielectric Targets Part I: By Scattering Parameters, Part II: By Microwave Time Delay Spectroscopy," *Med. Phys.*, Vol. 6, pp. 500–513, Nov./Dec., 1978.

33. —— "Microwve Scattering Parameter Imagery of an Isolated Canine Kidney," *Med. Phys.* Vol. 6, Sept./Oct., 1979.

34. N. H. Farhat, "Longwave Holography (Acoustic and Microwave), in *Holography in Medicine*, P. Greguss (Ed.), IPC Science and Technology Press, Richmond, Surrey, England, pp. 17–25, 1975.

APPENDIX I

The Angular Spectrum Representation of Scalar Wavefields

For convenience a simplified treatment of the angular spectrum representation of a one-dimensional field distribution is presented here. More detailed treatments can be found in the literature as for example in reference [10]. Consider a uniform plane wave of amplitude A propagating in the direction of the wavevector $\bar{k}$ as shown in Fig. 11 with $k = 2\pi/\lambda$, λ being the wavelength. Then for the two-dimensional geometry shown,

$$\left.\begin{array}{l}\bar{k} = k_x\bar{I}_x + k_z\bar{I}_z \\[1em] k_x = k\sin\theta \\[0.5em] k_z = k\cos\theta\end{array}\right\} \quad \text{(A.1)}$$

where,

and where $\bar{I}_x$ and $\bar{I}_z$ are the Cartesian unit vectors. Since the position vector of a field point P on the x-axis is $\bar{r} = x\,\bar{I}_x$, we find that $\bar{k}\cdot\bar{r} = k\sin\theta\,x$. Therefore, the CFA along x will be

$$U(x) = Ae^{-j\bar{k}\cdot\bar{r}} = Ae^{-jk_x x} \quad \text{(A.2)}$$

It is seen that the phase of $U(x)$ is linearly dependent on x just as the expression $e^{j\omega t}$ in the time domain represents a time signal whose phase varies linearly with time. The quantity $f_x = k_x/2\pi = \sin\theta/\lambda$ can be regarded therefore, analogously, as a spatial frequency just as $f = \omega/2\pi$ is the temporal frequency. The unit of the spatial frequency is seen to be inverse length $[\text{m}^{-1}]$.

Referring now to the geometry of Fig. 11 and to Eq. (A.1) we see that

$$\bar{k}\cdot\bar{k} = k^2 = k_x^2 + k_z^2$$

therefore

$$k_z = \sqrt{k^2 - k_x^2} \quad \text{(A.3)}$$

Thus, since $k = 2\pi/\lambda$ is a known quantity because λ is known, knowledge of k_x is sufficient to determine k_z and specify fully the direction of propagation of the plane wave.

Imagine now a spectrum of such plane waves of the same wavelength λ each with its own complex amplitude propagating in a distinct direction specified by different angles θ in Fig. 11. To distinguish between the complex amplitudes

of the various plane waves we designate this amplitude $\frac{1}{2}\pi A(k_x)$. Then according to the discussion in the previous section, the CFA produced along the x coordinate by any one component of this "spectrum" may be expressed as

$$\frac{1}{2\pi} A(k_x)e^{-jk_x x} \quad \text{(A.4)}$$

Assuming that the direction of polarization of the electric field in all the component plane waves is identical, or equivalently that we are dealing with scalar fields, the net CFA produced along x by the entire set of waves comprising the spectrum can be shown to be given by the superposition integral,

$$f(x) = \frac{1}{2\pi}\int_{-\infty}^{\infty} A(k_x)e^{-jk_x x}\,dk_x \quad \text{(A.5)}$$

which indicates that the CFA along x is a Fourier transform of $A(k_x)$. It follows then by the inverse Fourier transform that,

$$A(k_x) = \int_{-\infty}^{\infty} f(x)e^{jk_z x}\,dx \quad \text{(A.6)}$$

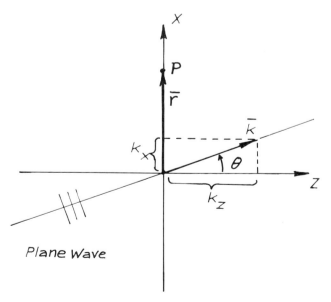

Plane Wave

Fig. 11. Plane wave geometry for development of angular spectrum representation of scalar monochromatic wavefields.

Thus f(x) and $A(k_x)$ form a so-called Fourier transform pair, a condition abbreviated symbolically by $f(x) \leftrightarrow A(k_x)$. The complex quantity $A(k_x)$ is called the "angular spectrum" associated with the CFA f(x). Equations (A.5) and (A.6) suggest, subject to the existence of a Fourier transform, that any CFA distribution f(x) along the x coordinate can be synthesized through the superposition of the CFA's produced by the component waves of an angular spectrum of plane waves with specific complex amplitude weighting $A(k_x)$ given by Eq. (A.6). The analogy to the synthesis of time functions by a continuous superposition of a spectrum of sinusoidal time functions in Fourier analysis is evident.

The two-dimensional generalizations of Eqs. (A.5) and (A.6), can be shown to be,

$$f(x, y) = \frac{1}{(2\pi)^2} \int_{-\infty}^{\infty} A(k_x, k_y)e^{-j(k_x x + k_y y)} \, dk_x \, dk_y \qquad (A.7)$$

and

$$A(k_x, k_y) = \int_{-\infty}^{\infty} f(x, y)e^{j(k_x x + k_y y)} \, dx \, dy \qquad (A.8)$$

where f(x, y) is the CFA produced in the x-y plane by the angular spectrum of plane waves $A(k_x, k_y)$. Since a Fourier transform is involved, necessary and sufficient conditions for the existance of an angular spectrum representation of a given CFA distribution f are basically those for the existance of a Fourier transform of a function, namely that f is continuous with finite number of finite discontinuities and is square integrable, that is $(\int_{\infty}^{\infty} |f(x)|^2 \, dx)$ is finite.

Examination of Video Pulse Radar Systems as Potential Biological Exploratory Tools

J. D. Young and L. Peters, Jr.

Possible video pulse applications to biological structures are to be discussed. The pertinent parts of the current state-of-the-art of our underground video pulse radar and target identification systems is briefly reviewed. Suggestions made for extending these concepts to biological structures are given. These extensions could possibly include a) detection, b) classification, c) interaction (or heating), and d) measurement of electrical parameters "in vivo."

1. INTRODUCTION

For more than ten years, considerable research and development efforts at Ohio State University and elsewhere has been devoted to the complex but important problem of using radar for underground sensing. Much has been discovered about the electrical characteristics of the ground medium, and the characteristics and signatures of desired underground targets. Using this knowledge, several systems have been developed for applications where radar seems particularly suited, and new, more difficult problems are now being attacked.

All of these systems may be called video pulse radars, because they make use of an impulsive transmitted signal whose spectrum covers at least a 10:1 frequency bandwidth.

Concurrently, other researchers have made important discoveries in the area of microwave interaction with biological tissues. We have not made an exhaustive literature search, but are aware of significant results in two broad areas: 1) biological effects relating to safety precautions, design requirements, and operating restrictions on all microwave hardware, and 2) possible medical applications of microwave instruments.

In reviewing the goals, problem areas, and systems involved in these two disciplines, many similarities become apparent. This paper discusses the goals and problems in underground video pulse radar applications and shows their relevance to biological sensing. Second, it describes several systems which are used in buried utility location, target location and identification, and underground mapping. Fi-

nally, it discusses the possible application of video pulse radars to microwave interrogation of biological targets.

2. GOALS AND PROBLEMS IN UNDERGROUND RADAR

The problem of sensing underground objects with radar systems is summarized in Fig. 1. A broadband signal generator produces energy which is coupled into a lossy, dispersive medium by means of an antenna placed on the interface. Within the medium are a wide variety of regions whose electrical constitutive parameters ϵ (permittivity), σ (conductivity), and/or μ (permeability), are different than the surrounding material, and thus alter the propagation to a measurable extent. A receiving antenna senses a portion of the radiated energy which has been affected by the scattering of these "targets." A receiver/processor is used to record the received signals as a function of time, frequency and/or antenna position. The existence, location, and nature of specific targets of interest are to be extracted from the above information.

Note that the above description is precisely applicable to the problem of microwave interrogation of biological targets. Additional information on biological dielectrics and propagational interfaces is presented elsewhere in this volume by Lin as well as by Burdette *et al.* Since videopulse has never been applied to biosystems, the experimental portion of this paper will be in geophysical settings. The theory is, however, applicable to biomedical targets.

The electrical characteristics of the media play a very important role in the design and potential performance of underground radar sensing systems. Experimental studies have been made involving the propagation of electromagnetic signals through soil overburden [1] and through deeper geological strata [2]. Results show that the dielectric constant and conductivity are frequently dependent, and vary con-

The Ohio State University ElectroScience Laboratory, Department of Electrical Engineering, Columbus, Ohio 43212.

The work reported in this paper was sponsored in part by Contracts DAAG53-76-C-0178 and DAAK-77-C-0174 with U.S. Army Mobility Equipment Research and Development Command; Contract RP 7856-1 with Electric Power Research Institute, Inc.; and Contract 5015-353-0234 with Gas Research Institute.

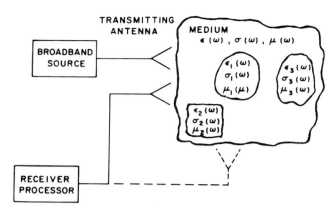

Fig. 1. Video pulse radar system for detecting objects in a material medium.

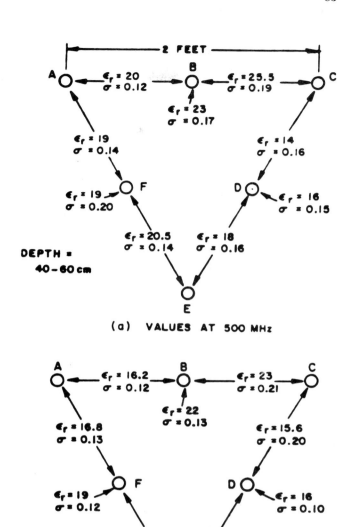

(a) VALUES AT 500 MHz

(b) TIME-DOMAIN VALUES

Fig. 3. (a) Parameter values of soil at 500 MHz using two methods in cross-verification study #1. The monopole probe values point to locations B, D, and F. The dipole values are given between the holes. (b) Parameter values from the same set of measurements using simple time-domain computations.

siderably as a function of soil (or rock) chemical composition, which is in turn a function of geographical location, season, position, depth, and moisture content.

One of the more recent and most accurate techniques for *in situ* measurement of soil constitutive parameters have been developed by Hayes [1]. Figure 2 shows a diagram of the experiment. A steel probe with a small transmitting antenna at its tip is inserted into the soil, and a calibrated

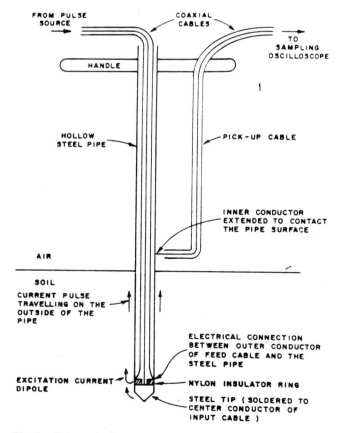

Fig. 2. Schematic diagram of the monopole probe. The pick-up cable should not touch the ground at the contact point.

transient pulse is radiated. The received signal from a small stub antenna at the ground surface is recorded as a function of probe tip depth. For the particular transient pulse used, this technique produces information on relative dielectric constant ϵ_r and conductivity σ vs frequency and depth within ranges of ~50 MHz to 1.3 GHz and 30 cm to 150 cm. Probes can also be used in pairs to obtain lateral propagation information from one probe to another. Figure 3 shows a set of data for six probe positions spaced 1 ft. apart in a triangle, located in a mowed lawn on the Ohio State University. Fig-

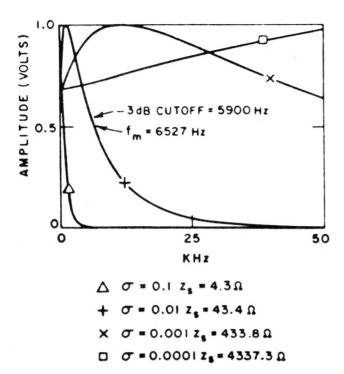

Fig. 4. Normalized frequency responses for transmission between two parallel 10m long dipoles 240 m apart in homogeneous media of various conductivities. The difference between the −3 dB cutoff frequency (with respect to the amplitude at zero frequency) of the system of two dipoles is compared to f_m for $\sigma = 10^{-2}$ mhos/ m.

ure 3(b) shows data averaged over the frequency spectrum of the transient pulse, Fig. 3(a) shows the data for a particular frequency (500 MHz). Note that this soil which seems uniform as far as physical appearance on its surface is concerned is remarkably inhomogeneous. Other measurements using similar techniques at a specific site discovered variations of more than 40% in ϵ and 100% in σ for soil before and after a rain.

In summary, the soil is correctly characterized as a lossy, dispersive, inhomogeneous medium, characterized by significant conductivity which is strongly related to moisture content. This statement also applies to human tissue!

3. THE BIOLOGICAL MEDIA

3.1 Introduction

The discussion given by Johnson and Guy [3] are used in this paper to establish a basis for the electromagnetic properties of human tissue. The basic parameters of interest to us here are listed in Tables I and II of Ref. [3]. Specifically, for tissues with high water content, the frequency band of interest could range from 0.200 to 10 GHz, the relative permittivity from 56 to 40 and the conductivity from 1.28 to 10.3 mhos/m. For tissues with low water content, the relative permittivity over the same frequency band ranges from 6.0 to 4.5 and the conductivity from 0.25 to .500 mhos/m.

Observe that the experiments performed to obtain the constitutive parameters of the earth could also be performed in "vivo" for biological tissue using a carefully designed hypodermic needle for the probe in this case using small semi-rigid cable as feed cables as has been done by Burdette et al. The reader is directed to the papers of Burdette in this volume and in a prior paper [4], [5]. It is observed that Burdette presented new values for the electrical properties of tissue at the Electromagnetic Dosimetric Imagery Symposium but these values have not been incorporated in this present work. There are several distinct differences in the two approaches. Burdette is measuring self impedance of a probe quite similar to that of Fig. 2. Whereas the measurements we have made are mutual impedances between probes Burdette's approach leads to results from a more localized sample in the form of local reflections. This technique has the advantage of being sensitive to only a small tissue volume but has the disadvantage of reduced accuracy in loss tangent measurements for low loss tissues. Our measurements include a propagation path and inherently represents a more accurate measure of conductivity for low loss tissues.

This is quite similar to the tomographic techniques used by Lyttle at Lawrence Livermore [6]. They have recently designated this technique as "Geotomography" [7] and it usually has been applied to isolate anomalies in the earth in a manner similar to the image development of Larsen and Jacobi in this volume.

A second difference lies in our use of the transient signals which makes it possible to "gate out" scattering from adjacent anomalies. We would recommend that both procedures be combined in any future work.

An analysis of propagation through lossy dielectric media has been performed. It has been shown that, depending upon constitutive parameters and total energy path through the medium, there may be two "windows" in the frequency spectrum where operation is feasible. These results may be applied to biological materials also.

If the antenna is in electrical contact with the medium, then "propagation" in the so-called Low Frequency Window (LFW) is possible in a conducting media for frequencies below the critical frequency

$$f_c = \frac{3.76 \times 10^6}{\sigma r^2} \qquad (1)$$

where r is the total propagation path in the media [8], [9]. Here f_c is the frequency at which the fields at the distance r have decayed 3 dB below the field at f = 0. This is illustrated in Fig. 4 for an example pertinent to an earth medium.

This low frequency window becomes significant primarily because of its dispersionless characteristics. The source of these characteristics is readily seen by expanding the fields of a short dipole. There are two significant factors given by

$$e^{-\gamma r}[1 + \sqrt{r} + (\sqrt{r})^2]$$

Expanding the first factor in a binomial expansion and retaining the first two terms for the product gives

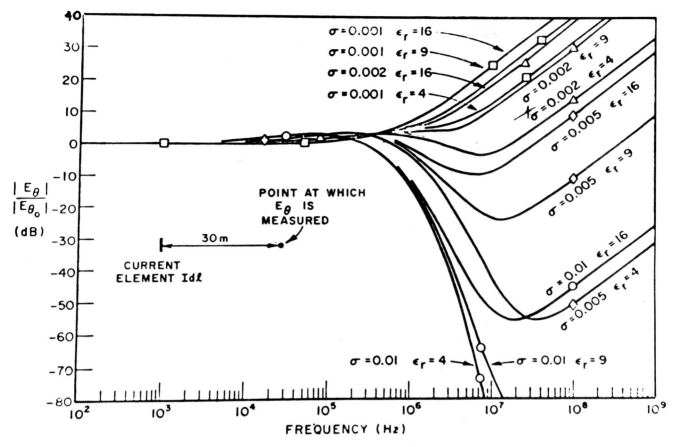

Fig. 5.   Normalized electric field strength 30 m from current element for lossy media.

$$1 + \frac{(\gamma r)^2}{2} ----$$

One observes that the linear term does not appear and thus the media appears to have a deeper range of non-dispersionless propagation when the radiator is a small dipole in contrast to plane wave propagation.

There is also a High Frequency Window (HFW). This now has the form of a high pass filter for infinitessimal dipoles. The critical frequency for the HFW occurs when the fields again become equal to the DC value (as frequency increases). See Fig. 5 as an example. A universal set of curves is shown in Fig. 6 so that the biological media parameters can readily be introduced. For finite length antenna elements the monotonic increasing nature of the fields shown in Fig. 6 is not valid after the frequency is increased beyond the resonant frequency of the antenna is seen in Fig. 7.

We observe also that Eq. (1) defines the critical frequency for the LFW to be independent of the dielectric constant since it is based on the assumption of a conducting medium. The more general case could be obtained from Fig. 6 for any medium. It would appear that operation in some biological media for short ranges is worthy of consideration provided the target is a significant scatterer in this frequency band.

For the depths of interest and the electrical parameters of the high water content media, only the low values of $[r\sigma/2]$ $(\mu/\epsilon)]$ are significant.

The tissues with high water content represent a conducting medium for frequencies below 400 MHz. In this case, the propagation factor

$$\gamma \simeq \alpha + j\beta \cong \sqrt{\frac{\omega\mu\sigma}{2}} + j\sqrt{\frac{\omega\mu\sigma}{2}} \qquad (2)$$

We observe that the dielectric constant has little real significance in this case.

The tissues with high water content for frequencies above 400 MHz and all tissues reported in Ref. [3] with low water content are lossy dielectrics and

$$\gamma = \alpha + j\beta = \frac{\sigma}{2}\sqrt{\frac{\mu}{\epsilon}} + j\omega\sqrt{\mu\epsilon} \qquad (3)$$

Note that the dividing frequency of 400 MHz was selected here as that frequency where $\sigma/\omega\epsilon = 1$. Of course, this is not an absolute boundary and in its vicinity

$$\gamma = \sqrt{-\omega^2\mu\epsilon + j\omega\mu\sigma} \qquad (4)$$

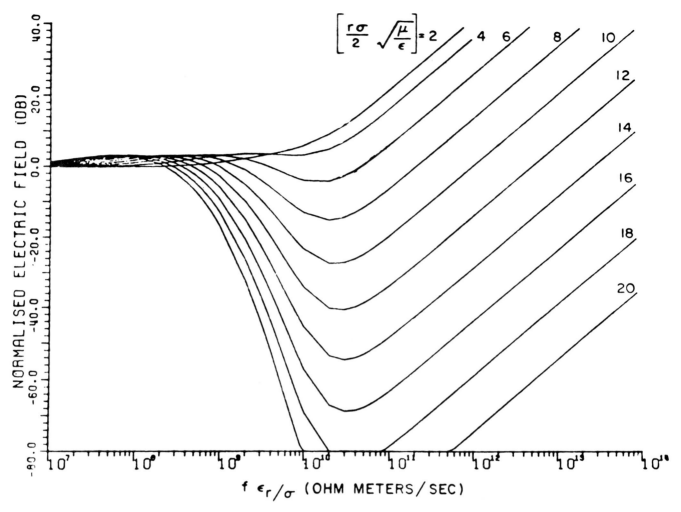

Fig. 6. Universal curves for the normalized electric field of a current element in lossy media (due to Gabillard et al., 1971).

so that neither of the above approximations [Eqs. (2) and (3)] are valid. If these equations are not properly interpreted, then results obtained from experimental data can be misleading. As we have indicated, we have not returned to the original sources of the data and our comments are based on the data given in Tables I and II.

These media in general are not dispersionless, i.e., a waveform transmitted through such a medium becomes distorted. However, the degree of dispersion depends on the path length in the medium.

The electrical values given in Ref. [1] for high moisture content would indicate small dispersion would occur at over the band quoted for one cm depths but that it would be severe at 10 cm depths, primarily because of the loss mechanism. For video pulses, the launcher (antenna) would dictate the lower frequency limit whereas some multiple of skin depth would fix the high frequency limit. As we shall see, it will be desirable to include the high frequency content in the pulse even though it will not be present in the received signal.

4. "TARGET" CHARACTERISTICS

The next factor of importance is the size and nature of the "target" of interest. For purposes of this discussion we must define "target" in both an electromagnetic and a diagnostic sense. Let us concentrate on some organ, bone or growth in the body which is a distinct object of interest, and which has at least one contrasting electromagnetic parameter, thus altering an electromagnetic field propagation in the surrounding medium. It is already known that there are significant contrasts in the electromagnetic parameters of the different body tissues. The size of the targets relates to the required interrogating signal characteristics, the amount of disturbance (response, echo) to be expected, and the amount of information which can be extracted from the electromagnetic scattering of the target.

Objects whose size is less than $1/10$ wavelength (in the biological medium) are so-called Rayleigh scatterers, appearing to the radar (see also Skolnik elsewhere in this volume) as point scatterers having an echo proportional to the

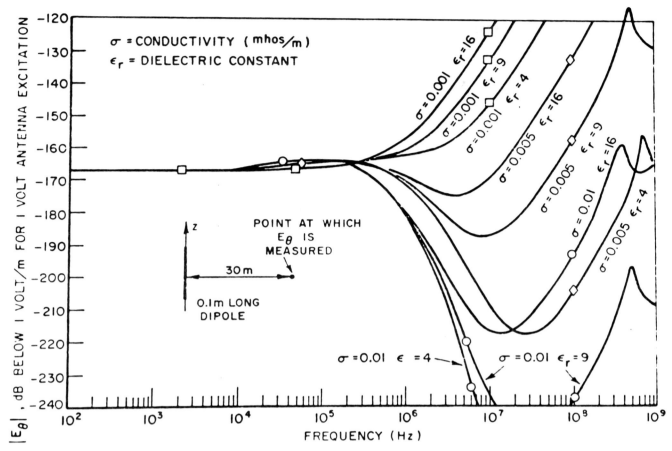

Fig. 7. Magnitude of electric field strength of 0.1 m long, 0.0001 m radius perfectly conducting dipole in homogeneous media. The distance to the measurement point is 30 m.

product of target volume and constitutive parameter contrast.

Objects which are quite large in terms of wavelengths (1 to 100 wavelengths in major dimension) give not only much larger echoes, but have sufficient spatial variation in scattering vs angle and/or position to make imaging appropriate, as Larsen and Jacobi have demonstrated [10].

What about the objects in between ($\lambda/10$ to 10λ), or even those portions of target images where improved resolution is desired within the constraints of interrogating wavelength and antenna size? This is the area where video pulse radar techniques may be of most interest.

First, it has been shown that every scattering target has a unique impulse response. Second, and more importantly, it seems that the resonance frequency and low frequency region ($\lambda/10$ to 10λ) of the impulse response is highly correlated to target shape (see also Skolnik's discussion of Mei scattering elsewhere in this volume). A typical result is shown in Fig. 8. A preliminary evaluation of imaging concepts has been discussed in Ref. [8] and indicates that there are much more stringent requirements for generating images. At this time, this approach [11] has not been used to generate images of targets immersed in a lossy media. However, target

identification [12] has been achieved for buried targets in a lossy earth making use of target resonances. This will be discussed later.

The problem encountered for creating images from backscattered signals is that the loss mechanism causes the signal to decay as it propagates over the target. This attenuation distorts the image that would be created from backscattered data. This loss must be known and should be removed from the radar data before the image processing. Observe that this does not represent a problem for imaging when forward scattered waveforms are used as discussed by Larsen and Jacobi in this volume.

Natural resonances (represented by poles in the complex frequency plane) are very important for the identification of scatterers. The impulse excites the resonances of a scatterer just as an acoustic impulse might excite the resonances of a tuning fork. The lowest electromagnetic frequency resonances are most distinct, and most clearly correlated to object size and shape. Furthermore, the total echo response of the object may vary drastically depending on the angle of incidence of the interrogating energy and on interactions with other nearby objects, but the location of the poles is invariant.

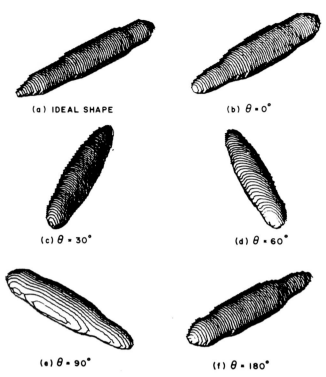

Fig. 8.   Multiple-frustrum cylinder images.

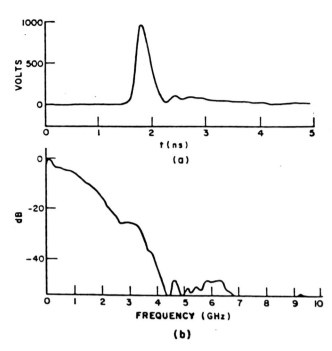

Fig. 9.   Characteristics of the impulse source in time and frequency domain.

In summary, it is seen that the electrical characteristics of the soil and biological media have much in common. Also, the targets in both cases are comprised of regions having ϵ and σ differences, and whose scattering depends on size and shape.

There are, of course, some differences between the systems generally used for the two applications. First, it is usually impossible to place a receiving antenna on the "other side" of the medium to detect the forward scattering affect of a buried object. Second, the target sizes and depths combined with soil parameters result in a lower frequency band of operation. For shallow targets, frequencies between 50 MHz and 2 GHz are used. For targets beyond 10 feet deep, the spectrum used ranges lower and lower, until frequencies of 1 to 100 KHz are necessary for detection of layers at depths to 1 KM. A third difference is that general physiological information constitutes an important body of a priori information for biological sensing applications. The locations and shapes of the internal organs are known, and diagnosis of abnormalities, or other specific information is desired. For underground radar, the detection of an object (such as a pipe) of unknown location takes precedence over diagnostic studies of its characteristics.

4.1 State of the Art in Video Pulse Radars

The characteristics of the video pulse which is used for underground sensing and other applications have such an important impact on the design and configuration of the other components that it is appropriate to call the whole system a video pulse radar.

The primary feature of a video pulse is its lack of any center or "carrier" frequency. The time domain waveform, and frequency spectrum of a typical generator is shown in Fig. 9, which also includes the characteristics of a "normal" radar pulse, which is really a pulse burst of some specific carrier frequency. The pulse shape can be controlled by proper design of the generator. Video pulse generators having widths of 180 psec, and a resulting spectrum ±4 dB from 100 MHz to 4 GHz are in use for laboratory measurements.* A small, portable, battery-powered generator having a pulse width of 0.5 nsec, 1 kV peak amplitude, and average power of less than 1/4 m watt has been very useful for underground radar systems. For deeper work, devices with up to a joule of transmitted energy have been proposed. All of these generators produce coherent energy over exceedingly wide bandwidths (at least 100:1), from a fairly simple design. By using a fast fourier transform, the time domain signals in these systems can be transformed into frequency domain information equivalent to that obtained by a swept frequency source operating over the same bandwidth. For our applications, the video pulse generator has been found to be simpler, cheaper, and less power consuming than a 100:1 bandwidth swept frequency oscillator.

In order to illustrate the rest of the system, a photo and block diagram of the Terrascan underground utility locator

---

* This 1000 volt pulse does not represent a health hazard because of the small power involved. The duration of the pulse is too small to excite any thermoelastic expansion.

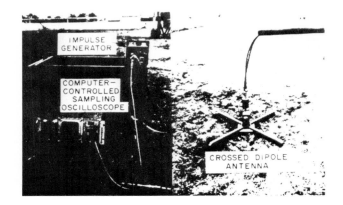

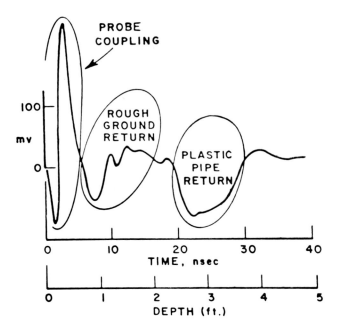

Fig. 11.   Typical underground radar received waveform.

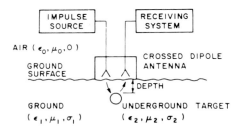

Fig. 10.   The subsurface pulse radar and its block diagram.

is shown in Fig. 10. This commercial device is a rather simple portable battery-powered radar for locating non-metallic as well as metallic utility lines (pipes, cables, conduits, drain titles) to depths of 10 ft.

The pulse is launched into the medium via a resistively loaded dipole resting on the ground surface. An echo is produced by a dielectric constant contrast (plastic pipe) or a conductivity contrast, (metal pipe), and received on a similar dipole antenna positioned orthogonal to the transmitting antenna. The receiver is similar to a sampling oscilloscope interfaced to a microprocessor for processing and display of the time domain waveform.

The resistive loading of the dipoles damps out their normal resonant characteristics, giving a relatively flat spectrum at the expense of antenna efficiency. The orthogonal arrangement of the two antennas permits a relatively compact transmit receive pair, focused over the same spot, and yet with theoretially perfect transmit-receive isolation independent of frequency. Also, the orthogonal structure is relatively insensitive extended to horizontal layers, an undesired target for this application. The system would detect any target not rotationally symmetrical about the antenna axis.

A typical received time-domain waveform is shown in Fig. 11. The initial residual antenna coupling spike serves as a time reference for the waveform. This is followed by some random clutter, due mainly to tufts of grass and other surface irregularities beneath the antenna. Finally, a pipe return echo is seen. By moving and rotating the antenna to maxi-

mize this signal, the location of the pipe is found with an accuracy of ±6 in. in location and ±5° in direction. The time delay of the echo yields target depth accuracy of ±15%. Of course, the antenna system and the video pulse required for exploration of biological tissues should both be much shorter with increased resolution in both depth and position.

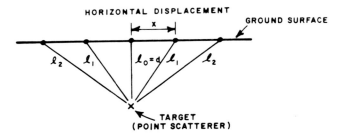

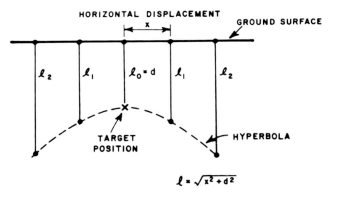

$$\ell = \sqrt{x^2 + d^2}$$

Fig. 12.   Derivation of target signature in cross section.

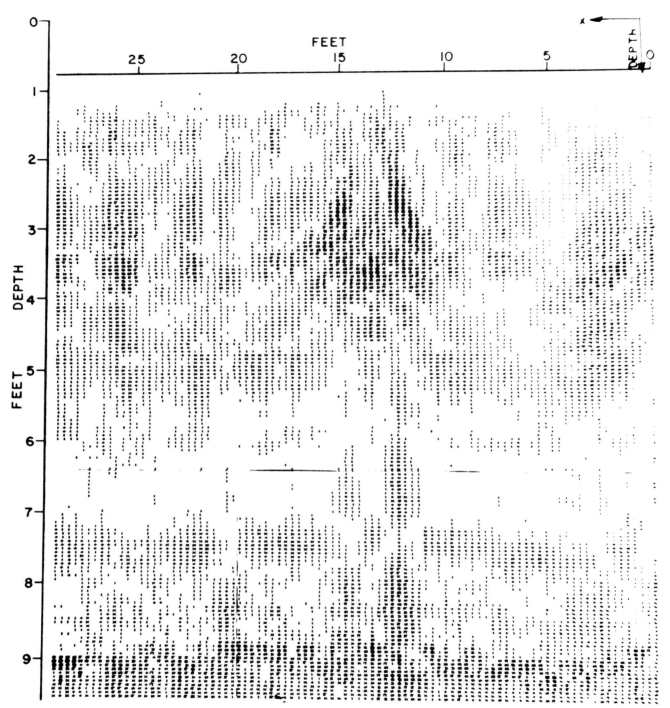

Fig. 13. Grey level map prepared using raw data.

In cluttered environments additional echos may be seen at random depths. However, the trained operator can recognize the characteristic pipe echo shape in the midst of the random clutter as he shifts the antenna in small steps.

Two state-of-the-art systems using the same basic radar combined with more sophisticated processing are now described: 1) mapping or (imaging) and 2) transient signature target identification. Some results for a combination of the two techniques are presented and briefly discussed.

A. Underground Mapping

If the sensing antenna is moved in raster fashion over an area of interest, and the time domain echo waveforms vs position

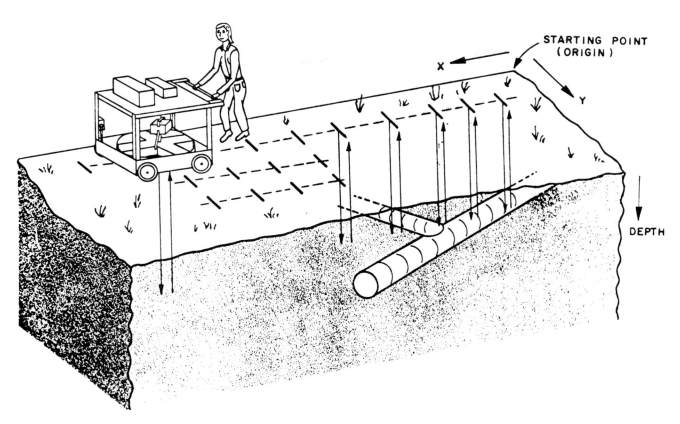

Fig. 14.   Mobile radar system over pipe system.

are recorded and stored, then this information is sufficient to produce a 3-dimensional map of the volume beneath the scanned area. In common with other imaging processes, the resolution of this process is limited to the wavelength of the highest frequency in the received echos, and also by antenna size (if antenna inverse transfer function processing is not done). Thus, our maps shows location and direction of buried pipes, but not their size and shape. In addition, some interesting ground inhomogeneities show up.

Processing in this case involved several different steps now being investigated. The two most significant ones included a focusing where a minicomputer was used to eliminate differences caused by different propagation paths into the ground is illustrated in Fig. 12.

It is first assumed that the target is at the position shown. The time waveform is recorded for each position (x_i) of the radar on the surface above the target. The time of arrival of each such waveform $t_i = 1/v$ is adjusted to be equal to $t_0 = d/v$ where v is velocity of the wave in the earth. After repeating this process for every target position a grey level map is prepared.

The second step involved statistical processing making use of the variance of the received signal. Fig. 13 shows a particular grey level map obtained without any special processing from raw data generated by moving the radar over region containing a plastic pipe buried at a depth of 30 inches as shown in Fig. 14. One sees a blackened area that occurs over the main pipe and several other darkened areas. After

applying the statistical processing to the same raw data, a second grey scale map was developed and is shown in Fig. 15. This clearly shows the three darkened areas. The one labeled trench is from the walls of the trench in which the pipe was buried. This is still detectable even though the trench was filled with the dirt that was removed and compacted almost a decade ago. The pipe and the bottom of the trench causes the blackened region at the bottom of the trench. The blackened area labeled side pipe is cauded by a pipe that is part of T junction in the pipe system. The hazy darkened area labeled rock was actually a rock that was subsequently dug up to identify it. The introduction of the statistical process clearly lifts these targets out of the background. Introducing the focusing further sharpens the picture and also tends to eliminate the trench walls as is shown in Fig. 16.

A second type of map has also been prepared for the purpose of detecting pipe. This is a plan view in which all data is included for depths of 0 to 0.7 m and 0.7 to 1.5 m and are shown in Figs. 17 and 18, respectively. The plan view for the shallower pipe vaguely shows the outline of the pipe system. However, the T is clearly evident for targets contained in 0.7 to 1.5m depths.

In concluding this mapping discussion, we observe that proper processing does indeed lift the target out of the background clutter. Known targets have been detected and unknown targets such as the rock have been detected and later confirmed.

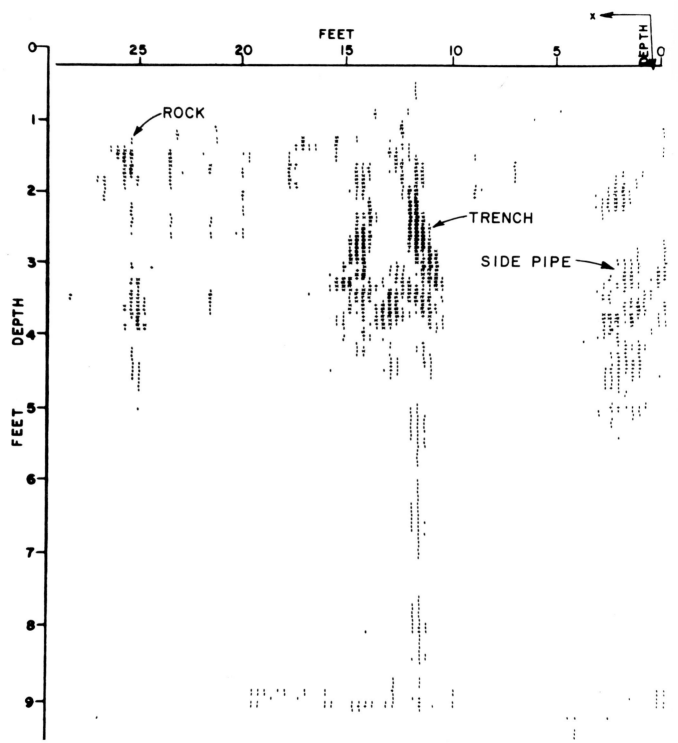

Fig. 15. Grey level map obtained after applying statistical processing to raw data.

B. Target Resonances and Target Identification

For some underground radar applications, the targets are small enough that they can be detected, but not identified by a mapping system. However, these objects possess unique transient signatures, so that identification based solely on the characteristics of a single return echo may be attempted.

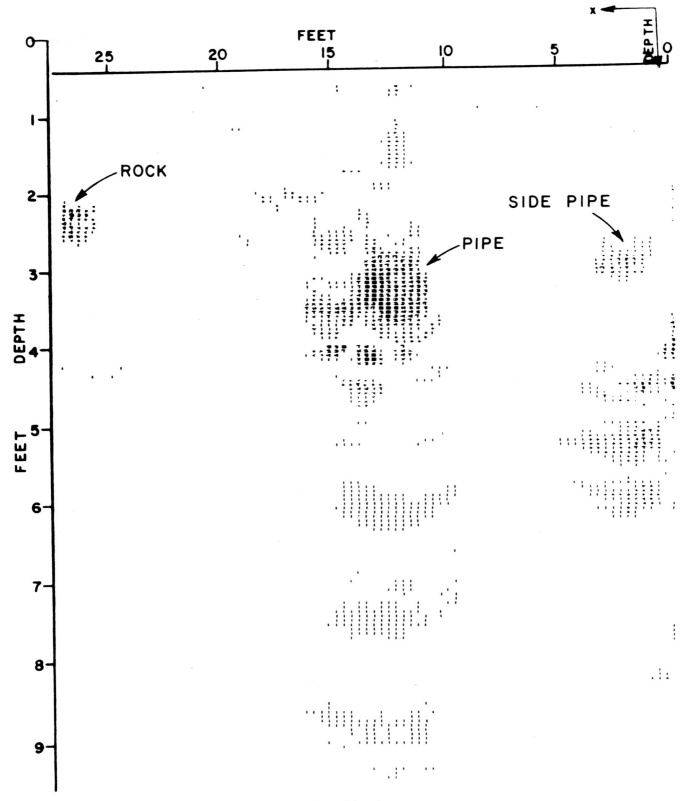

Fig. 16.   Grey level map obtained after applying both statistical and focusing.

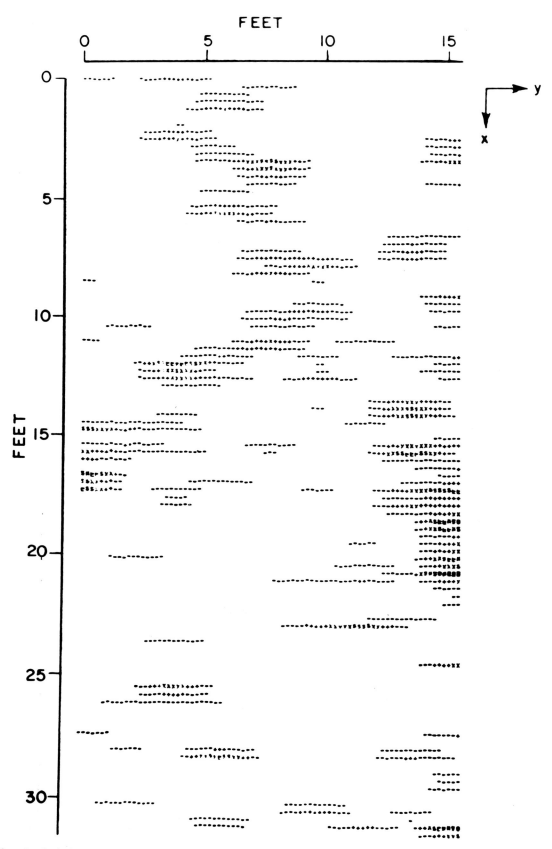

Fig. 17.   Plan view including data. Data from 0 to 0.7m depths.

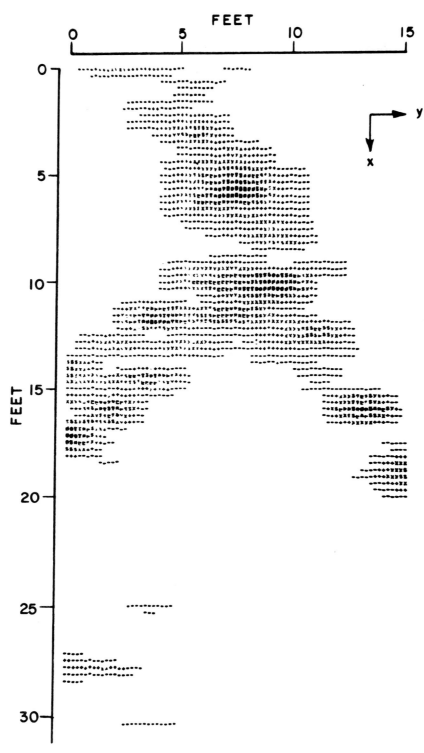

Fig. 18.   Plan view including data from 0.7 to 1.5m depths.

The broad spectrum of the video pulse is necessary to excite these different transient signatures. The set of targets shown in Fig. 19 were buried to a shallow depth in a lossy earth (ϵ_r $\simeq$ 20, $\sigma \approx 0.03$ s/m). A series of measurements were made over these targets using a radar system with a 500 ps, 1000 V periodic pulser. The received time domain signals are

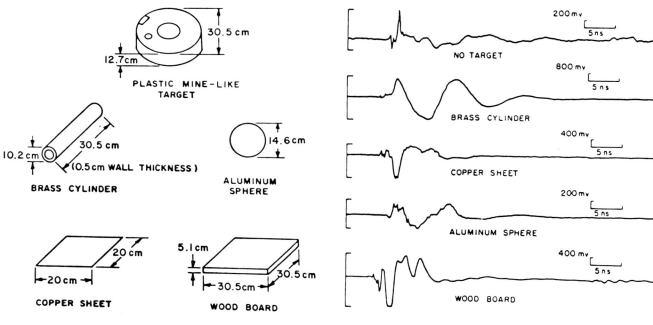

Fig. 19. Physical characteristics of the subsurface targets.

Fig. 20. Processed waveforms from the subsurface targets.

shown in Figs. 20 and 21. Clearly the no target response of Fig. 20 shows a maximum of the order of 100 mv after the initial coupling spike. Thus we appear to be able to observe scattered signals of order of 80 dB below the transmitted pulse. Cable losses, etc., would reduce this to about 70 dB.

These waveforms and their received spectrum are dependent on the relative positions of the antenna and the target. However, the natural resonances (poles) of the target are not dependent on the excitation, except for the magnitude of the excitation. There is a procedure known as Prony by which

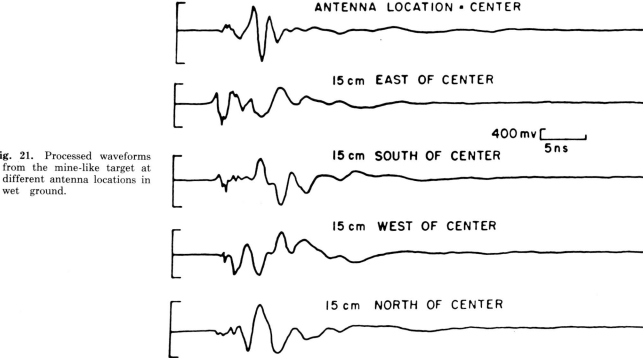

Fig. 21. Processed waveforms from the mine-like target at different antenna locations in wet ground.

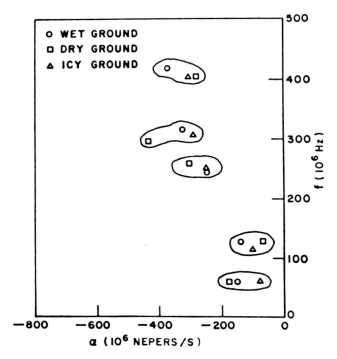

Fig. 22. Locations of the average extracted resonances of the mine-like target in different ground conditions.

such poles can be obtained from the time domain data [13].

In simples terms, this is an expansion of a function in terms of a set of decaying oscillatory functions. These are represented as a set of natural resonances of the form $A_i e^{-\alpha_i t} e^{-j2\pi f_i t}$. The LaPlace transform of this function is

$$A_i/S + (\alpha_i + j2\pi f_i) \qquad (5)$$

where A_i is the residues of the ith function and $\alpha \pm j\,2\pi f_i$ are the poles or natural resonances. Natural resonances, of course, occur in many physical systems such as the traditional stretched string, drum, organ, and electrical circuits.

For a given target the natural resonances are independent of the excitation including the angle of incidence. The residues on the other hand are not excitation invariant. Thus, the shape of the received waveform, its spectral distribution etc are not reliable signatures. However, any system that uses only the natural resonances represents an excellent candidate for obtaining signatures.

Some typical results are shown in Figs. 22 and 23. The mine-like target was a penetrable body and we observe that the resonances are independent of ground conditions (or electrical parameters), whereas the brass cylinder was a conducting body and we observe the resonances to be highly dependent on ground conditions.

Even though the waveforms of Fig. 21 are quite different, the poles extracted from these waveforms are clustered in a manner quite similar to those of Fig. 22.

The generation of these natural resonances from the raw data via the Prony procedure is a lengthy process and requires substantial computation time on a modern machine. Once determined, however, these resonances can be used to generate a predictor-correlator equation that is extremely

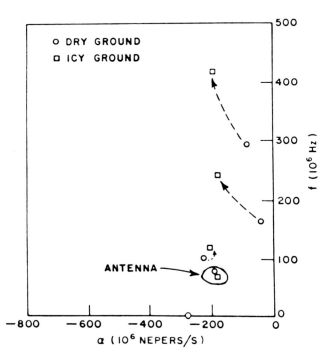

Fig. 23. Average extracted resonances of the brass cylinder in different ground conditions.

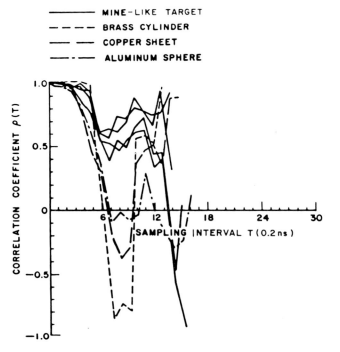

Fig. 24. Typical $\rho(T)$ curves for the identification of the mine-like target in dry ground.

rapid even on a microcomputer. The output of this predictor correlator is shown in Fig. 24. Since the abscissa is completely under the control of the operator, then the mine-like target can easily be separated from the other targets if one sets a threshold $\rho(\tau) \sim 0.3$. This scheme has proven to be extremely successful for a large number of measurements.

To shorten computation time three other constraints were introduced in the target identification algorithms. This included maximum and minimum peak signal levels, total energy content and limits on the arrival time of the maximum. On this basis, we have achieved 100% identification with a single look false alarm rate of the order of a few per cent, including reflections from a number of unknown false targets in a dirt, debris filled road outside this laboratory. Fifty-three waveforms were measured over these sites whose level exceeded the minimum value.

C. Combination of Mapping and Signature Processing

If a (or set of) natural resonance(s) can be attributed to a desired target, then it would appear desirable to focus attention on that (those) natural resonance(s). For example, the antenna resonances contribute little or no information concerning the identity of the target. Further, the residue associated with that resonance may be relatively large so that the identification algorithms are relatively insensitive to changes in the target natural resonances. Recent studies [9] have led to techniques for removing or filtering these unwanted resonances from the data using a difference equation technique. For cases where desired target echos are submerged in clutter, the combination of filtering with mapping produces improved results.

A geometry considered for this investigation was a tunnel in the Rocky Mountains near Gold Hill, Colorado, at a site known as the Hazel A. mine.

A modified version of the Terrascan system was used to obtain a sequence of waveforms for positions 20 ft. above a known tunnel site. The radar system was moved along a line transverse to the axis of the tunnel. This position is the horizontal scale in the maps to follow and the vertical scale will with the arrival time of the signal corresponding to depth in ground. The surface over which the radar was operating was rough and strong clutter signals existed in the time range where the tunnel response would be expected. A grey level map was created by digitizing the signal magnitude and then assigning darker characters to the stronger signals. Figure 25 shows a typical grey level map obtained using only relatively simple data processing steps. In this figure, the symbol T indicates the position of the radar when it was vertically above the tunnel. The ground was not horizontal so the minimum distance from the surface to the tunnel would occur at about the position designated as S-6–T-8. One can see little evidence of the tunnel in this grey level map. Several more complex data processing steps were introduced with some improvement. However, the most dramatic improvement resulted where natural resonances not associated with

the tunnel were removed from the raw waveform via a difference equation approach described in a thesis by Volakis [14]. The result is shown in Fig. 26. Clearly the tunnel is properly located. One observes a "premonition" of the tunel in the map of Fig. 26. This has been created by data processing but the onset of the signal reflection from the tunnel can be clearly observed in processed waveforms as is discussed by Volakis [14]. The tunnel has thus been clearly detected and also identified as an object with a particular natural resonance.

5. APPLICABILITY OF VIDEO PULSE CONCEPTS TO RESEARCH FOR BIOLOGICAL STRUCTURES

To our knowledge, no transient radar system has yet been applied to biological remote sensing. However, systems measuring coherent scattering for microwave frequency bands of at least 2:1 have been built. Such a frequency spread is approaching the equivalent information contained in transient scattering signatures. Thus, it is appropriate to consider several basic concepts used in transient radar even when discussing advances in frequency-domain measurement techniques.

First, it is important to state that dispersion caused by the medium is important when considering any target imaging or identification processing with broadband (or baseband video pulse) radar. With the recent advances in soil *in situ* constitutive parameter techniques, the need for still more measurements and a deeper insight into soil electromagnetic propagation mechanisms vs soil mechanics has become evident. If such a need also exists in biological media, then it is feasible to adapt the single-probe technique described in this paper into a hypodermic needle assembly for in vivo tests of constitutive parameters. This has been achieved by Burdette.

Obviously, the universal curves showing the low-frequency and high-frequency "windows" can be applied to biological tissue. First consider the low-frequency window.

A small dipole immersed in a conducting media radiates a field that is 3 dB below the d.c. field of that dipole at the critical frequency given by

$$f_c = \frac{3.76 \times 10^6}{r^2 \sigma}$$

where r is the distance to the observation point. For example, assume r = 10.0 cm, σ = 1.0 S/m

$$f_c = 3.76 \times 10^8 \text{ Hz}$$

For all frequencies below this value and for conductivities below σ = 1.0 S/m, propagation is essentially dispersionless. Figure 27 gives the magnitude of the receive dpulse after propagation of a range of 2 d (d = distance to a target). This data was obtained by scaling computed values for different sets of parameters [9].

The signal is launched and received by a dipole antenna of length (21). The transmitted pulse for the case of Fig. 28

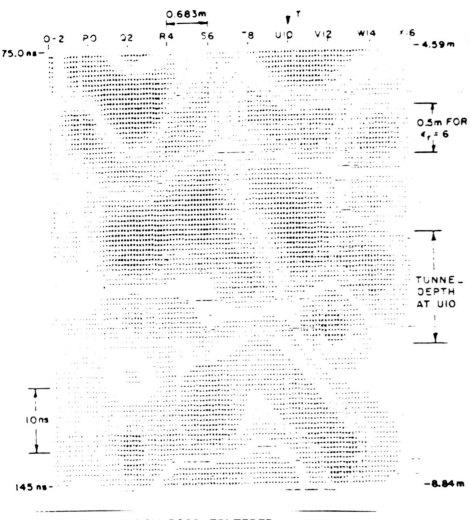

Fig. 25. Mapping of the top traverse over the tunnel, as given by Stapp.

LOW-PASS FILTERED

Long Box Antenna   Clipped at 1.91
NXDIV=12   NYDIV=0   Fold Last
LIX=3   LIY=1

has a duration of 10 ns. Clearly, the received signal would overlap the timing of the transmitted pulse. If, however, the transmitted pulse is shortened to a one ns pulse, the relative receive and transmitted pulse are shown in Fig. 28. Clearly, the received pulse arrives at a time well after the transmitted pulse has decayed to zero. The maximum frequency contained in the pulse of Fig. 28 is of the order of 100 MHz. The wavelength in the conducting medium is of the order of 30 cm. Using the rules of thumb established earlier for penetrable targets that a target of the order of $\lambda/10$ would give an appreciable scatter, then a 3 cm target might be observable. The skin depth is about 3 cm, in high water content media, so the resonance might be observed. This is in the absence of scatter, etc. This would represent a difficult target to detect. However, the antenna could be lengthened and the

depth shortened. A 20cm antenna and a propagation path of 5 cm would produce a pulse attenuation factor of the order of ±2 dB. The pulse length would be shortened for the purpose illustrated in Fig. 28 yielding another 10–15 dB or a net of the order of 55 dB. This would produce additional high frequency content and result in a situation where the target could be detected.

The high frequency window begins at about 1.5 to 2.5 GHz, depending upon target depth and the electrical parameters of the localized medium. As has been stated elsewhere, losses get prohibitive in biological media for submerged targets at frequencies much above 3 GHz. For imaging, this high frequency cut-off puts a limit on possible resolution, and on the minimum size of a few mm for targets which can be effectively imaged.

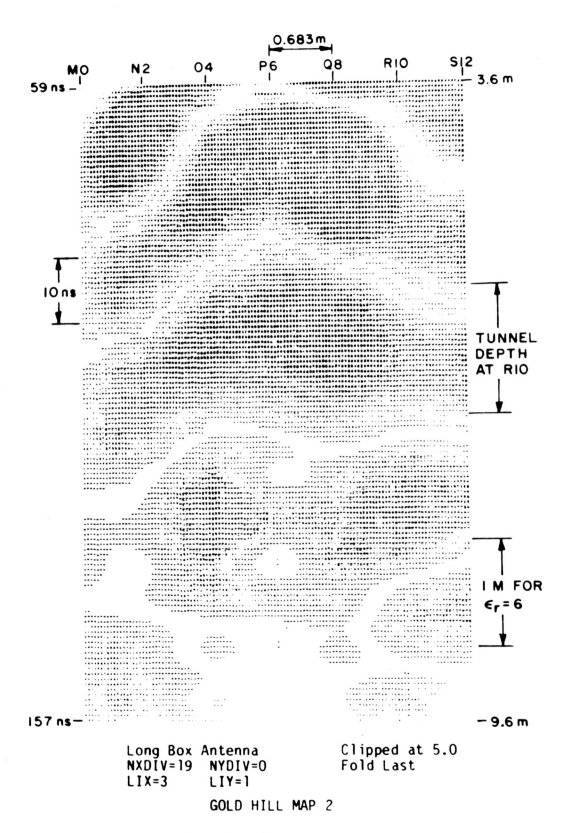

Fig. 26. Mapping of the lower traverse over the tunnel.

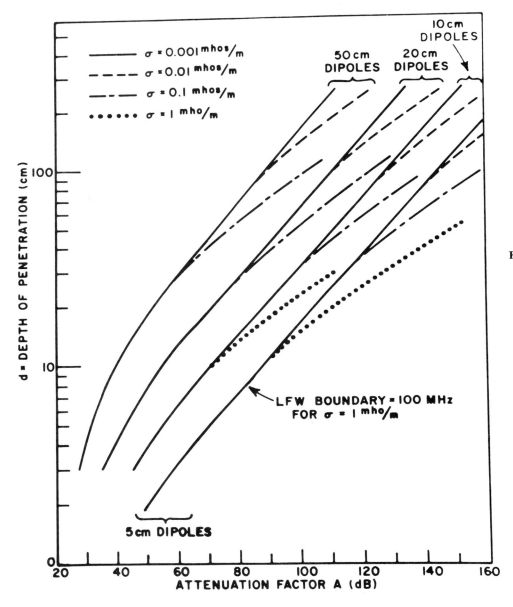

Fig. 27. LFW radar attenuation factor curves for 5cm, 10cm, 20cm and 50cm dipoles. the data was computed for a 10ns Gaussian input pulse which is bandlimited at about 100 MHz. The LFW breakpoints therefore are for a frequency of 100 MHz.

A most important concept from video pulse radar practice is the transient signature properties of the targets. This is particularly true when the detection of targets too small to be effectively imaged is desired. For example, one might be able to monitor the growth of a tumor.

It is our understanding that even the infrared radiometry techniques tend to look for differences. Breast cancer diagnosis for example compares results obtained from the right and left breasts as is discussed by Barrett and Myers elsewhere in this volume. One would also be interested then in the time history development of a tumor. Thus a system that would monitor changes could make a significant contribution to diagnosis.

Here one would set up the difference-equation correlator from measurements on the tumor. Then the patient would be monitored. Any changes in the size of the tumor would lead to a false target reading. If one has an interest in the size of an organ of known shape, etc., a difference equation could be devised for this purpose. For example, Hill [15] has shown that metallic spheres of different size can be discriminated using the difference equation-correlator scheme. Fig. 29 shows the correlation function he obtained as a function of normalized sphere radius a/a_0. Where a_0 is the sphere radius for which the correlator was set. The parameter $\tau = 2a/c$ is the transit time for a wave traveling the diameter of the sphere. For optimum discrimination the sampling interval $\Delta t/\tau = 0.8$ should be used.

One very important number can also be obtained from the results of Fig. 22. The lowest target resonance has a real frequency of the order of 150 MHz. This corresponds to a free space wavelength of 6 m. Assuming a dielectric constant of the mine-like target (unknown) to be of the order of 2.5,

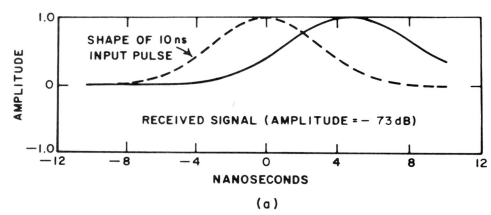

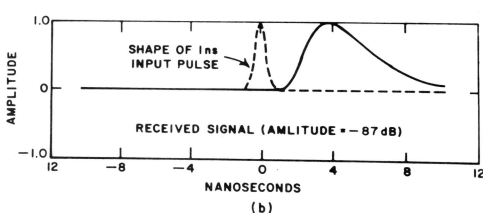

Fig. 28. Relative locations of received signal and (a) 10ns input noise and (b) 1ns Gaussian input pulse for 10cm dipoles spaced 2d = 20 cm in a medium of $\sigma = 1.0$ mhos/m.

then the wavelength in the target medium is 3.8 m. The maximum dimension of the target, then, is of the order of $\lambda_\tau/10$. The stability of these resonances under changing ground conditions indicates that the first value $\lambda_\tau/10$ is

significant. However, it is at this time not clear just what mechanism is creating this resonant behavior. Even though it has been substantiated by subsequent measurements, this resonance does not appear to be attributable to internal

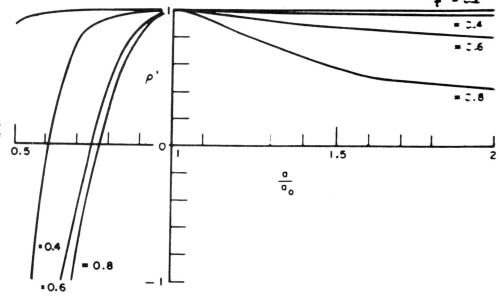

Fig. 29. Discrimination against spheres of different sizes by the use of the difference equation.

mechanisms. The resonance (pole) whose imaginary part is approximately 250 MHz could possibly be introduced by internal reflections associated primarily with the 30.5 cm dimension.

All tissues classified as lossy dielectrics are assumed to be penetrable bodies. This must be qualified to the extent that if the propagation paths in the particular body substantially exceed the skin depth, then the body is indeed inpenetrable. It is significant, then, that the resonances of penetrable bodies are not strongly dependent on the external media. Thus identification can be achieved under a variety of circumstances.

It may even be possible to make use of these small target resonances to achieve microwave dielectric heating with much better selectivity and revolution than could be obtained by antenna focusing techniques. Heating would be most efficiently achieved at the set of target resonance frequencies. Thus, a sophisticated approach would be to evaluate the resonances of the obstacle to be heated and select a set of frequency bands appropriate for that obstacle. Quite possibly the residues would be equally important. One would desire to tailor the heating signal to the resonances that are most strongly excited. As a long range goal, an investigation could be directed using the dispersive properties of the media. It may prove possible to design transient signals so that these natural resonances would tend to line up to achieve a peak power at the desired position in the medium. Thus, it may be possible to achieve a maximum instantaneous (short time) power density and thus maximum heating at this position and minimize the short time power density elsewhere. Even evaluating the shape of this transient signal would be difficult. Achieving it in practice would be an order of magnitude more difficult. Its potential, however, may make it worth the effort.

Finally, all of the techniques developed for our video pulse radars are needed in order to obtain relatively clutter free data needed to exploit the three potential applications outlined above. This implies a substantial research and development effort is necessary.

6. VIDEO PULSE RADAR SYSTEMS FOR BIOLOGICAL APPLICATIONS

Assuming that a video pulse radar system for direct application of the concepts just discussed is a desirable goal, this section looks at the modifications which might be necessary, and some approaches which might be promising.

Considering the three system blocks (transmitter, antenna(s), receiver), it can be immediately stated that no modifications whatsoever are needed in the receiver. Present sampling oscilloscopes possess tangential sensitivity of a few millivolts, and flat frequency performance from a few MHz to 12.4 GHz. This is plenty of frequency response and sensitivity for any biological sensing applications. Thus the real discussion here concerns the development of appropriate pulse generators and antennas.

The present, commercially available pulse generator puts out an average power of $\simeq 1$ mw, with a flat spectrum to $\simeq 1.2$ GHz. This pulser would be appropriate for "low-frequency window" applications. Note also that its total radiated power is within the ANSI C95 standard, (assuming at least 1 cm^2 antenna aperture), and could be decreased further by lowering the pulser repetition rate. For use of video pulse techniques with higher frequencies, more exotic solid state devices are required. A unit which is flat to 4 GHz has been demonstrated. Actually, good results to frequencies of 7.5 GHz have been obtained with this advanced pulser. Therefore, it is feasible that in the near future, reasonably priced impulse generators covering the total appropriate spectrum for biological sensing would become available.

It is in the area of antenna design where most work is needed. The physical constraints combined with the requirements on very broadband performance combined with minimum time delay dispersion all point to new antenna designs for this special application.

Much work has been done on linear dipole antennas for underground applications, so the adaptation of these to biological applications deserves study.

7. ANTENNA PROPERTIES

The impedance of the linear dipole antenna in electrical contact with a conducting medium has the very nice property that its input impedance is nearly real and constant when operation is in the Low Frequency Window. This means that a good impedance match can be achieved over a rather wide

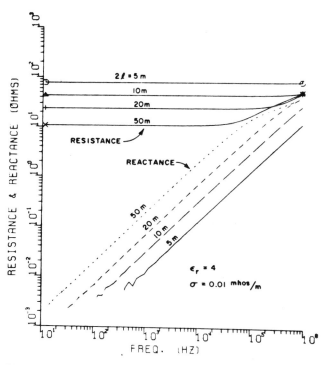

Fig. 30. Input impedance of 0.002m radius bare copper wire dipole antennas in an infinite homogeneous medium.

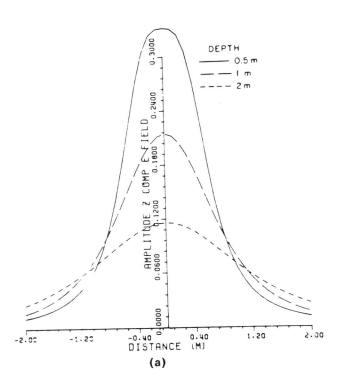

(a)

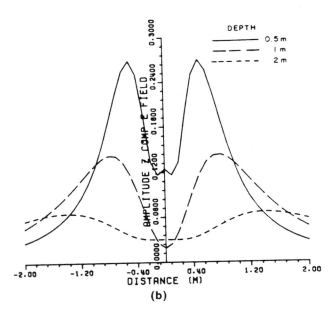

(b)

Fig. 31. (a) Electric field (E_z) at a constant depth while scanning along the z axis below the antenna at 100 MHz. (b) Electric field (E_z) at a constant depth at 140 MHz.

spectrum. This is illustrated in Fig. 30. This is not as fortunate for operation in the High Frequency Window where impedance matching has been achieved primarily by lossy loading of the antenna. There is also the problem of bifurcation of the antenna beam as frequency increases [16] for near zone patterns. A typical result is shown in Figs. 31. These patterns were computed for a model of the current Terrascan antenna with length of one meter immersed in a lossy homogeneous earth type medium. The patterns at 100 MHz show the usual antenna beam whereas at 140 MHz, there is a minimum on axis. This is a result of different paths for energy radiating from the feed section and for energy radiating from the ends of the antenna. Similar near zone bifurcation can occur for aperture antennas where the geometrical optics radiation and the edge diffracted radiation create an interference pattern.

The major constraint on scaling the present antennas for operation at higher frequencies or shorter pulses consists of the balun transformer. However, if shorter probes are used in conjunction with shorter pulses, then the balun transformer could be eliminated. Typical broad band baluns are as used on such antennas as the spiral could be used. This could provide balanced operation over a time interval such that the desired received signal has been recorded.

Another interesting approach for antennas operating in the HFW would be the use of broad band spiral antennas. Since these devices are dispersive in themselves some data processing steps are required if they are to be used in a video pulse radar.

Another basic antenna type worthy of consideration is the dielectric-loaded TEM horn antenna. Studies have shown good broadband performance from these types, although their low frequency sensitivity is generally less than dipole types.

Considering all of the above, it seems feasible at the present time to put together a transient radar system which would operate much like present time delay spectroscopy systems, yielding scattering vs time information directly, but with a greater signal bandwidth than is currently used. In the past, studies of underground radar have made use of both time-domain video pulse and swept-frequency systems. With the use of the fast Fourier transform, the best of both time domain and frequency domain imaging and identification algorithms may be applied to raw data from either type of system.

All combinations are finding particular applications where they are most suitable. In the same sense, video pulse radar techniques seem worthy of further serious investigation as one more tool available for improved results or wider applicability of microwave biological sensing.

References

1. P. K. Hayes, "An On-Site Method for Measuring the Dielectric Constant and Conductivity of Soils over a One Gigahertz Bandwidth," Report 528X-1, Aug. 1979, The Ohio State University ElectroScience Laboratory, Department of Electrical Engineering, Columbus, Ohio. Prepared for Columbia Gas System Service Corp.

2. C. W. Davis III, "A Computational Model for Subsurface Propagation and Scattering for Antennas in the Presence of a Conducting Half Space," Report 479X-7, Oct., 1979, The Ohio State University ElectroScience Laboratory, Department of Electrical Engineering, Columbus, Ohio. Prepared for Columbia Gas System Service Corp.

3. C. C. Johnson and A. W. Guy, "Nonionizing Electromagnetic Waves in Biological Materials and Systems, *Proc. IEEE*, Vol. 60, No. 6, pp. 692–718, June 1972.

4. E. C. Burdette, F. L. Cam, and J. Seal, "In Vivo Probe Measurement Technique for Determining Dielectric Properties at VHF Through Microwave Frequencies," *IEEE Trans. Microwave Theory Tech.*, Vol MTT-28, No. 4, pp. 414–428.

5. E. C. Burdette, R. L. Seaman, J. Seals, and F. L. Cain, "In Vivo Techniques for Measuring Electrical Properties of Tissues," Engineering Properties of Tissues," Engineering Experiment Station, Georgia Institute of Technology, July, 1979.

6. K. A. Davis and R. J. Lytle, "Computerized Geophysical Tomography," *Proc. IEEE*, Vol. 67, No. 7, pp. 1065–1073, July, 1979.

7. R. J. Lytle, J. T. Okada, and E. E. Laine, "Geotomography Applied to Nuclear Waste Repository Site Assessment," Paper presented at the International Union of Radio Science [U.R.S.I.] Symposium, Quebec, Canada, June 2–6, 1980.

8. R. Gabillard P. DeGauge, and J. R. Wait, "Subsurface Electromagnetic Telecommunications—A Review," *IEEE Trans. Commun. Technol.*, Vol. COM 19, pp. 1217–1228, 1971.

9. G. A. Burrell and L. Peters, Jr., "Pulse Propagation in Lossy Media Using the Low-Frequency Window for Video Pulse Radar Application," *Proc. IEEE*, Vol. 67, No. 7, pp. 981–990, July, 1979.

10. L. E. Larsen and J. H. Jacobi, "Microwave Scattering Parameter Imagery of an Isolated Canine Kidney," *Med. Phys.*, Vol. 6, No. 5, pp. 394–403, Sept./Oct. 1979.

11. J. D. Young, "Target Imaging from Multiple-Frequency Radar Returns," Report 2768-6, June, 1971, The Ohio State University ElectroScience Laboratory, Department of Electrical Engineering, Columbus, Ohio. Prepared under Grant No. AFOSR-69-1710 for Dept. of the Air Force, Arlington, Va. (AD 728235).

12. L. C. Chan, D. L. Moffatt, and L. Peters, Jr., "A Characterization of Sub-Surface Radar Targets," *Proc. IEEE*, Vol. 67, pp. 91–1000, July, 1979.

13. R. Prony, "Essai expérimental et analytique sur les lois de la dilatabilité de fluides elastiques et sur celles del la force expansive de la vapeur de l'alkool, a differentes témperatures," *J. l'Ecole Polytech*, (Paris), Vol. 1, No. 2, 1795, pp. 24–76.

14. I. L. Volakis, "Improved Identification of Underground Targets Using Video-Pulse Radars by Elimination of Undesired Natural Resonances," Report No. 710816, Oct., 1979, The Ohio State University ElectroScience Laboratory, Department of Electrical Engineering, Columbus, Ohio. Prepared under Contract N00014-78-C-0049 for Dept. of the Navy, Office of Naval Research.

15. D. A. Hill, "Electromagnetic Scattering Concepts Applied to the Detection of Targets Near the Ground," Report 2971-1, Sept., 1970, The Ohio State University Electroscience Laboratory, Department of Electrical Engineering, Columbus, Ohio. Prepared under Contract F19628-70-C-0125 (AF CRL 70-0250) for the Air Force Systems Command. (AD 875889).

Medical Imaging Using Electrical Impedance

Yongmin Kim* and John G. Webster

The basic principles of image formation by measurement of electrical impedance at ca 100 KHz in biomedical systems are presented. The forward problem is described in terms of known voltage drives and conductivity distribution from which the current density is to be determined. The inverse problem is presented as known voltage drives and current densities with unknown conductivity distribution. Frontal plane and tomographic systems are discussed along with iterative reconstruction methods.

1. INTRODUCTION

Conventional techniques for forming images of the structures within the body use: (1) x rays, (2) radioisotopes, and (3) ultrasound. X-ray photographic (radiographic) techniques are most used and usually result in frontal and sagittal plane images. Recently the Computerized-Tomography (CT) scanner has been developed to reconstruct a fine detail transverse plane image based on the absorption characteristics in the plane. However, x rays are a hazard to the patient. The hazard increases with repeated usage and x rays to pregnant women and children should be minimized.

Ultrasound is widely used with the advantage that it can outline some organs not successfullly imaged by x rays and it does not pose any hazard to the patient. There is ultrasonic imaging similar to that of the CT scanner [1]. It provides separate images of propagation velocity and attenuation. However, reflections from tissue-to-air interfaces are so large that it is impossible to penetrate the lung. Ultrasonic waves do not travel in straight lines within the body. Hence the reconstructed images are distorted.

Impedance imaging is a technique that can form images of the body structures noninvasively. It utilizes the fact that voltages applied to the regions of different conductivity conduct different currents. From these current measurements, it should be possible to reconstruct the impedance image of the body for the frontal plane, the transverse plane, and eventually three dimensions.

Although much progress has been made in the last decade, the science of impedance imaging is still in its infancy. Before we can use it clinically, we need to make technical improvements, increase understanding of the algorithms, and establish its validity.

Department of Electrical and Computer Engineering, University of Wisconsin-Madison, Madison, Wisconsin 53706
* Present address: Department of Electrical Engineering, University of Washington, Seattle, Washington 98195

2. VOLTAGE AND CURRENT DISTRIBUTION INSIDE THE BODY

The human body is electrically inhomogeneous and anisotropic. Body conductivity changes by a factor of more than 10 from region to region. Intracardiac blood is most con-

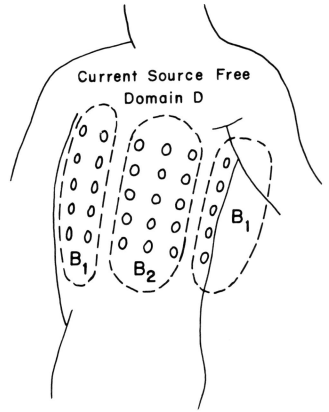

Fig. 1. A possible configuration of electrodes on the thorax. On boundary B_1 we specify the voltages. On boundary B_2 the normal derivative of electric potential vanishes.

ductive while lungs and bone are least conductive. Skeletal muscle exhibits different conductivities in different directions. Furthermore, torso boundaries are irregular and individual differences can be significant.

Figure 1 shows a possible configuration of electrodes on the thorax. At high frequencies, the thorax is source free. The steady-state governing equation for the voltage distribution within the inhomogeneous and anisotropic body is given by

$$\frac{\partial}{\partial x}\left(\sigma_x \frac{\partial v}{\partial x}\right) + \frac{\partial}{\partial y}\left(\sigma_y \frac{\partial v}{\partial y}\right) + \frac{\partial}{\partial z}\left(\sigma_z \frac{\partial v}{\partial z}\right) = 0 \qquad (1)$$

where v is the scalar electric potential, and σ_x, σ_y, and σ_z are the electric conductivities in the three directions for the electric conduction problem. For the electrostatic field problem, v is the electric field intensity, and σ_x, σ_y, and σ_z are the permittivities in three directions. However, in impedance imaging, we usually consider the conductivity profile. Therefore, we will only consider the electric conduction problem. Equation (1) is subject to two sets of boundary conditions. On boundary B_1, we specify the voltage values. On boundary B_2, the normal derivative of electric potential must vanish, i.e.,

$$\sigma_x \frac{\partial v}{\partial x} n_x + \sigma_y \frac{\partial v}{\partial y} n_y + \sigma_z \frac{\partial v}{\partial z} n_z = 0 \qquad (2)$$

where n_x, n_y, and n_z are the direction cosines of the outward normal of the surface. The union of B_1 and B_2 forms the complete boundary. For a homogeneous medium, σ_x, σ_y, and σ_z are constant throughout the domain and for an isotropic medium, $\sigma_x = \sigma_y = \sigma_z = \sigma$.

We do not necessarily require the solution of this equation to reconstruct the impedance image. However, it does show the voltage distribution and current paths so that we can better understand how current flows, especially through an inhomogeneous medium. Furthermore, some reconstruction techniques require the solution of Eq. (1). Analytic solutions exist only for very simple configurations. When we encounter either irregular boundaries or inhomogeneity, analytic solutions are simply out of the question. Frank [2] built a homogeneous electrolytic tank to model the torso and measured the voltage at the torso surface when he excited a dipole in the heart region. Burger and Van Milaan [3] studied an inhomogeneous electrolytic tank model using sand bags to simulate the lungs. To model the brain tissue and scalp, Lifshitz [4] employed a conductive sheet of paper (Teledeltos paper) with resistive paint for the skull and silver paint for the electrodes. He formed a two-dimensional model of the brain and plotted the equipotential lines and current paths. Rush [5] built a simulated thorax made of a resistive network and an electrolytic tank. He simulated the inhomogeneity of the body with a matrix of interlocking plastic rods. These controlled current flow to satisfy known conductivity values of each region. Price [6] modeled a plane as a linear network of lumped impedances. He applied current to one pair at a time, used all possible pairs of peripheral nodes, and measured potentials at all other surface nodes. None of these models has been used extensively because of inflexibility, lack of resolution, and poor accuracy.

Numerical methods are becoming widely used for the following reasons: (1) it is easy to change parameter values or boundary conditions, (2) we can control resolution and accuracy, (3) the availability of high-speed digital computers has increased rapidly in last few years, and (4) the cost of computing has declined continuously. We assign an element to each portion of the body. An element is the smallest unit in which we know conductivity, voltage, and current distribution. By assigning a different conductivity to each element, we can construct a piecewise homogeneous model. The accuracy of the model and the solution depend on the number of elements. While a large number of elements improves the solution, it also increases the computing cost.

If elements have a regular geometry (rectangular or regular hexahedral element), we call this numerical technique the finite difference method. If elements can have mixed shapes (triangular, quadrilateral, tetrahedral, prism or an general hexahedral element), we call it the finite element method [7], [8]. Thus the finite difference method is a special case of the finite element method. The finite difference method has been used in solving Laplace's and Poisson's equations. The finite element method works better for irregular body boundaries, inhomogeneous internal structure, and anisotropy, but has increased computing time. When using these numerical techniques, we should make reasonable approximations and simplifications before attempting a computer solution. We model the internal structures as piecewise homogeneous, and approximate their conductivity values from published data [9], [10]. We model the relative locations, and orientations of the limited number of elements from standard cross-sections of gross anatomy [11]. To simplify the governing equation, most models assume isotropy and solve only the two-dimensional (transverse plane) problem. This makes Eq. (1) a two-dimensional Laplace's equation. For each element,

$$\frac{\partial^2 v}{\partial x^2} + \frac{\partial^2 v}{\partial y^2} = 0 \qquad (3)$$

However, to model only the thorax of the body can lead to significant error, because truncation of the cephalic and caudal parts of the body disturbs the overall potential distribution.

Silvester and Tymchyshyn [12] and Demers et al. [13] built crude three-dimensional torso models based on the finite element method with less than 300 elements. They did not present any results. However, to be useful for impedance image reconstruction, the model must have enough picture elements (pixels) to show relatively fine detail. Kinnen [14] used the finite difference method with thousands of elements to model a three-dimensional thorax. He calculated the current distribution through blood, muscle, lungs, and fat tissue and mapped it on transverse planes. Guha et al. [15] used the finite difference method to model a thoracic cross section and calculated the equipotential lines. Also, Lytle and Dines [16] and Tasto and Schomberg [17] used the finite difference method to determine the voltages, current densities, and current paths which are required to reconstruct the impedance image. Natarajan and Seshadri [18] used the

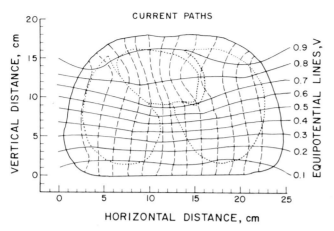

Fig. 2.   Equipotential lines and current paths inside the thoracic cross-section (from [19]).

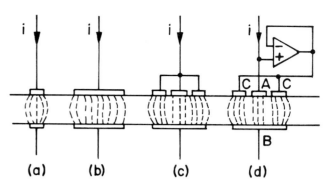

Fig. 3.   Current paths with (a) small electrodes, (b) large electrodes, (c) split electrodes, (d) guard electrodes (from [20]).

finite element method to model a cross section using 94 elements. Tjandrasa [19] also used the finite element method with 265 triangular elements. She used a resistivity of 1200 $\Omega\cdot$cm for the lung, 150 $\Omega\cdot$cm for the heart, 600 $\Omega\cdot$cm for the thoracic wall, and 300 $\Omega\cdot$cm for the interstitial fluid between the lungs and the heart. Boundary conditions were 1 V applied to the chest and grounded current-sensing electrodes at the back. She solved Laplace's equation. Figure 2 shows the equipotential lines and current paths. Current paths are concentrated in the heart region because current takes the path of least resistance.

After the voltage distribution is obtained, we can easily calculate the current densities by

$$\mathbf{J} = \sigma\mathbf{E} = -\sigma\nabla v$$

where J is the current density, σ is the conductivity of an element and v is voltage at the edges of the element. However, the real objective of impedance imaging is just the opposite of this forward problem. We wish to solve the inverse problem: given a black box, applied voltages and current measurements on the boundary, determine the resistivity image of the black box. We will discuss this inverse problem subsequently.

3.   GUARDING

The biggest problem with impedance imaging is that current does not travel in straight lines. Thus the reconstructed image is distorted. It helps to make current paths as straight as possible. Within a homogeneous medium, a guarding electrode can force current to flow in straight lines. In an inhomogeneous medium, currents take paths of least resistance, so no matter how good the guard electrodes are, current flows in curved paths. But guarding helps to minimize the nonlinearity of these paths.

Webster [20] compares the current paths of various electrodes with and without guarding. Figure 3 shows four situations. For small electrodes [Fig. 3(a)] some current flows

outside the region of the electrodes, which causes a fringing effect. Also the measurement emphasizes impedance changes in the high-current-density region near the electrodes. Figure 3(b) shows that this effect can be reduced by using large-area electrodes. The current paths are linear near the center of the electrode. By splitting the electrodes [Fig. 3(c)], fringing is eliminated at the central electrode. The surrounding annular electrode is known as the guard electrode. In Fig. 3(d) a unity gain follower amplifier delivers

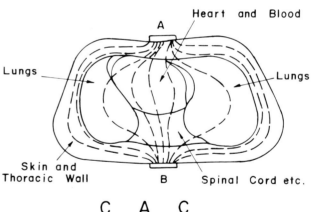

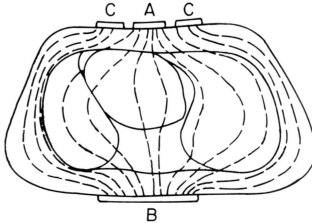

Fig. 4.   Current paths inside the thoracic cross section. (a) Without guarding. (b) With guarding.

current to the guard separately. This forces the center electrode to deliver current only to the region of interest. The center electrode and the guard are maintained at the same potential.

Figure 4(a) qualitatively shows the current distribution without guarding in a body cross section. The current passes from electrode A to electrode B. Most of current passes through the high-conductivity regions and very little passes through the lungs. Figure 4(b) shows the current distribution with guarding. A circular guard ring electrode C around electrode A supplies most of current to the surrounding high-conductivity region (thoracic wall). Electrode A supplies the current to the inner region, which includes the low-conductivity lungs and the high-conductivity heart.

Cooley and Longini [21] found the impedance of the thorax to be 120 Ω without the guard and 600 Ω to 1000 Ω with the guard. Henderson and Webster [22] used guard electrodes in frontal plane imaging. For transverse plane current density measurements, current applied in one plane can deviate from the plane of interest and travel to another plane along low-resistivity paths because it takes the paths of least resistance. Guard electrode rings around the body just above and below the plane of interest help to keep current paths in one plane. Lifshitz [4] and Cooley and Longini [21] found that without the guard, most of the potential drop occurs very near the electrode and the measurement is more sensitive to impedance change near the electrode. In a guarded system, the measured impedance is derived from approximately evenly weighted increments of volume in the column under the center electrode. Thus the guard electrodes make the specificity more uniform and constrain it to the central region.

4.   FRONTAL PLANE IMAGING

The first frontal plane impedance imaging technique was proposed by Swanson [23]. Henderson et al. [24] constructed the first equipment and obtained the first images. Henderson and Webster [22] described this "Impedance camera" which yielded a frontal plane image of the thorax. We now describe the principle and the design of the impedance camera in detail.

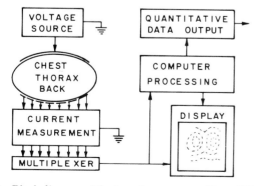

Fig. 5.   Block diagram of the impedance camera (from [22]).

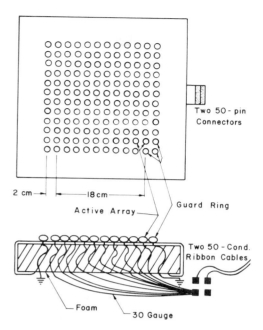

Fig. 6.   Electrode array for the impedance camera (from [22]).

Figure 5 shows a block diagram of the impedance camera. The impedance camera uses 144 mutually guarding electrodes to make 100 spatially specific measurements of the thorax. The electrode array consists of 100 active current-sensing electrodes connected to input amplifiers and 44 guard electrodes connected to ground. The electrode array is made of coat snaps fastened to rubber bed sheeting and is laid out on a urethane-foam pad (Fig. 6). Aquasonic gel is used to make contact between the skin and the electrodes. The ground electrode establishes an equipotential surface on the chest. The electrode consists of a flexible copper sheet mounted on an open-cell foam sheet saturated with a highly conductive electrode paste.

The impedance camera measures admittance at each electrode. A constant voltage source at 100 kHz is transformer coupled and drives the chest electrode. The use of the 100-kHz voltage source produces a relatively low electrode-through-skin impedance and prevents any possible electric shock hazard to the subject. Figure 7 shows one input stage of the 100 current-to-voltage converters. After dc blocking, ten of the 100 ac voltages each representing input current from an electrode are connected via the ten level-one multiplexers to the ten demodulators. Each demodulator is allowed sufficient time to settle for each electrode measurement. Meanwhile the nine other demodulators are sampled. After a demodulator is sampled, its multiplexer latch is strobed, which selects a new electrode. A second level of multiplexing samples the ten demodulators to provide all 100 values as one analog signal.

Images are made in two ways. The first and simpler method provides an intensity-modulated image on a CRT display. Figure 8 shows that 100 electrodes are displayed on

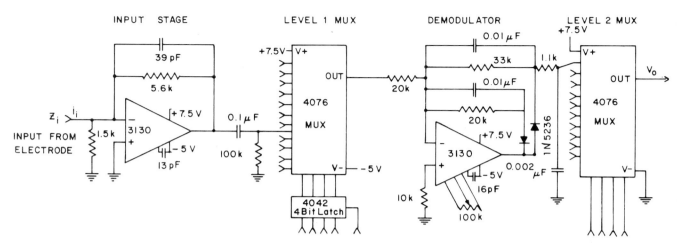

Fig. 7. Partial schematic diagram of the impedance camera (from [22]).

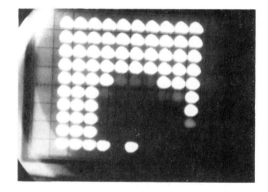

Fig. 8. Intensity modulated image of 100 spots on a CRT display.

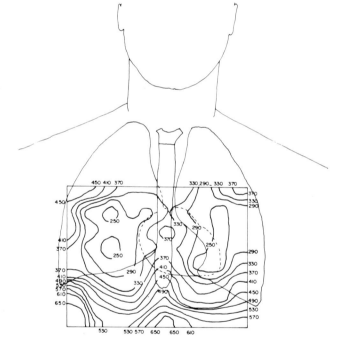

Fig. 9. Isoadmittance contour map of human thorax. Admittance values are in units of microsiemens per electrode (from [22]).

the CRT each as an intensity-modulated spot. The analog signal is fed to the z axis to modulate the spot corresponding to the respective electrode. A point-by-point raster scan is provided by the two 4-bit level-one and level-two multiplexer-select signals. These are D-to-A converted, which generates horizontal and vertical scan signals respectively. Eight frame rates from 32 frames per second (FPS) down to ¼ FPS are provided. High frame rates provide a flicker-free display. 16 FPS allows quantitative FM tape recordings of the data. The slower frame rates allow direct strip-chart recording of the data.

The intensity-modulated display has the advantage that it provides an on-line image on the CRT. Because only 100 electrodes are used, there is a very poor spatial resolution. Having poor spatial resolution demands an effective use of amplitude resolution to obtain maximum information from the impedance image. The human eye is unable to make quantitative use of the amplitude information if presented as intensity modulation. Alternately, isoadmittance curves can be plotted. The analog data recorded, on FM tape or on-line, are converted to digital values and fed to a computer.

A computer program plots isoadmittance curves as shown in Fig. 9. This provides a more understandable picture. Some information such as that of pulmonary edema, represented by high admittance contours may also be available. The impedance camera demonstrates the ability to form a frontal-plane image of the thorax. However, this technique does not use the reconstruction techniques developed for CT. Due to the bending of current paths, the admittance value of one electrode site does not necessarily represent the admittance of a path straight down below that electrode even if 143

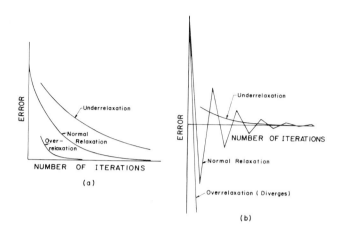

Fig. 10. Overrelaxation and underrelaxation techniques. (a) When the approximated solution converges to the exact solution from one direction. (b) When the sign of error alternates on every iteration.

guard electrodes are used. Because the current paths are not linear, the isoadmittance plot has spatial distortion. Work needs to be done to correlate data with pathologies and present them in a meaningful form.

5. TRANSVERSE PLANE IMAGING BY RECONSTRUCTION

Transverse plane imaging is a technique for displaying the image of an anatomical plane that sections the body. Such an image is also called a tomogram, which means a picture of a slice. The use of x-ray tomography is now wide-spread and to a lesser extent radioisotopic and ultrasonic imaging are finding acceptance. It is impossible to form an impedance image in the transverse plane by just putting electrodes surrounding the torso and displaying the data. As with the CT scanner, we must use a computer to manipulate the data and reconstruct the image.

The reconstruction of an object from projections results in a two-dimensional slice of a three-dimensional object. We construct it from a line integral along a current path S in the unknown resistivity distribution function, $r(x,y)$. Mathematically, along a specific current path the projection is,

$$p = \int_{\text{path } S} r(x,y) ds.$$

We can numerically approximate this integral equation by a set of equations. For each projection,

$$p_j = \sum_{i=1}^{n} w_{ij} r_i; \; j = 1, \ldots, m,$$

or in a matrix form

$$P = WR,$$

where w_{ij} is a weighting coefficient that represents the contribution of the ith element to the jth projection, r_i is the

resistivity of the ith element, n is the number of elements, and m is the number of current paths. X-ray beams travel in straight lines, thus W is independent of the resistivity distribution vector R. Therefore R can be calculated by solving a system of linear equations, usually using an algebraic reconstruction technique (ART) [25], [26]. The ART is an iterative method for solving linear equations and very popular in reconstructive tomography.

Unfortunately, in impedance imaging, the current paths depend on the resistivity distribution, i.e., the solution. Thus, W is a function of R and we have to solve a nonlinear reconstruction problem. To solve this problem properly, we must know current paths for each projection. But how do we obtain the current paths if we do not know the resistivity profile? All iterative numerical methods proceed from some initial guess, R^0. *A priori* information of rough regional conductivities can save the number of iterations and minimize the ambiguities. In body imaging, we can make use of this readily available information. However, if we have no prior knowledge of the resistivity profile—the black box—then we have to start *de novo*. Ambiguities in image reconstruction are likely to persist [27] and it takes many more iterations to have the solution converge.

Whether our initial guess is good or bad, the solution of Laplace's equation with specified boundary conditions gives the voltage distribution and, finally, current paths. Along each of these curved current paths, we calculate the resistance contribution of each element to the current path (w_{ij}) and sum the total computed resistance. We back-project the difference between the computed and measured resistance and apply an equal resistance correction to each element along this path. Mathematically,

$$r_i^n = r_i^{n-1} + k \sum_{j=1}^{m} \Delta r_{ij}^n$$

where Δr_{ij}^n is the correction term after the nth iteration to the previous ith element's resistivity value (r_i^{n-1}) for the jth current path. k is a constant that sets the rate of convergence of the solution. Figure 10(a) shows that the approximated solution may converge to the exact solution from the initial guess without changing the sign of error ($\sum_{j=1}^{m} \Delta r_{ij}^n$). In this case, the overrelaxation technique ($k > 1$) increases the rate of convergence, while the underrelaxation technique ($0 < k < 1$) slows it down. However, if the sign of the error alternates on every iteration as shown in Fig. 10(b), the underrelaxation technique is better. Which course a solution will take cannot be predicted. It is strongly dependent on the initial guess and the properties of the problem. Even if we knew the course, it would be difficult to determine the optimum value of k theoretically. If we choose k too large or too small, the process may not converge or sometimes it may diverge. Proper use of the relaxation technique makes computing shorter and less expensive.

Figure 11 shows the flowchart of the reconstruction procedure. We select one projection angle, perform the computations, and test for degree of convergence. If sufficient convergence is not obtained, we index to the next projection angle and iterate. There is no definite stopping rule to de-

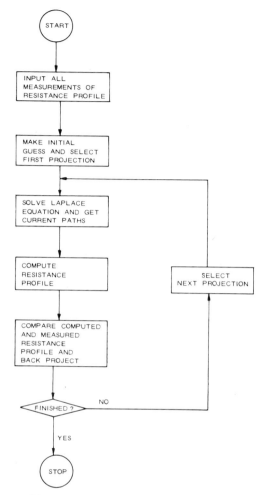

Fig. 11. The simplified flowchart for reconstructing images in the transverse plane.

Tjandrasa [19] made four different projections by solving the two-dimensional Laplace equation based on the finite element method. She rotated the electrode location (boundary conditions) around a body cross section for each different projection and plotted equipotential lines and current paths (Fig. 2). She started with a postulated initial guess of the resistivity distribution, calculated the resistance contribution of each element along current paths, back-projected the difference until satisfactory agreement with measured resistance profile was obtained, and produced the reconstructed image shown in Fig. 12. The image easily shows the low-resistivity heart region and high-resistivity lung regions. She started with uniform lung resistivity of 1200 Ω·cm. After computation, the highest resistivity in the image occurred in the lower lung regions because the proximity of the low-resistivity heart interfered in imaging the high-resistivity upper lung portions. How much the initial guess influenced the final image is not clear. To evaluate the effect of the initial guess on reconstructing body images, we would need to obtain images from different resistance profiles, but using the same initial guess.

Lytle and Dines [16] simulated a synthetic square core sample implemented in a computer program with easy and difficult resistance profiles. First, they solved the Laplace equation using the finite difference method. They calculated the current entering or leaving each simulated electrode for many current angles. They controlled the current angle by the different voltage values applied to each electrode. In reconstructing the image, they started with no *a priori* knowledge about the resistivity distribution inside the core sample. For each current angle, they solved a new Laplace equation and computed a new resistance profile by back projection. Ten iterations computed a good reconstruction of a simple profile (a square with conductivity twice as high as the surrounding low-conductivity core sample). Figure 13 shows inaccuracies after 10 iterations of a difficult resistance profile. The low-conductivity black region in the middle surrounded by a high-conductivity white region present in the ideal synthetic profile is falsely interpreted as a high-conductivity region.

Tasto and Schomberg [17] present more convincing results. They simulated a square core with a computer program. They interpolated their initial 400 elements to 1600 elements, and also tried easy and difficult resistance profiles. They calculated the voltage distribution for each projection angle using the two-dimensional finite difference method and the overrelaxation technique. Without any assumed resistivity distribution, they determined current paths and the computed resistance profile. After comparing the computed and measured (simulated by computer) resistance profiles, they performed the back projection. They used the underrelaxation technique to obtain successive resistance profiles and intermediate smoothing of the reconstructed resistance profile. Figure 14 shows the improvement in reconstructed image quality of a rotationally symmetric hat object as the number of projection angles increases from 1 to 432. Figure 15 shows the original and reconstructed images of a more complex profile with no rotational symmetry. Some

termine when we are "finished" as shown in Fig. 11, because we do not have information about the behavior of this solution beforehand. Two methods are widely used and if they are combined, they produce an effective and efficient stopping rule. One method uses a preselected constant for the maximum number of iterations and protects us from excessive computing time due to poor convergence or error. The other method requires $|r_i^n - r_i^{n-1}|$ to be less than a prespecified error range and allows an early termination after the convergent solution is obtained. We select the error range depending on the initial guess and the accuracy desired. After running the program many times, we are able to choose good values for the two constants. An alternative method is an interactive computing approach. After every iteration, the computer displays the resistance profile and correction terms from the previous iteration. Upon viewing these, we decide whether to have another iteration or not. In this way, we can see the image improvement for successive iterations and have better control on when to stop the reconstruction procedure.

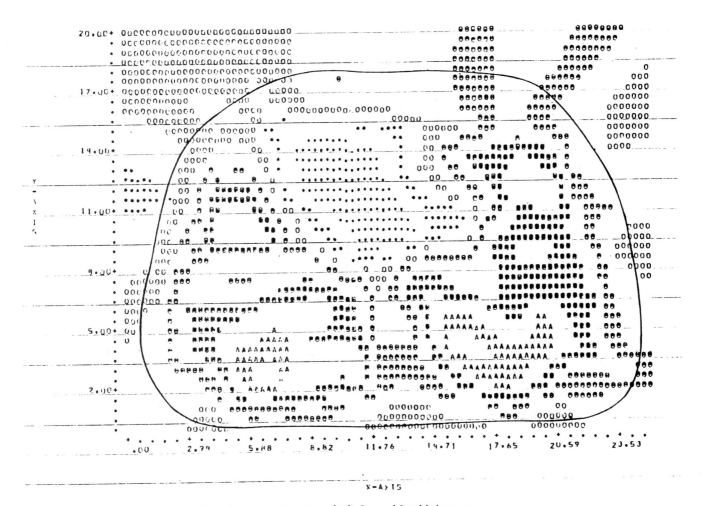

Fig. 12. Impedance tomogram of a thoracic cross section (from [19]). Legend for this image:

100–200 Ω·cm	,	700–800 Ω·cm	θ
300–400 Ω·cm	*	900–1000 Ω·cm	θ
500–600 Ω·cm	0	1100–1200 Ω·cm	A
All others	space	Missing values	M

blurring and local overshoots are apparent in their reconstructed images. The results are promising in that they used only 400 elements to represent an object.

Kim [28] reconstructed impedance images using measured current densities at the current-sensing electrodes. He did not determine the current paths. Instead, before reconstructing the images, he calculated the sensitivity of each element inside a computer model to each current-sensing electrode by changing the resistivity of one element at-a-time and observing the current density changes in all current-sensing electrodes. He calculated the current densities based on the most recently modified conductivity profile by solving Laplace's equation for every iteration. During image reconstruction, he back-projected differences between true and computed current densities according to each element's sensitivity to each electrode. He used a computer model with 332 elements based on the finite element method and 4 projection angles. Figure 16 shows the reconstructed image after 200 iterations without using any *a priori* information.

With an intelligent initial guess (20% different from the true object's resistivity), he could reconstruct an impedance image that is almost the same as the original one in fewer than 12 iterations.

6.   OTHER IMAGING TECHNIQUES

Benabid et al. [29] constructed a linear array of 128 electrodes encircled by a guard electrode. In a saline-filled plexiglas tank they successively measured magnitude and phase of each electrode while maintaining all other electrodes at the guard potential. They detected and localized Plexiglas and aluminum cylindrical test objects even behind a plastic screen (which simulated the high-resistivity skull). They proposed a rotational displacement of the electrode matrix and an algorithm similar to that used in CT scanners to solve for the impedance profile.

Schmitt [30] has proposed using "mutual impedivity

Low conductivities are difficult to electrically sense

Ideal   Interpreted

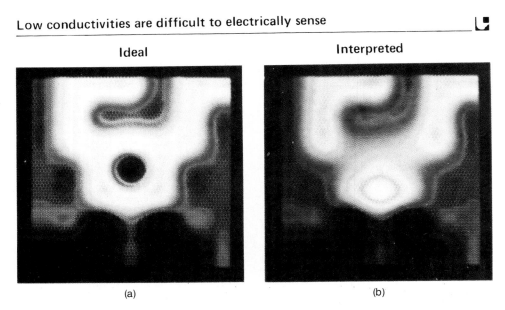

(a)   (b)

Fig. 13. Image reconstruction of a difficult resistivity profile within synthetic square core sample. Conductivity for white: $\sigma = 0.015$ S/m, for black: $\sigma = 0.005$ S/m. (a) Ideal image, (b) after ten iterations (from [16]).

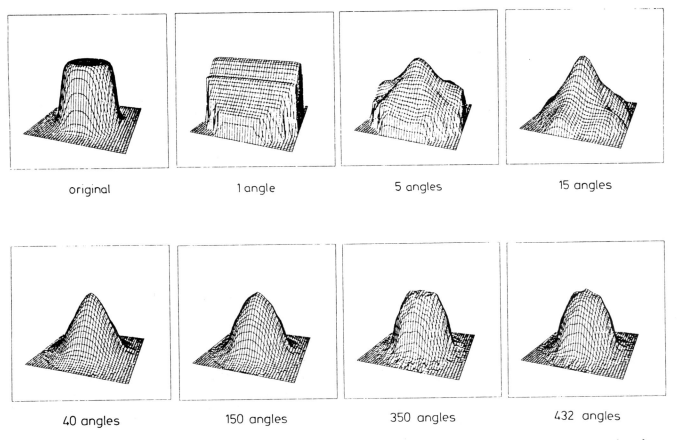

original   1 angle   5 angles   15 angles

40 angles   150 angles   350 angles   432 angles

Fig. 14. Sequence of iterations during tomographic reconstruction of a rotationally symmetric object. The vertical axis represents impedance magnitude (from [17]).

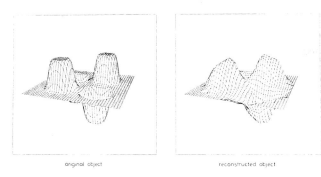

Fig. 15. Original and reconstructed object with no rotational symmetry (from [17]).

spectrometry," to create an analog of the CT scanner. It would determine impedance arrays at each of many frequencies for example from 100 to 500,000 Hz. From an array of a few dozen electrodes, input driving current in one pair of electrodes would yield output voltage on another pair of electrodes. By using many different combinations he would develop the CT reconstruction. He does not explain

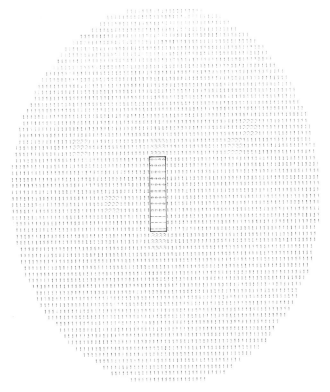

Fig. 16. The image reconstructed without using any a priori information, with a rectangle representing the location of the phantom object. The original image has 500 Ω-cm inside this rectangle and 100 Ω-cm elsewhere.

Legend for this image:

Range	Symbol	Range	Symbol
90–130 Ω-cm	1	400–460 Ω-cm	–
130–160 Ω-cm	2	460–540 Ω-cm	*
160–220 Ω-cm	3	(from [28]).	

the method, suggest the reconstruction rule, or provide results of pilot studies.

Price [6] modeled a tomographic slice into pixels made up of lumped impedances. The data that would be obtained from such an impedance array were computed by the computer and images were reconstructed using these data. Except for the peripheral ring, the results were very poor because the guarding technique was not used. He suggested that a computer simulation using guarded electrodes should yield better results.

Frei and Sollish have developed a mammo-scanner for detecting breast tumors [31]. An 8×8 array of 7.5-mm plates passes 100 μA/cm^2 into the breast. By measuring the current flowing to each of the 64 plates, it calculates the tissue conductivity and dielectric constant. The mammo-scanner yields a picture of one quadrant of the breast. Squares with abnormal conductivity and dielectric constant appear brighter, indicating suspicious areas.

Bates et al. [32] investigate the simple two-dimensional case of a circle which has an axial circular core of one conductivity and the remaining annular ring of a second conductivity. For simple circumferential voltage distributions, it is not possible to deduce the radius of the core or the conductivities from measurements of current density flowing across the circumference. By using additional circumferential voltage distributions (modes), it is possible to deduce any circularly symmetric (onion-like) distribution of conductivity. However, biological resistivity distributions are complex and it is not possible to use analytical techniques. They concluded that if the problem of impedance imaging has a unique solution, in general, algorithms for the solution must be iterative.

7.   THE FUTURE

The science of impedance imaging looks promising but unproven, as are most other impedance applications. The data from humans have not been used in reconstructions. Most reconstructions have been on nonhuman synthetic objects. Image reconstructions of the frontal or transverse plane similar to x-ray tomography from real resistance measurements have not been attempted. Most of the cross sections used in transverse imaging were simple ones (squares) or had a fixed geometry due to the limitations of the computer models. Because electrodes must contact the body, it will be difficult to account for individual body geometry differences. Problems associated with anisotropy and dielectric effects have been ignored. Tissue or disease specificity of impedance imaging is unknown. In spite of all these deficiencies and uncertainties, we feel that we can obtain fairly good images in the frontal and transverse planes as well as in three dimensions.

We must improve the guarding technique for the frontal and transverse planes. We must use reconstruction methods to improve frontal plane images. We must establish the specificity to various diseases. We can minimize the ambiguities in transverse plane imaging by building a three-dimensional girdle of electrodes that encircles the torso. We can force current to flow in the plane of interest by controlling the voltage values to nonplanar electrodes. This should minimize the ambiguities in the central region, that might exist for a difficult profile similar to that of Fig. 13. Also, out-of-plane measurements and in-plane measurements should help to reconstruct the transverse plane image accurately. We can improve spatial resolution by carefully increasing the number of electrodes while maintaining each electrode isolated from all other electrodes and increasing the number of elements in the computer body model.

We can program the computer to control many voltage sequences applied to the electrodes for measurement of the resistance profiles. The individual's body geometry and electrode locations are prior information. The computer model must be interactive according to the individual differences and contain enough elements to produce good resolution. Key factors to be considered are the computer model in reconstructing the image and the computing time. Finite element models are much more elaborate and flexible than finite difference models especially in modeling irregular body boundaries. Finite element models have been implemented in only medium to large-scale computers if the number of elements exceeds about 500. The computer memory required is approximately proportional to n^2, where n is the number of elements in the model. Kim et al. [33] present a three-dimensional finite element model implemented in a mini-computer. Kim et al. [34] and Kim [35] also describe an efficient finite element algorithm whose memory requirement increases by about $n \ln n$, which gives substantial memory space saving when n is large. Even if this memory requirement is resolved, the computing time can be exorbitantly large, and may take hours. The time depends on number of projections and accuracy desired, because for each projection, Laplace's equation based on the old resistivity profile must be solved. Jordan [36], [37] proposed a special parallel computer just for finite element analysis with 1024 identical computers. Today a small prototype exists. For computerized impedance tomography (CIT), there should be a dedicated special purpose computer that controls data acquisition and reconstructs images, with array processors or fast floating point execution units similar to those in a CT scanner. It will take more computing effort to reconstruct the impedance image than for x-ray tomography unless more effective computer models and more efficient numerical methods are developed.

Kim et al. [38] present reconstructed impedance images with 8 projection angles and different projection methods using computer-generated phantom objects. They have evaluated the effects of the number of projection angles employed using *a priori* resistivity distribution information. The use of an available approximately correct initial guess reduces the computing time and ambiguities significantly, while the wrong initial guess does not deteriorate the final reconstructed impedance image. Generally, the fan-beam projection method results in better reconstructed impedance images than the parallel beam projection method especially in imaging the central region, while the required computing time is always higher for the fan-beam projection than the parallel beam projection. They also report that the accuracy in reconstructed impedance images is lower in the central region than the periphery, because the sensitivity distribution is found to be shaped like an upside-down bell.

Barber and Brown [39] review electrical impedance imaging in general. They discuss various techniques proposed and implemented for solving the forward problem and for reconstructing two-dimensional impedance images. They also show reconstructed *in vivo* impedance images by applying current between a pair of a set of 16 electrodes connected to the boundary of the forearm and measuring the potential difference between all adjacent pairs of electrodes. This four-electrode system minimizes the errors caused by electrode contact impedances and stray capacitances. The difference between the measured potential and computed potential is back-projected between the equipotentials ending on the source electrode pair. This process is repeated for many source electrode configurations. The reconstructed images show the bones and fat in the forearm, but are limited in resolution mainly due to the small number of electrodes and limited number of independent measurements using 16 electrodes.

Kim et al. [40] discuss an impedance imaging system under development, which integrates the hardware components, and software models and algorithms necessary in electrical impedance imaging into one system. The impedance imaging system consists of a two-dimensional electrode array, a microcomputer-based electrode array controller, a host computer based on a powerful microcomputer system with large main memory, fast floating-point processing capability, high-speed secondary storage devices, and a computer human body model developed with the finite element method and a group of scanning and reconstruction algorithms implemented in the host computer.

We think that CIT images in two or three dimensions are a definite possibility in the future.

References

1. J. F. Greenleaf and R. C. Bahn, "Clinical Imaging with Transmissive Ultrasonic Computerized Tomography," *IEEE Trans. Biomed. Eng.*, Vol. BME-28, pp. 177–185, 1981.
2. E. Frank, "Spread of Current in Volume Conductor of Finite Extent," *Ann. N.Y. Acad. Sci.*, Vol. 65, pp. 980–1002, 1957.
3. H. C. Burger and J. B. Van Milaan, "Heart Vector and Leads," *Br. Heart J.*, Vol. 8, pp. 157–161, 1946.
4. K. Lifshitz, "Electrode Impedance Cephalography, Electrode Guarding, and Analog Studies," *Ann. N.Y. Acad. Sci.*, Vol. 170, pp. 532–549, 1970.
5. S. Rush, "An Inhomogeneous Anisotropic Model of the Human Torso for Electrocardiographic Studies," *Med. Biol. Eng.*, Vol. 9, pp. 201–211, 1971.
6. L. R. Price, "Electrical Impedance Computed Tomography (ICT): A New CT Imaging Technique," *IEEE Trans. Nucl. Sci.*, Vol. NS-26, pp. 2736–2739, 1979.
7. O. C. Zienkiewicz, *The Finite Element Method in Engineering Science*, McGraw-Hill, London, 1971.
8. K. H. Hubner, *The Finite Element Method for Engineers*, Wiley, New York, 1975.

9. H. P. Schwan, "Determination of Biological Impedances," in W. L. Nastuk (Ed.), *Physical Techniques in Biological Research*, Vol. 6, Academic Press, New York, 1963.

10. L. A. Geddes and L. E. Baker, "The Specific Resistance of Biological Material—A Compendium of Data for the Biomedical Engineer and Physiologist," *Med. Biol. Eng.*, Vol. 5, pp. 271–293, 1967.

11. A. C. Eycleshymer and D. M. Shoemaker, "A Cross-Section Anatomy," Appleton Century Crofts, New York, 1911.

12. P. Silvester and S. Tymchyshyn, "Finite-Element Modeling of the Inhomogeneous Human Thorax," in S. Rush and E. Lepeschkin, (Eds.), *Advances in Cardiology*, Vol. 10, S. Karger, Basel, pp. 46–50, 1974.

13. R. Demers, S. Freiwald, and P. Silvester, "Construction of a Finite Element Model of Thorax Impedance," *Proc. Annu. Conf. Eng. Med. Biol.*, Vol. 18, p. 320, 1976.

14. E. Kinnen, "Determining Electric Current Flow Patterns in the Thorax, *IEEE Reg. Six Conf. Rec.*, pp. 379–381, 1966.

15. S. K. Guha, S. N. Tandon, and M. R. Khan, "Electrical Field Plethysmography," *Biomed. Eng.*, Vol. 9, pp. 510–518, 1974.

16. R. J. Lytle and K. A. Dines, "An Impedance Camera: A System for Determining the Spatial Variation of Electrical Conductivity," Lawrence Livermore Lab. Rept., UCRL-52413, Livermore, CA, 1978.

17. M. Tasto and H. Schomberg, "Object Reconstruction from Projections and Some Non-linear Extensions," in C. H. Chen (Ed.), *Pattern Recognition and Signal Processing*, Sijthoff & Noordhoff, Alphen aan den Rijn—The Netherlands, pp. 485–503, 1978.

18. R. Natarajan and V. Seshadri, "Electric-Field Distribution in the Human Body Using Finite Element Method," *Med. Biol. Eng.*, Vol. 14, pp. 489–493, 1976.

19. H. Tjandrasa, "Two-Dimensional Analysis of the Electrical Field Distribution in the Thorax and Reconstruction of a Cross-Section from its Projections," M. S. Thesis, Dept. of Elec. and Comp. Eng., Univ. of Wisconsin, Madison, WI 53706, 1978.

20. J. G. Webster (Ed.), *Medical Instrumentation: Application and Design*, Houghton Mifflin, Boston, 1978.

21. W. A. Cooley and R. L. Longini, "A New Design for an Impedance Pneumograph," *J. Appl. Physiol.*, Vol. 25, pp. 429–432, 1968.

22. R. P. Henderson and J. G. Webster, "An Impedance Camera for Spatially Specific Measurements of the Thorax," *IEEE Trans. Biomed. Eng.*, Vol. BME-25, pp. 250–254, 1978.

23. D. K. Swanson, "Measurement Errors and the Origin of Electrical Impedance Changes in the Limb," Ph.D. Dissertation, Dept. of Elec. & Comp. Eng., Univ. of Wis., Madison, WI 53706, 1976.

24. R. P. Henderson, J. G. Webster, and D. K. Swanson," A thoracic Electrical Impedance Camera," *Proc. Annu. Conf. Eng. Med. Biol.*, Vol. 18, p. 322, 1976.

25. R. Gordon, "A Tutorial on ART (Algebraic Reconstruction Techniques)," *IEEE Trans. Nucl. Sci.*, Vol. NS-21, pp. 78–93, 1974.

26. R. A. Brooks and G. DiChiro, "Theory of Image Reconstruction in Computed Tomography," *Radiology*, Vol. 117, pp. 561–572, 1975.

27. R. Plonsey and R. Collins, "Electrode Guarding in Electrical Impedance Measurement of Physiological Systems—A Critique," *Med. Biol. Eng. Comput.*, Vol. 15, pp. 519–527, 1977.

28. Y. Kim, W. J. Tompkins, and J. G. Webster, "Medical Body Imaging Using Electrical Impedence and Nonlinear Reconstruction," Northeast Bioeng. Conf., Vol. 10, pp. 298–303, 1982.

29. A. L. Benabid, L. Balme, J. C. Persat, M. Belleville, J. P. Chirossel, M. Buyle-Bodin, J. de Rougemont, and C. Poupot, "Electrical Impedance Brain Scanner: Principles and Preliminary Results of Simulation," *T.-I.-T. J. Life Sci.*, Vol. 8, pp. 59–68, 1978.

30. O. H. Schmitt, "Mutual Impedivity Spectrometry and the Feasibility of its Incorporation into Tissue-Diagnostic Anatomical Reconstitution and Multivariate Time-Coherent Physiological Measurements," in K. Preston, Jr., K. J. W. Taylor, S. A. Johnson, and W. R. Ayers (Eds.), *Medical Imaging Techniques—A Comparison*, Plenum, New York, 1979.

31. A. Kemelman, "Abnormal Electrical Properties Betray Breast Tumors to Scanner," *Electronics*, Vol. 53(1), pp. 68–70, 1980.

32. R. H. T. Bates, G. C. McKinnon, and A. D. Seager, "A Limitation on Systems for Imaging Electrical Conductivity Distributions," *IEEE Trans. Biomed. Eng.*, Vol. BME-27, pp. 418–420, 1980.

33. Y. Kim, W. J. Tompkins, and J. G. Webster, "A Three Dimensional Modifiable Body Model for Biomedical Application," *IEEE-EMBS Frontiers Eng. Health Care*, Vol. 3, pp. 8–12 1981.

34. Y. Kim, W. J. Tompkins, and J. G. Webster, "An Efficient Finite Element Algorithm," *Frontiers Comput. Med.*, Vol. 1, pp. 34–37, 1981.

35. Y. Kim, "A Three-Dimensional Modifiable Computer Body Model and Its Applications," Ph.D. Dissertation, Dept. of Elec. & Comp. Eng., Univ. of Wis., Madison, WI 53706, 1982.

36. H. F. Jordan, "A Special Purpose Architecture for Finite Element Analysis," in G. J. Lipovski (Ed.), *Proc. Int. Conf. Parallel Processing*, IEEE Computer Society, Long Beach, CA, pp. 263–266, 1978.

37. H. F. Jordan (Ed.), The Finite Element Machine: Programmer's Reference Manual, Dept. of Elec. Eng., Univ. of Colorado, Boulder, CO 80309, 1979.

38. Y. Kim, J. G. Webster, and W. J. Tompkins, "Electrical Impedance Imaging of Thorax," J. Microwave Power, Vol. 18, pp. 245–257, 1983.

39. D. C. Barber and B. H. Brown, "Applied Potential Tomography," *J. Physics E: Sci. Instrum.*, Vol. 17, pp. 723–733, 1984.

40. Y. Kim, T. J. Brooks, and S. O. Elliott, "Electrical Impedance Technique in Medical Body Imaging," Proc. Annu. Conf. Eng. Med. Biol., Vol. 26, p. 19, 1984.

Methods of Active Microwave Imagery for Dosimetric Applications

Lawrence E. Larsen and John H. Jacobi

The topic of active microwave imagery is described from the molecular, microscopic and macroscopic perspectives for the canine kidney. Design considerations for data acquisition systems are discussed in terms of antenna design and operation, the role of scattering geometry, the choice of frequency of operation, and use of the polarization scattering matrix. Image processing for raster scanned (nontomographic) systems and image interpretation based on renal regional specialization are presented.

INTRODUCTION

The spatial distribution of microwave energy transmitted from an incident field through a biosystem to a receiving antenna depends upon features of both the biosystem and the incident energy. With respect to the incident field, features such as frequency, polarization, and mode are important parameters of such a radiative transfer. With respect to the biosystem, the spatial distribution of dielectric properties and their time as well as frequency dependencies are the important parameters. Frequency dependencies are intrinsic to the particular molecular structure involved, whereas time dependencies often represent physiologic sources of variation. Further, to the extent that the incident flux density is sufficient to heat the biosystem, power density may itself effect the spatial distribution of absorbed microwave energy. This is a consequence of two facts: first the microwave constitutive parameters of biological dielectrics are temperature dependent *per se;* secondly, *in situ* thermoregulatory mechanisms may substantially alter the spatial distribution of complex permittivity as a result of vasomotor activity.

The basic motivation for diagnostic applications of microwave imagery is improved physiologic and pathophysiologic correlation, especially in soft tissue. This expectation is based on the molecular (dielectric) rather than atomic (density) based interactions of the radiation with the target when compared with x-ray imagery. Furthermore, much greater contrast is available with microwave imagery. The range of microwave constitutive parameters in soft tissue is ca. 20 to one compared to the few percent range of densities in soft tissue. This suggests not only improved contrast but also better tissue characterization.

The following sections of this paper deal with characteristics of the incident radiation and properties of biosystems as media which are important for the propagation of microwave energy. After a discussion of dielectric properties and various biological considerations, electromagnetic wave

descriptors and system design criteria will be considered. This will be followed by a description of actual data collection systems, subsequent image processing and, lastly, image interpretation keyed to known regional specialization in the canine kidney.

The Nature of Dielectrics

Since microwave images are formed on the basis of the spatial distribution of complex permittivity, the nature of biological dielectrics and their role as media for propagation of microwave energy are items of necessary background [1], [2], [3]. Electromagnetic waves in the microwave region of the spectrum (typically defined as 300 MHz to 300 GHz in frequency, although a much smaller range of ca. 1 to 10 GHz is suitable for diagnostic imagery) propagate in uniform dielectrics according to

$$E = E_o \exp(-\gamma z)$$

where E_o is the electric field at the origin, E is the scalar instantaneous electric field in the dielectric at a distance z from the origin and γ is the complex propagation coefficient. The complex propagation coefficient is defined as

$$\gamma = \alpha + j\beta = j\omega\sqrt{\epsilon^*\mu^*}$$

where α is the attenuation factor, β is the phase factor, ϵ^* is the complex permittivity, μ^* is the complex permeability and j has its usual significance. The factors α and β are related to the microwave constituative properties of the medium as follows:

$$\alpha = \lambda\omega^2/4\pi \ (\epsilon'\mu'' + \epsilon''\mu')$$

and

$$\beta = \frac{2\pi}{\lambda} = \omega\left[\frac{\epsilon'\mu' - \epsilon'\mu''}{2} \cdot \sqrt{1 + \frac{\epsilon'\mu'' + \epsilon''\mu'}{\epsilon'\mu' - \epsilon'\mu''} + 1}\right]^{1/2}$$

where λ is the wavelength, ω is the angular frequency, ie., $\omega = 2\pi f$, and f is the frequency in Hertz of the electromagnetic wave. The constitutive parameters of μ^* and ϵ^* refer to the magnetic (complex permeability) and dielectric (complex

———
Department of Microwave Research, Walter Reed Army Institute of Research, Walter Reed Army Medical Center, Washington, D.C. 20012

permittivity) properties of the medium. The magnetic properties are assumed to be those of free space, i.e., purely real with no attenuation or phase shift. The dielectric properties are complex as the result of the fact the biological dielectrics (other than air) contain both conduction and displacement currents when polarized by an electric field or when immersed in a time harmonic electric field. The conduction currents represent current flow that is in phase with the applied voltage whereas displacement currents are in phase quadrature with the applied voltage. The complex nomenclature is applied as follows:

$$\epsilon^* = \epsilon' - j\epsilon''$$

where ϵ' is the real part of the complex permittivity, known as the dielectric constant, and ϵ'' is the imaginary part known as the dielectric loss factor. It is important to understand that the loss in question is dielectric, that is, such loss exists in the absence of D.C. conductivity although ionic conduction may also contribute. The real and imaginary parts of the complex permittivity represent the complementary processes of energy storage and energy dissipation, respectively. Since heat production is related to the product of frequency and the dielectric loss factor, these are often combined in the quality known as dielectric conductivity, σ

$$\sigma = \omega\epsilon''$$

As a matter of convenience, the complex permittivity is often normalized to that of free space. That is,

$$\epsilon_r^* = \epsilon^*/\epsilon_o$$

where ϵ_o is the permittivity of free space and ϵ_r^* is the relative complex permittivity. Likewise, the real and imaginary parts are normalized as follows

$$\epsilon_r^* = \epsilon_r' - j\epsilon_r'' = \epsilon'/\epsilon_o - j\epsilon''/\epsilon_o$$

where ϵ_r' is the relative dielectric constant and ϵ_r'' is the relative loss factor.

Both ϵ_r' and ϵ_r'' are important for diagnostic imagery of biological targets in the decimeter wavelengths where the experiments to be later described were conducted. Most biological dielectrics are water dominated. Water has a rather high relative dielectric constant of about 80 at middle microwave frequencies. This accomplishes a significant reduction in wavelength as compared to air since the phase velocity in lossless dielectrics is

$$dz/dt = \nu = \omega/\beta \text{ and } \lambda_{air}/\lambda_{diel} = c/\nu \cong 1/\sqrt{\epsilon_r'}$$

where β is approximately $c\sqrt{\epsilon_r'}$ and c is the velocity of light in a vacuum. Thus, the retarded phase velocity results in wavelength contraction to about $\frac{1}{9}$ of its value in air. This property is very useful in microwave imaging systems when water serves as a coupling and loading medium for waveguide antennas to probe the fields scattered by organs under study. Water is also rather lossy. At a frequency of 3 GHz, pure water has an alpha or dissipation factor such that a propagating wave is attenuated at a rate of about 3.82 dB/cm. In water coupled targets, this provides for effective attenuation of wave propagation along the first dielectric interface which

suppresses a major component of multipath in systems with forward scatter geometries. With respect to biologic dielectrics *in situ*, the measured insertion loss for thorax and abdomen at 2 GHz agree rather well for that predicted on the basis of bulk equivalent muscle at about 50 Nepers/meter or about 85 dB loss for a 20 cm path length [4]. At 3 GHz, attenuation is about 60 Np/m or about 104 dB for the same path length. At 4 GHz the loss would be about 134 dB. Because of additional "loss" due to the phase factor, 4 GHz is about the highest practical frequency in abdomen and thorax when using state of the art receivers (noise floor at ca. -140 dBm) and modest amplification prior to the transmitting antenna. In the case of breast scanners, however, appreciably higher frequencies (ca. 10 GHz) may be employed and in the case of head scanners intermediate frequencies (ca. 6 GHz) may be useful in transmission systems as a result of the appreciably lower attenuations of these tissues and the shorter path lengths involved. Physiologic variations in dielectric properties of biosystems is described in the chapters by Burdette et al and Lin elsewhere in this volume.

Dielectric properties at microwave frequencies represent the electrical behaviour of biological materials at wavelengths short enough that useful spatial resolution may be obtained and images can be produced. Dielectric properties are relatable to molecular structure by way of charge asymmetries and rotational mobilities. These factors are expressed in large part by the dipole moment. Net charge imbalance along one axis may be reduced to an equivalent dipole whereupon two charges equal to plus and minus one electron charge when separated by one nanometer represent about one half Debye. In these terms, water has a dipole moment of about 2 Debyes. This may be deduced from a knowledge of the static (low frequency) dielectric constant of water. Large molecules may have charge asymmetries that cannot be resolved along a single axis. Under these conditions, higher order moments may result. Also, a number of dispersion mechanisms exist which involve various irrotational states of water, as well as side groups and end groups of large molecules. Alterations in conformational state also effect dielectric properties in large molecules. Numerous small molecules such as amino-acids, oligopeptides, etc. also contribute to dielectric properties in the microwave region.

Dielectric properties at microwave frequencies of various tissues *in situ* are a relatively new area of study with respect to physiologic and pathophysiologic correlations [5]. However, it may be stated with certainty that various regional specializations in brain (e.g. white matter as opposed to grey matter) and kidney (medulla as opposed to cortex) are represented in both ϵ_r' and σ. Furthermore, antemortem/postmortem comparisons in brain and changes in renal permittivity with blood flow do not limit interpretation to simple changes in blood volume. Certainly, the images to be presented later in this paper attest to the fact that spatial variations in complex permittivity are relatable to known regional specialization in the canine kidney.

A brief comparison of microwave interactions in distinction with x-ray interactions in biosystems may be useful at this point. These distinctions arise because of the difference in photon energy between microwaves and x-rays. This is a consequence of the ca. 10^9 times higher frequency of photons in the x-ray region of the electromagnetic spectrum. That is,

$$e = hf$$

where h is Planck's constant and e the photon energy. Diagnostic x-rays typically have photon energies in the 100 KeV (Kilo-electron-volt) range where as microwave photons at decimeter wavelengths have photon energies in the 10^{-4}eV range. X-ray absorption mechanisms in this range are chiefly K-capture and Compton scattering, both of which may be modeled by exponential functions of the atomic number (s) of the constituent atoms [6]. Microwave absorption is based on rotation of molecules or component side chains, end groups, etc. The central notion is that the photon energy of the incident radiation determines which work functions in the target will be addressed. K-capture requires the incident photon to supply the correct energy to dislocate a K-band electron. This is a vastly more energetic process than overcoming the weak inter- and intra-molecular forces in rotatable dipoles. Inasmuch as atomic composition is not uniquely associated with biological function, the biological relevance of x-ray interrogation may be expected to be chiefly structural. This is evidenced by the popularity and often necessity of contrast enhancement procedures in diagnostic radiology. On the other hand, there is reason to believe that the molecular level (especially with respect to secondary and tertiary structure) offers greater promise for biological relevance to health and disease. Examples of this assertion include the importance of secondary and tertiary structure for enzyme substrate interactions, ligand-receptor interactions and the role of water in the structural stability of biopolymers. It is these hopeful expectations that motivate attempts to form images on the basis of dielectric properties. Since the information addressed by x-ray and microwave interrogation are not redundant, the combination of properties will provide a more complete description of the target than either one alone. Furthermore, the low photon energies of microwave radiation are of little added risk when compared to the ionization produced by x-ray photons [7].

At this juncture, the next item of necessary background is the choice of the target organ for microwave imagery demonstration and its relationship to dielectric properties. After this, methods of description for electromagnetic waves will continue. Of course, this has begun in the present section, but the item remaining—polarization—deserves special treatment after the biological and dielectric preliminaries are concluded.

Choice of Target Organ

The choice of a target for demonstrations of microwave imagery in biosystems is an important topic. Since water is an important constituent molecule in biological dielectrics, an organ with specialization for water transport would be a

good choice. Further, since tomographic reconstruction of organ images has not yet been achieved, an organ with symmetry in a plane perpendicular to the direction of propagation would be desirable. Finally, an organ with regional specialization and a fibrous capsule to maintain its shape *in vitro* would be helpful. All of these considerations suggest that the kidney is a good choice for imagery demonstation.

Another set of considerations apply to the choice of *in vitro* rather than *in situ* imagery. Obviously, *in situ* imagery is the desired goal, but imagery demonstrations *in vitro* limit the number of problems which require simultaneous solution. Imagery *in vivo* is complicated by several important but temporarily subsidiary problems. Among these, three are cf prime importance: *In vivo* imagery will be contaminated by significant movement artifact even in abdominal organs due to diaphragm contraction during respiration. Since data collection times are so long (ca. 4.5 hr), this movement would seriously degrade the image and make it difficult to assess the fundamental problem of biological relevance/interpretation of microwave imagery in biosystems. Secondly, the images would be contaminated by greater refractive error when intervening heterogeneous dielectrics are interposed between the antennas and organ of interest. This problem requires solution, but only after the questions of biological relevance and pathophysiological correlation confirm the motivation for use of microwave energy as the interrogating radiation. Thirdly, the present receiver could not accommodate the insertion loss commensurate with the long path lengths in canine kidneys *in situ*. The solution to this problem requires no new technologies for S band receivers; rather, it is a statement of the relative insensitivity of the existing receiver (noise floor of ca. −80 dBm) compared to state-of-the-art S band receiver (ca. −140 dBm).

Polarization Concepts

Microwave energy in free space is often described as a transverse electric and magnetic wave which is characterized by its amplitude, frequency and initial phase. This description is incomplete at least to the extent that the direction of the electric and magnetic components are only required to be mutually perpendicular and perpendicular to the direction of propagation. That is to say, polarization is not described and the vector nature of the radiation is not fully specified.

A reference co-ordinate system for polarization description is needed. Such a reference is established by projection of the electric field component into a plane which contains all possible directions of the electric field vector within the constraint that the electric field must be perpendicular to the direction of propagation. The electric field projection plane may be resolved into two orthogonal directions. In general, the electric field may occupy any position on that plane. This condition represents random polarization. However, random polarization is rarely encountered in radar or communication systems insofar as propagation and power transfer is substantially affected by polarization. This is

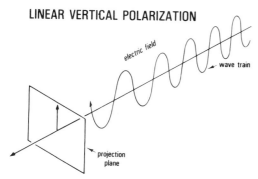

Fig. 1.   Linear polarization of an electromagnetic wave.

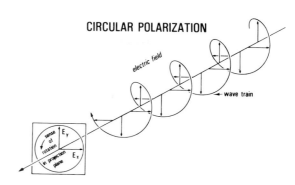

Fig. 2.   Circular polarization of an electromagnetic wave.

especially true when asymmetrical conductors and/or dielectrics are interposed in the propagation path [8].

The conventional definitions of polarization are referenced to the locus of the electric field vector as projected onto the plane viewed from the direction of propagation as illustrated by Fig. 1. A co-ordinate system is then applied to the two components of the projection plane. The IEEE standard provides a basis for this co-ordinate system which is an orthogonal pair parallel and perpendicular to the horizon. The simplest locus of polarization is linear wherein the oscillations of the electric field are confined to a single line when viewed from the direction of propagation. When the electric field component of the electromagnetic wave is colinear with either basis, the plane of oscillation is either perpendicular to the horizon (linear, vertical) or parallel to the horizon (linear, horizontal). On the projection plane, these are vertical and horizontal lines, respectively. Slant linear polarizations represent rotation of the plane of polarization from these bases. Another common condition is circular polarization wherein the electric field rotates in space such that on the projection plane a circle is traced by the tip of the electric vector for each cycle. As a matter of convention, the time harmonics are suppressed in the projection. Circular polarization of the electromagnetic wave train is then often described as a cork-screw where the direction of propagation is along the axis of the cylinder and the sense of polarization may be either clockwise when viewed from the direction of propagation (left hand circular) or counter-clockwise (right hand circular). Note that the sense of rotation is switched on the projection plane from that in the wavetrain, as shown in the Fig. 2. Circular polarization is especially attractive in that the waveguide aperture is symmetric, i.e., either round or square. This may provide some advantage as will be apparent later in this paper.

Since all positions on the projection plane of the electric field may be described by an orthogonal bivariate quantity, the most general polarization is elliptical whereby the tip of the electric field vector traces an ellipse on the projection plane. In this way, linear and circular polarization represent specializations of elliptical polarization: namely, when the two orthogonal components are equal, circular polarization results; and when one or the other orthogonal component is zero, linear polarization results. Slant linear polarization may

result from mixtures of left and right-hand circular polarization with differences in initial phase.

Power transfer between two dipole antennas nicely illustrates the importance of polarization. In this case, linearly polarized fields produced by the transmitting antenna couple energy into the receiving antenna according to a cosine law. That is, when the plane of the transmitted electric field is parallel to the plane of the receiving dipole (i.e., the transmitter dipole and receiver dipole are co-planar and parallel), the polarizations are matched and power transfer is maximized. In other words, the angle between them is zero and the cosine is unity. As the angle approaches 90°, the power transfer approaches zero.

With respect to the general elliptic polarization, several parameters of description are pertinent. The orthogonal field components are

$$E_x = |E_x| \exp(i\theta_x)$$

and

$$E_y = |E_y| \exp(i\theta_y)$$

where the magnitudes are for any orthogonal pair with phases of θ_x and θ_y. The time harmonics are suppressed. The amplitude of an elliptically polarized wave is

$$E^2 = |E_x|^2 + |E_y|^2$$

The axial ratio, r, is the ratio of the minor to the major axes of the polarization ellipse as viewed on the projection plane. The angle the major axis of the ellipse makes to the reference coordinate frame is known as the orientation angle, β. The ellipticity angle, α is related to the axial ratio as follows:

$$\alpha = \pm \arctan r$$

where the sign represents the sense of rotation. These definitions are illustrated in Fig. 3. The complex polarization ratio is defined as,

$$P = |E_y|/|E_x| \angle \theta_y - \theta_x$$

That is, the magnitude of the complex polarization ratio is the ratio of the magnitudes of the orthogonal component magnitudes and its phase is the difference of the phases of the orthogonal components.

In more general terms, the polarization state may be represented as a point on the surface of a sphere as shown

POLARIZATION ELLIPSE

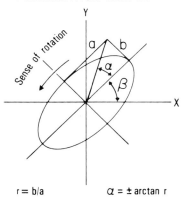

$$r = b/a \qquad \alpha = \pm \arctan r$$

Fig. 3. Parameters of elliptical polarization.

in Fig. 4. The point is located by a longitude of 2β and a latitude of 2α. One half of the sphere represents right sensed polarization (positive α) and the other half represents left sensed polarization (negative α). The poles represent circular polarization and the equator represents linear polarization states. This representation of polarization on the surface of a unit sphere was first described for optical radiation by Poincare. The sphere is known as the Poincare sphere [9].

Polarization is more conveniently shown with a projection of the pertinent hemisphere onto a plane which is shown in Fig. 5. Note that horizontal linear polarization is at the right hand perimeter whereas vertical polarization is at the left hand perimeter. Further, the radial distance to a polarization state on the chart is 2α. This polarization representation is very useful for the concept of power transfer. If the polarization of each antenna is defined as the polarization of the wave it would produce in the far field of its radiation pattern when energized, and each polarization is represented as a point on the Poincare sphere, then the efficiency of power transfer is proportional to the cosine of the polar angle, Δ, between the two points as shown in Fig. 6. This polar angle,

POLARIZATION SPHERE

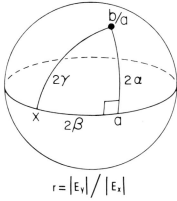

$$r = |E_Y| \Big/ |E_x|$$

Fig. 4. Polarization state on the Poincare sphere.

POLARIZATION CHART

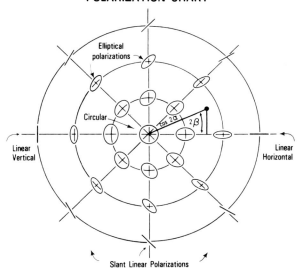

Fig. 5. Polarization hemisphere projected onto a plane.

Δ, is twice the difference between the orientation angles of the major axis of the two elliptical polarizations.

$$\Delta = 2|\beta_1 - \beta_2|$$

The functional form of the power transfer efficiency, Γ, depends upon the polarization states involved. In the case of linear polarization.

$$\Gamma = 1 + \frac{\cos \Delta}{2}$$

In terms of axial ratios, the power transfer efficiency is

$$\Gamma = \frac{(1 \pm r_1 r_2)^2 + (r_1 \pm r_2)^2 + (1 - r_1^2)(1 - r_2^2) \cos \Delta}{(1 + r_1^2)(1 + r_2^2)}$$

where the sign indicates sense of rotation and Δ is the polar angle previously defined. Power transfer is maximized when the axial ratios and direction of rotation are matched. Transmission and reflection at dielectric interfaces as a function of polarization presented by Lin elsewhere in this volume.

POWER TRANSFER

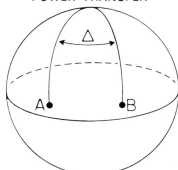

Fig. 6. Power transfer between two polarizations.

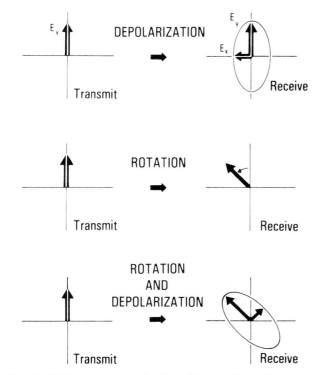

Fig. 7. Various examples of polarization transformation.

Polarization Transformation

The polarization of the electromagnetic wave launched by a transmitting antenna may be modified by the spatial distribution of propagation constants present in the media of propagation. To the extent that the polarization of the wave is different at the receiving antenna than that which was transmitted, the target has effected a polarization transformation upon the incident radiation. Such a polarization transformation may provide a useful description of the target.

Considerable theory exists for the case of backscattered microwave energy for a fixed aspect between the transceiver and target. The work of Kennaugh, Sinclair, Copeland, and Huynen are pertinent examples [10], [11], [12], [13]. Less work exists in the forward scatter case, but it is certainly well known that polarization transformation is possible. In the case of linearly polarized radiation, the transformation is depolarization since linear polarization is converted to elliptical upon reception. Some examples of polarization transformation are shown in Fig. 7. By analogy with the monostatic case, some physically reasonable assertions may be made for forward scatter, but in the case of imagery it is important to distinguish the additional factors of near field operation (whereby the fields have axial as well as transverse components) and the variable aspect of illumination. Also the effects of charges and magnetic fields (Faraday rotation) are ignored.

By analogy with the monostatic case, the polarization

properties of the target may be described for various bases. One convenient basis is linear polarization where the orthogonal pair is horizontal and vertical polarization. One useful representation is to define a target's polarization scattering matrix, T_π, for the bistatic angle of π (i.e. forward scatter),

$$\begin{bmatrix} E_{s_x} \\ E_{s_y} \end{bmatrix} = \begin{bmatrix} E_{i_x} \\ E_{i_y} \end{bmatrix} [T_\pi]$$

where T_π is a two by two symetric matrix of complex numbers which relate the incident electric field components, E_{i_x} and E_{i_y}, to the scattered electric field components, E_{s_x} and E_{s_y}.

The scattering matrix for the far field is:

$$[T_\pi] = \begin{bmatrix} |t_{11}| \exp(i\phi_{11}) & |t_{12}| \exp(i\phi_{12}) \\ |t_{21}| \exp(i\phi_{21}) & |t_{22}| \exp(i\phi_{22}) \end{bmatrix}$$

The elements of the matrix $[T_\pi]$ are magnitudes of the forward scattered fields, $|t_{m,n}|$, and their associated phase angles, $\phi_{m,n}$, for each combination of the two basis vectors, m and n, indicated as subscripts.

In the near field, the axial component requires an additional element to make a 3 by 3 matrix

$$\begin{bmatrix} E_{s_x} \\ E_{s_y} \\ E_{s_z} \end{bmatrix} = \begin{bmatrix} E_{i_x} \\ E_{i_y} \\ E_{i_z} \end{bmatrix} {}' [T_\pi]$$

where

$$[T_\pi] = \begin{bmatrix} |t_{11}| \exp(i\phi_{11}) & |t_{12}| \exp(i\phi_{12}) & |t_{13}| \exp(i\phi_{13}) \\ |t_{21}| \exp(i\phi_{21}) & |t_{22}| \exp(i\phi_{22}) & |t_{23}| \exp(i\phi_{23}) \\ |t_{13}| \exp(i\phi_{31}) & |t_{23}| \exp(i\phi_{23}) & |t_{33}| \exp(i\phi_{33}) \end{bmatrix}$$

In either case, the time harmonic notation is surpressed. The quantities actually measured are the magnitudes and phase angles at a single aspect for the Kurokawa scattering parameter S_{21} (i.e., coherent transmission or insertion loss) under all possible combinations of the basis polarizations [14]. In the far field case, the two polarizations may be vertical and horizontal. In the near field case, the three polarizations may be vertical, horizontal and axial. That is, for example, in the far field case with vertical and horizontal bases, the polarization transformation matrix may be written in shorthand notation as

$$[T_\pi] = \begin{bmatrix} HH & HV \\ VH & VV \end{bmatrix}$$

where the first letter (or subscript in the previous notation) applies to the polarization of the transmitting antenna and the second applies to the receiving antenna. It is understood that the matrix elements are complex quantities as shown in complete form above. The physical interpretation to be applied is, for the case of element HH, that the complex scattering parameter S_{21} is measured for one aspect of illumination and bistatic reception with both the transmitting and receiving antennas linearly polarized in the horizontal direction. In the case of element VV, both are vertically polarized. The off-diagonal elements are cross polarization states. For example, element HV is horizontal transmit and vertical receive. Frequency is an implicit variable.

In the case of an image, each element of the polarization transformation matrix becomes a two dimensional spatial series of complex numbers. Thus

$$[T_{\pi_i}] = \begin{bmatrix} HH_{mn} & HV_{mn} \\ VH_{mn} & VV_{mn} \end{bmatrix}$$

where the indices m and n are bistatic measurements at various positions azimuth and elevation.

In imagery systems, the magnitude and phase angles are represented by two real valued spatial series. In the case of the first copolarized images, t_{11} becomes indexed over azimuth and elevation to create the images; one is for phase, one is for magnitude. The magnitude image is known as the vertical-vertical or co-polarized (VV) image. The t_{12} component is similarly indexed over azimuth and elevation to create the t_{12} magnitude image. This image is known as the vertical-horizontal or cross polarized (VH) image. Similar procedures create the other possibilities in the 2×2 matrix T_π. Thus far, axial field components have not been measured and none of these elements (third row and/or third column) in the near field polarization transformation matrix have been used for imagery. Similarly, cross polarized phase images have not yet been utilized.

The various magnitudes in T_{π_i} may be combined to create a composite image. One of the possibilities in the so-called polarization invariant quantity [15], which for a pixel is

$$Q = |t_{11}|^2 + |t_{12}|^2 + |t_{21}|^2 + |t_{22}|^2$$

Another possibility is linear combinations of cannonical representations of the matrix T_π. Clearly, many fruitful avenues for research exist in this area; but in the present case only two magnitude images will be presented and these are from different kidneys. The reason for this slow start is the fact that data collection takes so long (ca. 4.5 hr). As a result, multiple images of a single target are not feasible with the existing scanner. The electronically scanned (cf Foti et al elsewhere in this volume) should improve the data acquisition time to the order of minutes rather than hours. The shorter data acquisition time will permit many polarization transformation studies (among them, pathophysiological correlations) which are not presently possible.

SYSTEM DESIGN CONSIDERATIONS

The advantages of the microwave region of the spectrum as a means for interrogation of biosystems have previously been frustrated by the problem of how to reconcile two contradictory requirements: adequate spatial resolution and managable propagation loss. These requirements are contradictory since shorter wavelengths improve spatial resolution, but simultaneously increase propagation loss. This dilemma may be mitigated by the use of high dielectric constant materials to fill waveguide based antennas as well as to provide a coupling medium between the antennas and the target. This technique has been presented elsewhere [16],

hence, only a summary will be provided here. The coupling/loading medium of choice is water. It provides high dielectric constant sufficient to contract the wavelength by a factor of approximately 9 (relative to air); it is sufficiently lossy as a coupling medium to attenuate multipath (chiefly at the first inerface of the target); it provides an easily implemented anechoic environment to prevent interference; and it provides a reasonably good, yet broad band, match to water dominated biological dielectrics.

The use of water as a loading/coupling medium, therefore, accomplishes wavelength contraction without the propagation loss penalty associated with increased frequency. Operation in the S band leaves the propagation losses managable at those of 2–4 GHz in water, but accomplishes resolution comparable to 18–36 GHz in air.

This technique has been used in the near field environment (i.e. when the target is in the near field of the antenna, or about one penetration depth away) where resolution is determined largely by aperture dimensions of the antenna. It is presently being generalized to far field operation by means of a water coupled, phased array antenna as described elsewhere in this volume by Foti et al and Guo et al. The array will offer several significant advantages: electronic beam steering will reduce data collection time; and the low f number of the array should provide 3-Dimensional resolution in the coupling medium. Far field operation does impose new requirements on the coupling medium due to the differentially increased loss of high spatial frequencies (cf Slaney et al elsewhere in this volume). A medium with more nearly real valued complex pemittivity would be advantageous.

Scattering Geometry

The system configuration as a forward scatter imager deserves some discussion. Forward scatter is, of course, uncommon in radar applications. This is largely because a simple forward scatter radar system can only provide information concerning detection of targets with no information concerning location or Doppler. In medical applications, location can be resolved in elevation and azimuth by scanning, but recovery of the range coordinate requires tomographic reconstruction. Range is easily obtained with a back scatter system, but the scattering cross section is often orders of magnitude smaller than in the case of forward scatter [17]. This is less true for small scatters (i.e., small relative to the wavelength in the medium) e.g., in the Rayleigh region. Resonances are a special case, but these are generally dampened in dissipative dielectrics. The behavior of forward scatter cross section as a function of the size of the scattering object relative to a wavelength in the medium (or, alternatively a fixed target object illuminated with increasing frequency) is a smoother function than in the case of backscatter. In the case of simple scatters such as spheres, the forward scatter cross section is a smooth, monotonic function of frequency/size; whereas for back scatter, the cross section is oscillatory and not monotonically related to frequency (size) as shown in Fig. 8. Polarization does not affect the scattering cross-section in the case of spherical objects. In

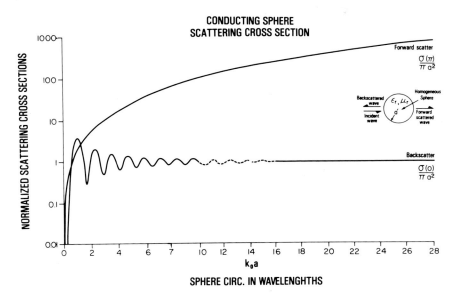

Fig. 8.   Scattering cross-section for conducting sphere (from Ruch, et al.).

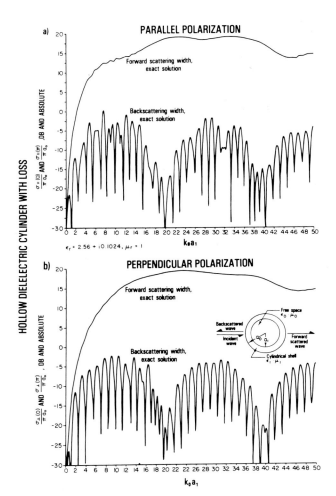

Fig. 9.   Scattering cross-section for heterogeneous dielectric cylinders with lossy walls (from Ruch *et al.*).

the case of double layered dielectric cylinders (see Fig. 9) it is apparent that the forward scatter cross-section exceeds the back-scatter cross-section. Further, the object size with respect to wavelength bears a nearly monotonic relationship to the cross-section in the case of forward scatter whereas this is conspicuously not true with back-scatter. Note also the effect of polarization.

Furthermore, spectroscopic methods for characterization of the medium in the propagation path are often more easily applied to forward scatter than to back scatter. These arguments were illustrated by improved sensitivity to small, concentric dielectrics with forward scatter rather than back scatter configurations as shown in Fig. 10.

The chief disadvantages to forward scatter are both consequences of the longer path length. These are the need for greater sensitivity of the receiver and more serious degradation of the image by diffraction effects.

On balance, these considerations appear to favor forward scatter systems, at least in the near field, although this may delay *in situ* operation. This is especially troublesome with tomographic systems. Reconstruction is certainly possible by conventional linear methods but the results are not of acceptable accuracy [18]. One special case does argue for back scatter systems. That is when the target of interest is superficial (e.g., a blood vessel) and much higher frequencies (e.g., 10 GHz) are needed for resolution.

IMPLEMENTATION

The present system configuration is a forward scatter (bistatic at 180°) design with one transmitting antenna and one receiving antenna. The antennas are water loaded and water coupled to the target which is interposed between the two antennas as shown in Fig. 11. The two antennas are rigidly

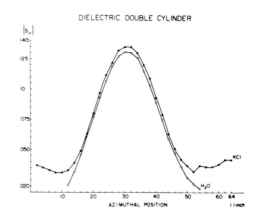

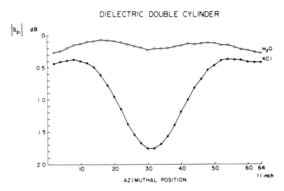

Fig. 10a,b. Glass tubing 6 mm OD and 3 mm ID in a water tank is filled with water and KCl solution, $|S_{11}|$ and $|S_{21}|$, respectively (Cf. reference 19).

interconnected and scanned as a pair in a raster pattern over the isolated canine kidney. The basic datum is the complex Kurokawa scattering parameter S_{21} for each of two combinations of a linear polarization [19], [20]. One case is copolarized (VV), the other is cross polarized (VH). The transmitting antenna is always vertically polarized whereas the receiving antenna is polarized alternately in the vertical and horizontal directions [21]. The data are converted into images by intensity modulation of a raster scanned display. Various image processing steps are applied in the amplitude and spatial frequency domain. Image interpretation is made on the basis of known regional specialization in the kidney.

Data Acquisition

The data presented later in this paper are two dimensional arrays of microwave transmission coefficients (magnitude and phase of S_{21}) which describe the relative insertion loss and relative phase shift of a 3.9 GHz signal as it propagates through the target of interest. This insertion loss and phase shift is affected by the path length, complex permittivity and geometry of the biological material, and the degree to which polarization transformation occurs as a signal is propagated

through the target. The image was collected as a square 64 × 64 array at 1.4mm sampling increments. The basic data are scattering parameters S_{21}, i.e., insertion loss and phase shift at each location on the raster. The block diagram of the data collection system is shown in Fig. 12. This diagram has been divided into three functional subsystems: control and recording, microwave stimulus/response, and the electromechanical scanner.

The electromechanical scanner consists of a water tank (.914 meters cubical tank), vertical and horizontal translation axes, independent drive motors for the axes, independant position transducers for the axes, and mounting frames upon which to connect the antennas and specimen. The linear position transducers were of the electro-optical type (Heidenhain Pos-Econ 1) with a total range of approximately 200 mm and an accuracy of ±0.002 mm (this accuracy is slightly degraded by the stepping motors). The position transducer readouts have +8421 BCD outputs which are interfaced to the Hewlett Packard (HP) 2100 mini computer using two HP 12604B Data Source Interface Cards. This interface required construction of an intermediate buffer that provided special signals such as flag and control logic required by the HP12604B.

The X and Y axes were driven by a Superior Electric M112-FJ25 synchronous stepper motor through a gear reduction box and worm gear. The motors were interfaced to the HP2100A computer through a SloSyn Translator Model BTR103RT and an HP sixteen bit duplex register in the computer. The computer controlled the X and Y coordinates of the positioner by sending pulses to the stepper motors under software control. This arrangement, in which the computer read the X and Y positions and controlled the motors, allowed closed loop control of the scanner position under software control. In this manner, the accuracy between image elements (pixels) was set to ±0.01 mm. Although higher accuracy was possible (±0.003 mm) it was not used because it considerably dilated the data acquisition time.

The cubical tank contained the water into which the antennas and specimen were immersed. The water was continuously filtered through ion-exchange columns and iodinated as a bacteriocidal measure.

The 3.9 GHz microwave interrogation signal was generated by a phase-locked HP8542A Automatic Network Analyzer. This signal was amplified to the level of approximately 1 watt in an HP491C Traveling Wave Tube amplifier, passed through an isolator to protect the TWT and sent to the transmitting antenna (American Electronics Laboratories Model H1561). The signal was received on an identical antenna; amplified by two Avantek low noise amplifiers in a series and passed to the receiver of the HP8542A. The uncorrected transmission coefficient (S_{21}) is generated in the HP8542A receiver, digitized, and stored in the HP2100A computer. This data is corrected for errors such as source mismatch, tracking errors, etc. by techniques described in the literature [22].

The corrected data was recorded on a disc file and later transferred to magnetic tape in a format compatible with the image processing system.

Fig. 11.   Canine kidney in microwave scanner with water coupled antennas.

Image Processing

Two types of digital image processing were performed: amplitude domain and spatial frequency domain. The processing sequence began in the amplitude domain. The first step was interpolation from the 64 × 64 data collection raster to a 256 × 256 raster. The data begins as 15 bits plus sign, but it is truncated to 7 bits plus sign. The interpolation function used was the cubic spline [23]. Interpolation properly consists of an ideal low pass filter applied to the data acquired on the scanner grid to regenerate the continuous, band limited (i.e., in spatial frequencies), two dimensional image of signal amplitudes. The reconstituted image is then in principle resampled at new grid positions. The interpolation kernal is convolved with the available data such that existing data values are not modified. The ideal kernal is the sin (x)/x function. This kernal is approximated by a cubic spline [24].

The cubic spline function has three ranges of values as shown below:

$$f(x) = \begin{cases} f_1(x) = a_1|x|^3 + b_1|x|^2 + c_1|x| + d_1; \ 0 \leq |x| \leq 1 \\ f_2(x) = a_2|x|^3 + b_2|x|^2 + c_2|x| + d_2; \ 1 \leq |x| \leq 2 \\ f_3(x) = 0; \ |x| \geq 2 \end{cases}$$

The function is evaluated for a range of ±4 pixels in azimuth and elevation to result in a 16 point kernal. The function approximates a sin(x)/x function in two dimensions and it is normalized to unity in magnitude. When the main lobe is coincident with existing data, the nulls are at 16 existing grid points. Thus, the function simply multiplies the value at the main lobe by one and adds zeros. At new locations, the interpolation error has a peak value of ca. 5% and an average error of ca. 2% [25].

The cubic spline is preferable to the two more common interpolation operators which are simple pixel replication and bilinear (two dimensional linear) interpolation. The former suffers from image blocking (checkering) and the latter substantially departs from the ideal low pass filtering. That is, spatial frequencies near but below the folding frequency are attenuated. Cubic spline interpolation also provides continuity in the first and second derivatives of the pixel values whereas bilinear interpolation provides continuity only in the displayed values. Pixel replication operates without regard to continuity of the displayed values. Of course, interpolation does not increase the number of "independant" pixels, neither does it compensate for any possible aliasing from the ca 1.4 mm sampling increment in azimuth and elevation. The objective is entirely to enhance the perceived image "quality" by simulating a continuous

DATA ACQUISITION SYSTEM

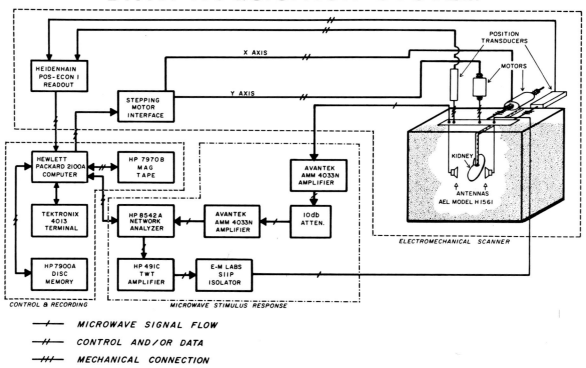

Fig. 12. Scattering parameter imaging system block diagram.

rather than a discrete representation of the data.

The next processing stage was also in the amplitude domain, but the objective was to adapt the dynamic range of the data to the dynamic range of the video display. The operation is known as grey scale mapping whereby the original scale of intensity values are mapped onto a different scale. The archetype of the operation is the ramp-intersect map which stretches some subset of the original range onto the full range of the new scale. This contrast stretcher is useful for part-range expansion, but it has the disadvantage that the upper intercept point maps all values at or above the upper intercept into one value at the new scale maximum and all values at or below the lower intercept are mapped onto the scale minimum. This, of course, is often an undesirable data compression. The map used in this study is different in that all values from the original scale are mapped into unique values on the new scale. The map actually used is a two-piece linear function with an interactively alterable hinge point. Since a null operator would be a 45° line through the origin, the expansion is proportional to the slope of the mapping function. Thus, the map expands the scale for values below the hinge and compresses those above the hinge when the hinge point is above the 45° line. The situation is reversed for hinge points below the 45° line. These grey scale operators are illustrated in the Fig. 13.

The last amplitude domain step is pseudocolor processing. This processing is often referred to as density slicing. This

nomenclature is an atavism from digital processing of photographic images whereby the optical density was assigned a color in a color map corresponding to the range of measured light transmission via a scanning desitometer. In the present context, the color map is applied to the scale of magnitudes of scattering parameter S_{21} as shown in Fig. 14. The color maps used in the present work are those from the NASA LANDSAT series.

Spatial frequency domain processing is accomplished by two-dimensional digital filters [26]. These are specified with radically symmetric transfer function magnitudes. The transfer functions are comprised of only even terms (cosines); thus, there is no detectable phase shift. The images are first Fourier transformed in one-dimension, transposed and Fourier transformed again in the same direction. Since the two-dimensional Fourier transform has a separable kernal, two one-dimensional transforms may be used to implement the two-dimensional transform. The Fourier processed image is then multiplied by the specified transfer function and the product is inverse transformed back to the spatial domain.

The form of the transfer function is a radically symmetric band-pass-filter. Such a transfer function is shown in perspective in Fig. 15. This class of filter is designed to reject spatial frequencies (i.e., spatial frequencies are the reciprocal of length in the same sense that ordinary frequency is the reciprocal of time) that are either above an upper "cut off"

GREY SCALE MAPPING

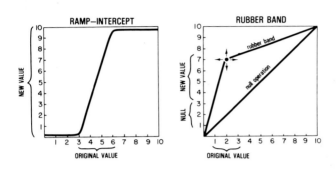

Fig. 13. Grey scale mapping operators.

or below a lower "cut off." The cut-off frequencies are half power points. The operation of the filter may be viewed as a cascade of a high pass filter set at a low cut-off frequency and a low pass filter set at a high cut-off frequency. The dampening factor is $\sqrt{2}$, and the asymptotic rate of attenuation is 12 dB/octave. These 2nd order filters were selected on the basis that higher order filters tend to introduce ringing in the image in the same way that the step function response of ordinary filters exhibit greater time domain overshoot as the order is raised. The dampening factor of $\sqrt{2}$ was selected to minimizes pass-band ripple.

The lower spatial frequency cut-off was selected with the aid of the two dimensional power spectrum of the image as shown in Fig. 16. Low spatial frequencies were rejected to remove effects of global changes in organ thickness and to

PSEUDOCOLOR DENSITY SLICING

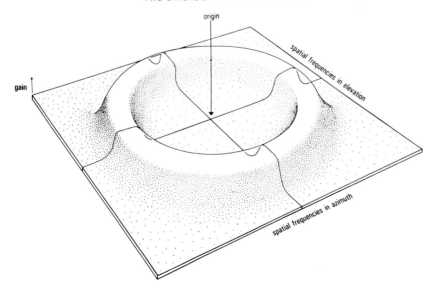

Fig. 14. Pseudocolor processing by density slicing.

TWO DIMENSIONAL BAND PASS FILTER

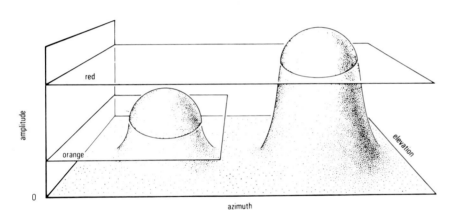

Fig. 15. Perspective view of the 2D band pass filter.

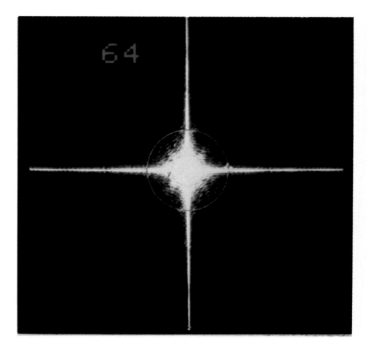

Fig. 16. Band pass filter overlaid on a power spectrum.

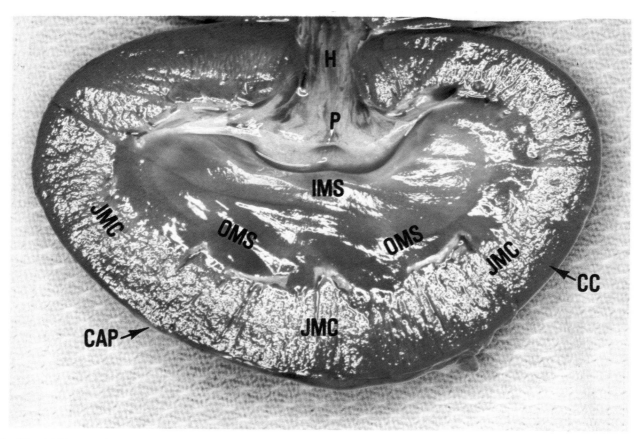

Fig. 17. A kidney specimen with dissection to demonstrate regional anatomy. Code for later use: CC is cortex corticis, CAP is capsule, JMC is juxtamedullary cortex, OMS is outer medullary stripe, IMS is inner medullary stripe, P is pelvis, H is hilus, F is sinus fat, R is support rod, A is artifact.

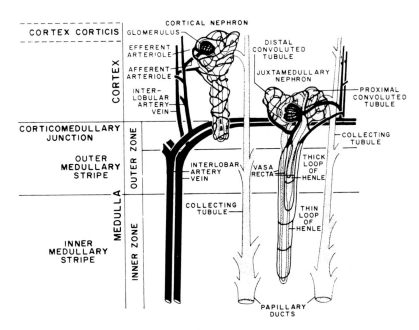

Fig. 18. A schematic view of the nephron.

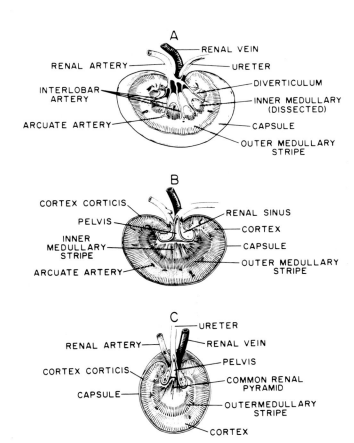

Fig. 19. Graphic illustration of canine renal anatomy.

ameliorate refractive effects in the image. High spatial frequencies were rejected to remove noise introduced by the cubic spline interpolation.

The convolution operator was used to enhance gradients within the image. This operation was implemented with a 3×3 kernal configured as a local high-pass-filter. It acts in a manner analogous to an isotropic three point moving average filter (the archetype being a differentiator) in two dimensions. The lack of directional sensitivity in such an isotropic filter requires symmetry in the kernal. That is, the generic form is

$$h(x,y) = \begin{bmatrix} A & B & A \\ B & 1 & B \\ A & B & A \end{bmatrix}$$

When both $|A|$ and $|B|$ are ≤ 1 and the coefficients are constrained to negative values for high pass filtering. Typically $|A| > |B|$ in order to avoid undue emphasis of the asymmetries due to the rectangular wave guide aperature. The operator is implemented by convolution of the kernal with the image as follows:

$$g(x,y) = f(x,y) * h(x,y)$$

where $g(x,y)$ is the gradient enhanced image, $f(x,y)$ is the original image and $h(x,y)$ is the kernal. In all cases, only mild high pass filtering was employed in order to provide contrast for photographic recording of the image.

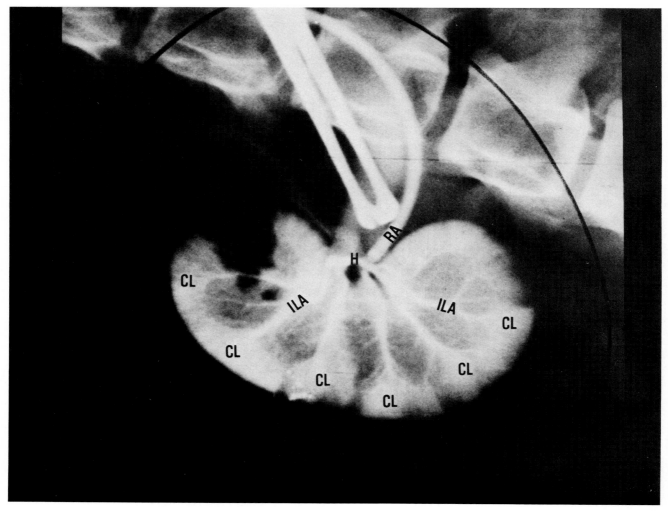

Fig. 20. X-ray angiogram of canine kidney. Note lobulations indicated by CL. Other structures are indicated as follows: ILA, interlobar artery; RA, renal artery; H, hilus.

CANINE RENAL ANATOMY AND PHYSIOLOGY

A brief discussion of canine renal anatomy and physiology is a necessary background for plausible interpretation of the microwave images that will be presented later in this chapter. These topics will be treated largely in terms of generalization since many references exist with detailed discussion and extensive bibliographies [27], [28], [29].

In very broad terms, the kidney consists of five regions (see Fig. 17). Of these, the central divisions are the cortex, medulla and pelvis. The other two regions are the outer margin, known as the capsule, and the medial structures of the hilus where blood vessels and the ureter have ingress and egress from the organ.

In similarly broad terms, the renal unit of function is the nephron (see Fig. 18). Each kidney contains ca 10^5 nephrons. The individual nephron functions in a series of sequential

processing steps. The first stage is ultra filtration of the blood at the glomerulus. Only blood cells and large protein molecules are retained; sugars, amino acids and electrolytes pass into the proximal nephron for further processing. The next processing step takes place in the proximal segment of the nephron where sugars, small protein molecules, amino acids and electrolytes are extracted from the ultrafiltrate and returned to the blood. The proximal segment also secretes p-aminohippurate into the partially processed ultrafiltrate. The next processing step is passive electrolyte and water movement out of the descending limb of the Loop of Henle. A counter current multiplier increases tonicity in the medulary interstitial spaces. Tonicity increases from a value of ca. 300 mosmols/L at the corticomedullary junction to a value of about 1200 mosmols/L at the tip of the medullary inner stripe. This is accomplished by sodium transport and the lack of water premeability in the ascending limb of the Loop of Henle. The vasa recta in combination with low me-

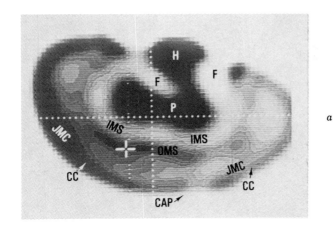

a

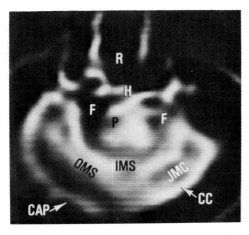

b

Fig. 21a,b.   Copolarized (21a) and cross-polarized (21b) images.

dullary blood flow apparently reduces dissipation of the osmotic gradients produced by the Loop of Henle. The collecting ducts later serve as an osmotic exchanger in the medulla. The distal convoluted tubule further reabsorbs and secretes electrolytes. Finally, the product is transferred to the collecting tubules where amonnia secretion and water transport take place chiefly by virtue of interstitial electrolyte concentrations produced via the Loop of Henle. The urine thereby becomes concentrated on route to the renal pelvis.

Canine renal anatomy differs from the more commonly known human anatomy in that the canine organ is nonpapillary. In other words, the medullary pyramids that are prominent in pylographic x-ray studies of humans are replaced by a single central pyramid. This is evident on both the transverse and longitudinal sections shown in Fig. 19. At the functional unit level (i.e. the nephron), human and canine kidneys are quite similar.

The outer margin of canine kidney is a thin fiberous capsule. Medial to the capsule is the cortex corticis. This is a sub-region of the cortex that is characterized by a relative paucity of glomeruli and high population density of proximal as well as distal convoluted tubules.

Medial to the corticis is the major portion of the cortex characterized by glomeruli, convoluted tubules and early urine collecting tubules. This region blends into the juxtamedullary cortex at the corticomedullary junction. The cortex is lobulated by the interlobar arteries in Fig. 19. These vascular patterns are further illustrated by angiography whereby x-ray contrast agent is injected into the renal artery. A canine renal angiogram is shown in Fig. 20.

The corticomedullary junction is formed by vascular patterns whereby the radially disposed lobar arteries produce a circumferential branch known as the arcuate artery. Small branches of the arcuate artery provide the lobular arteries, again in a radial pattern, from which the glomeruli are derived.

Medial to the corticomedullary junction is the first sub-region of the medulla, known as the outer medullary stripe. This is a region characterized by the thick portion of the loop of Henle and moderate size collecting tubules. The next sub-region is the inner medullary stripe. This is a region where the thin loops of Henle and the larger collecting ducts exist.

Medial to medulla is the pelvis where the collecting ducts empty the fully processed urine. The pelvis is a muscular cistern which is contiguous with the ureter in the hilar region of the organ. The other hilar structures are the renal artery and vein. Often a sinus or sinuses exists in the perihilar region which is occupied by fat.

CANINE RENAL IMAGES

The kidneys used for this study were prepared according to the methods in reference [20]. Important factors recounted here include the use of fresh specimens, maintenance of physiological blood volumes, filling of the ureter/pelvis with saline, and use of pentobarb anesthesia. The presentation of imagery is grouped according to the combination of polarization basis vectors used to generate the image and the image processing steps that have taken place. The first polarization basis is the co-polarized or VV pair. That is, the transmitting antenna is linearly polarized in the vertical direction and the receiving antenna is linearly polarized in the vertical directions. The image, therefore, represents the forward scattered radiation (i.e., insertion loss) at 4096 positions in a raster scan of 1.4 mm sampling increments under the condition that the received polarization is the same as transmitted polarization.

The first co-polarized (VV) image shown in monochrome (see Fig. 21a) is that obtained after only amplitude domain processing steps. These processing steps began with interpolation by a cubic spline from 64 × 64 × 15 bits deep to 256 × 256 × 7 bits deep. The interpolation step was followed by grey scale mapping with the rubber band operator. This was followed by the 3 × 3 convolution operator to enhance gradients. In the image of Fig. 21a, the pelvis and medullary inner strip are certainly visible as is the corticomedullary junction with juxtramedullary cortex and cortex corticis distal to the junction. Some suggestion of cortical lobulations is also visible.

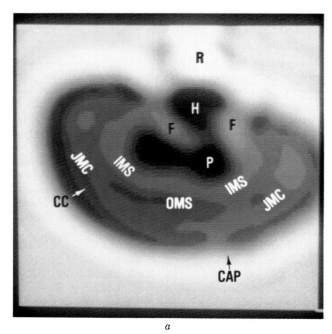

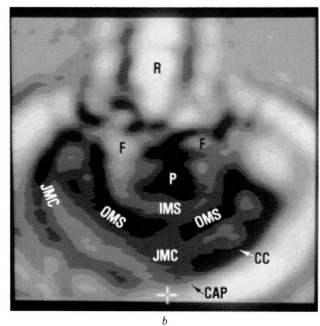

a b

Fig. 22a,b.   Pseudocolored image with only amplitude domain processing. Copolarized image at 22a, cross-polarized image at 22b.

The same sequence of image processing steps has been applied to the cross-polarized (VH) image with the addition that the grey scale is inverted. The image is formed by measurements of the forward scattered radiation under the condition that the transmitted radiation is depolarized by the organ and the component orthogonal to that radiated becomes the datum. It is evident in Fig. 21b that very different features of the organ are represented in this image when it is compared to the copolarized image. For example, the lamination between the medullary inner stripe and pelvis

is enhanced as is the lamination between the cortex and the medullary outer stripe. Also, the support rod is clearly represented in the VH image whereas only the end of rod was evident in the HH image. Pseudocolored versions of the amplitude domain processed images are shown in Fig. 22a, b. Of course, the same features are present; but the density slicing process enhances the distinctions between adjacent "grey" levels. The convolution operator was not used in the pseudocolored images. The differences between copolarized and cross polarized images are evident.

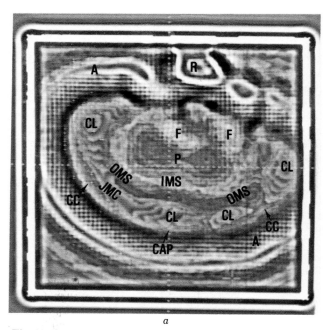

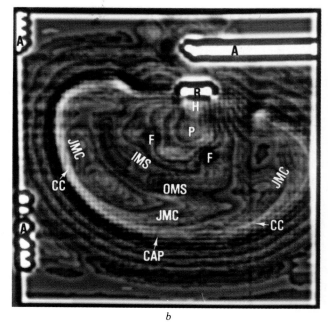

a b

Fig. 23a,b.   Fully processed images are shown for the magnitude and phase of scattering parameter S_{21} at 22a and 22b, respectively.

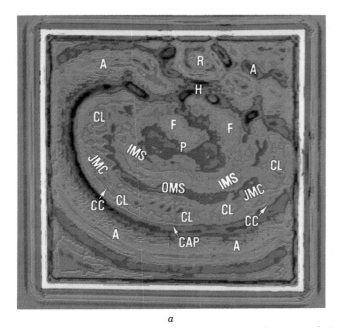
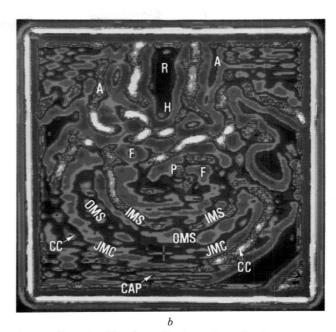

a   *b*

Fig. 24a, b. Copolarized and cross polarized magnitude images are shown after two dimensional band pass filtering and pseudocolor processing in 24a and 24b, respectively.

The next processing step was application of the two dimensional band pass filter followed by the rubber band and convolution operators. The copolarized magnitude image in Fig. 23a is now much improved with respect to enhancing the cortical lobulations as well as the medullary inner and outer stripes. Similarly, the cortex corticis is enhanced. The co-polarized phase of S_{21} image after the same image processing sequence is shown in Fig. 23b. The bright bars represent artifact. Once again, the cortex corticis, juxtamedullary cortex, medullary outer stripe, medullary inner stripe,

pelvis, hilus and fat in the sinus are distinguishable. Note that the magnitude and phase co-polarized images are from the same kidney specimen.

The co-polarized magnitude image in pseudocolor is shown in Fig. 24a, along with the cross-polarized magnitude image in Fig. 24b. With regard to the cross-polarized magnitude image, the region between the two lamina previously described is somewhat enhanced. It is also apparent in the band pass filtered, cross-polarized image that the effects of the waveguide aperture asymmetry are enhanced. This is an

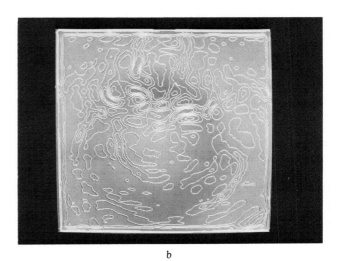

a   *b*

Fig. 25a, b. The fully processed data for the co-polarized and cross-polarized images are shown in contour format in 25a and 25b, respectively. The contour intervals are 10 and 50, respectively.

unfortunate consequence of the double ridged waneguide antenna in the cross-polarized configuration. Attempts to remove this effect by means of a horizontal notch filter were only partially successful. These factors argue for the use of a symmetrical aperture when band width is not a major concern.

The final presentations for both the copolarized (VV) and cross polarized (VH) scattering parameter images are in a contour format (see Figs. 25a and 25b) to permit quantitative comparison. Note that the contour intervals are not identical for the two images due to the need to preserve relatability to the grey scale and pseudocolor presentations.

DISCUSSION

The primary feature in the results is that both the copolarized and cross polarized images can be related to known regional specialization within the target organ. This reflects upon the choice of the kidney organ for microwave imagery demonstration wherein gross anatomy, in fact, corresponds rather closely to aggregate nephron morphology. Hence, aggregate physiology at the microscopic scale is correlated with regional specialization. The apparent correspondence between the microwave images and this regional specialization argues for the biological relevance of imagery based upon dielectric properties in the microwave region of the electromagnetic spectrum. We know of no other single form of radiation that offers comparable physiologic correlation in any soft tissue organ. The fact that four imagery methods yield self consistant images and interpretations is further source for optimism. In addition to a high level of biological relevance, microwave radiation is clearly less hazardous than x-ray imagery and no artificial contrast enhancement need be used. These properties have the potential advantage that invasive procedures may be avoided and early diagnosis of disease may be improved.

The non-invasive methods used herein do offer promise for providing spatial maps of the energy dissipation and energy storage in biologic dielectrics. Indeed, to the extent that insertion loss measurements can be made with resolution in three dimensions and the various scattering mechanisms can be discriminated, non-invasive dosimetry will become possible. The necessary intermediate step is to deduce the spatial distribution of the real and imaginary parts of the complex permittivity from scattering parameter measurements. For example, to the extent that higher insertion loss represents attenuation, the outer medullary stripe may be predicted to be subject to greater energy disposition than the adjacent juxtamedullary cortex and inner medullary stripe. This inference remains unproven at the present time.

The present work is encouraging, but several additional steps are required. The first of these is to decrease the data acquisition time by use of the phased array system. Present estimates suggest that spatial resolution comparable to that obtained in the near-field with a single element will be available to azimuth and elevation with data collection times

in the order of minutes (or less) rather than hours. In addition to electronic beam steering, the array will provide off-line focusing in range, which may be expected to improve image quality. Further details of this system are described by Foti *et al* elsewhere in this volume. At the present time, the array system data acquisition speed is largely determined by the electromechanical switches, which require about 45 msec per element. These switches serve as a multiplexer on the receive array and as beam steering on the transmit array. These switching times may be decreased to about 1% of their present value by the use of PIN diode switches instead of electromechanical switches. This implies that a high data rate may be achieved, perhaps in the order of 10–100 msec per frame. Even this is not fast by microwave standards. System configurations for maximum speed would be rather different from the one described here. It would appear to be technologically feasible to achieve video frame rates if this objective was included as a design goal. The predicted performance of additional array lattice structures and an element design more amenable to tape-controled-machining is presented in a companion paper by Guo et al. elsewhere in this volume. Similarly a method for array self focusing is described.

The isolated organ can be maintained on extracorporal circulation at 37°C. The use of physiologic temperature will decrease attenuation and pathophysiological correlations may be explored. Furthermore, the full use of polarization diversity may be explored without the inevitable specimen degradation that would otherwise take place.

In a speculative vein, it may be useful to consider what means are available to improve the spatial resolution of medical microwave imaging systems. Near field data is clearly the most valuable to the extent that the combined effects of dielectric loss and path length have conspired least to erode the high spatial frequency content of the scattered field. In the near field, resolution beyond the Rayleigh limit is possible by use of the evanescent fields excluded from the conventional optical (far field) analog. Indeed, evanescence is pandemic in water coupled imaging systems. In the near field case, then, resolution would seem to be largely a function of aperture dimensions and receiver noise floor in the context of vector measurement of the power wave scattering parameters. This would suggest that improvements in antenna loading dielectrics in the direction of higher K' and lower receiver noise floor (up to the limit of effective antenna noise temperature) would be worth while.

Far field imagery is influenced by these same factors with the addition that the coupling medium and the frequency of operation assume greater importance as a result of the longer path lengths involved. The far field is attractive since it permits the use of noncontacting antennas and antenna arrays. These are important steps to increase data acquisition rates. However, the retention of spatial resolution requires attention. The need for coupling media with lower dielectric loss assumes paramount importance because of the differential loss of high spatial frequencies in the angular spectrum (cf. Slaney et al. elsewhere in this volume). This also places new emphasis on lower frequency of operation if the receiver noise floor is inadequate to the task. Since microwave receivers have phase resolution far better than pi, it may be possible to exceed the Rayleigh limit by substituting phase resolution for higher carrier frequency [32]. This is not unlike the use of analytic continuation of a fixed aperture to improve transverse resolution and it is subject to the same limitations imposed by noise considerations.

Both near and far field imaging systems could benefit from multifrequency operation. This is useful not only as a means to suppress multipath [30], [31] (cf. Jacobi and Larsen elsewhere in this volume) but also a method to increase Fourier space acess in tomographic systems (cf. Farhat, and Slaney et al. elsewhere in this volume). Lastly, some improvement may be possible by the use of combined S_{11} and S_{21} measurements. This would offer greater Fourier space access than either one alone as described by Slaney et al. elsewhere in this volume.

These areas remain topics of future research, but the source of motivation must come from continued studies of the physiologic and pathophysiologic correlations between dielectric properties and health or disease. A significant fraction of this work can be accomplished by point probes as described by Burdette et al. elsewhere in this volume, but high speed, near field imagery would be of immense value in this application as well. Microwave imagery remains the best candidate technology for noninvasive dosimetric analysis, and it offers a view of biosystem function which is unparalled by other imaging modalities.

References

1. A. R. Von Hippel, *Dielectrics and Waves*, MIT Press, 1954.
2. J. B. Hasted, *Aqueous Dielectrics*, (Chapmand and Hall, 1973).
3. E. H. Grant, R. J. Sheppard, and G. P. South, *Dielectric Behavior of Biological Molecules in Solution*, Clarendon Press, 1978.
4. I. Yamaura, "Measurements of 1.8–2.7 GHz Microwave Attenuation in the Human Torso," *IEEE Trans. Microwave Theory Tech.*, MTT-25, 707–710, 1977.
5. E. C. Burdette, F. L. Cain, and J. Seals, "*In vivo* Probe Measurement Techniques for Determining Dielectric Properties at UHF through Microwave Frequencies," *IEEE Trans. Microwave Theory Tech.*, MTT-28, 414–427, 1980.
6. R. A. Rutherford, B. R. Pullan, and I. Isherwood, "Measurement of Effective Atomic Number and Electron Density using an EMI Scanner," *Neuroradiol.*, 11, 15–22, 1976.
7. E. C. Gregg, "Radiation Risks with Diagnostic x-Rays," *Radiology*, 123, 447–453, 1977.
8. K. H. Steinbach, "On the Polarization Transform Power of Radar Targets," in *Atmospheric Effects on Radar Target Identification and Imaging*, E. Jeski, Ed., 65–82, D. Reidel Co., 1976.
9. G. A. Deschamps, "Part II—Geometrical Representation of the Polarization of a Plane Electromagnetic Wave," *Proc. IEEE*, 39, 540–544, 1951.
10. E. M. Kennaugh, "Polarization Properties of Radar Reflectors," Ohio State University Antenna Laboratory, Project Report 389-12 (AD2494)RADC, AF28(099)-90, 1952.
11. G. Sinclair, "The transmission and Reception of Elliptically Polarized Waves," *Proc. IEEE*, 39, 535–540, 1951.
12. J. R. Copeland, "Radar Target Classification by Polarization Properties," *Proc. IRE*, 48, 1290–1296, 1960.
13. R. J. Huynen, *Phenomenological Theory of Radar Targets*, Drukkerij Bronder, 1970.
14. K. Kurokawa, "Power Waves and the Scattering Matrix," *IEEE Trans. Microwave Theory Tech.*, MTT-13, 194–202, 1965.
15. W. M. Boerner, M. B. El-Arini, C. Y. Chan, and P. M. Mastoris, "Polarization Dependence in Electromagnetic Inverse Problems," *IEEE Trans. Antennas Propag.*, AP-29, 262–270, 1981.
16. J. H. Jacobi, L. E. Larsen, and Hast, C. T. "Water Immersed Microwave Antennas and their Application to Microwave Interrogation of Biological Targets," *IEEE Trans. Microwave Theory Tech.*, MTT-27, 70–78, 1979.
17. G. T. Ruck, D. E. Burrick, W. D. Stuart, and C. K. Kirchbaum, *Radar Cross Section Handbook*, Vols. I and II Plenum, 1970.
18. S. P. Rao, K. Santosh, and E. C. Gregg, "Computer Tomography with Microwaves," *Radiology*, 135, 769–770, 1980.
19. L. E. Larsen and J. H. Jacobi, "Microwave Interrogation of Dielectric Targets, Part I: By Scattering Parameters," *Med. Phys.*, 5, 500–508, 1978.
20. L. E. Larsen and J. H. Jacobi, "Microwave Scattering Parameter Imagery of Isolated Canine Kidney," *Med. Phys.*, 6, 394–403, 1979.
21. L. E. Larsen and J. H. Jacobi, "The Use of Orthogonal Polarization in Microwave Imagery of Isolated Canine Kidney," *IEEE Trans. Nucl. Sci.*, NS-27, 1184–1191, 1980.
22. S. F. Adam, *Microwave Theory and Applications*, Prentice Hall, 1969.
23. R. Bernstein, "Digital Image Processing of Earth Observation Sensor Data." *IBM J. Res. and Develop.*, 20, 40–57, 1976.
24. S. S. Rifman and D. M. McKinum, "Evaluation of Digital Correction Techniques for ERTS Images," TRW Corporation Final Report, TRW 20634-6003-TU-00, NASA, GSFC, 1974.
25. K. W. Simon, "Digital Image Reconstruction and Resampling for Geometric Manipulation," in *Digital Image Processing for Remote Sensing*, R. Bernstein, Ed., 84–94, IEEE Press, 1978.
26. R. C. Gonzales and P. Wintz, *Digital Image Processing*, Addison-Wesley, 1977.
27. B. M. Brenner and F. C. Rector, Jr., *The Kidney*, Vols. I and II, Saunders, 1976.
28. R. F. Pitts, *Physiology of the Kidney and Body Fluids*, 3rd Ed., Yearbook Medical Publishers, 1974.
29. H. E. DeWardener, "The Control of Sodium Excretion," *Am. J. Physiol.*, 235, F163–F173, 1978.
30. J. H. Jacobi, and L. E. Larsen, "Microwave Interrogation of Dielectric Targets. Part II: By Microwave Time Delay Spectroscopy," *Med. Phys.*, 5, 509–513, 1978.
31. J. H. Jacobi, and L. E. Larsen, "Microwave Time Delay Spectroscopic Imagery of Isolated Canine Kidney," *Med. Phys.*, 7, 1–7, 1980.
32. H. O. Collins, T. J. Davis, L. J. Buse, R. P. Gribble, "Eddy current phasography" p 609–624 in *Acoustical Imaging*, Vol 11, [J. P. Powers Ed]. Plenum Press, 1981.

Linear FM Pulse Compression Radar Techniques Applied to Biological Imaging

John H. Jacobi and Lawrence E. Larsen

Forward scatter imaging systems are susceptable to multipath effects. The major method of multipath supression used up to this point has been system operation in a dissipative liquid dielectric. This paper presents another method more suitable for exterior multipath suppression. Time is encoded by frequency in a linear FM chirp signal from the transmitter. The results presented here are a hybrydization of differential propagation delay and amplitude since a $|S_{21}|$ maximum is selected within a defined range of arrival times.

INTRODUCTION

One of the potentially serious problems in non-invasive microwave imaging of the spatial distribution of complex permittivity in biological systems is that of multipath propagation. Microwave energy can travel from a transmitting antenna to a receiving antenna by a variety of paths other than the desired straight line. This is illustrated schematically in Fig. 1. Radiation can travel around the body or be reflected from objects external to the body. It can even travel by many paths through the body because of spatial variations in tissue permittivity which causes diffraction, refraction, and internal reflections. Radar engineers and engineers concerned with testing microwave circuits have developed many methods for discriminating against multipath propagation. In the case of stationary targets where Doppler shifts cannot be used to advantage (which is usually the case in biological systems) multipath discrimination techniques generally involve pulsed signals and "range gating." The term "range gating" means that the imaging system responds only to signals that arrive at the detector within a specified time window. Treatments of these various methods are well described in the literature [1]. The method described here to provide multipath discrimination is based on "chirp" radar or pulse compression radar (PCR). This approach was selected for a number of reasons:

A. It does not require generation of the very short duration, fast rise time pulses needed for good range resolution. Rather, conventional laboratory microwave oscillators can be used for generating the interrogating microwave signal.
B. One of the features of pulse compression as opposed to conventional pulse systems is that high peak power pulses are not required. This approach avoids

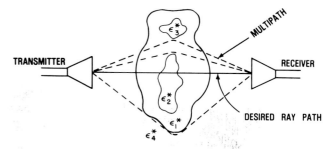

Fig. 1. Graphical illustration of multipath propagation.

the possibility that the system would be hazardous because of thermoacoustic expansion effects [2].
C. The raw data from the chirp system is easy to record using inexpensive, conventional techniques. This is useful for permanent records and for application of various experimental signal processing techniques. It is very difficult to record picosecond pulses.
D. The width and shape of the spectrum of the chirp signal is easy to vary. This is a potentially valuable attribute for biological systems since they are highly dispersive. For example, the attenuation of microwave signals in biological tissues is highly frequency dependent. It may be desirable to shape the spectrum of the interrogating signal in such a way as to compensate for this attenuation feature. Also, amplitude weighting may be used to reduce range sidelobes on the signal which are caused by the fact that the signal has finite duration [1], [3].
E. Pulse compression techniques offer high energy per pulse and high signal-to-noise ratios without high peak powers.

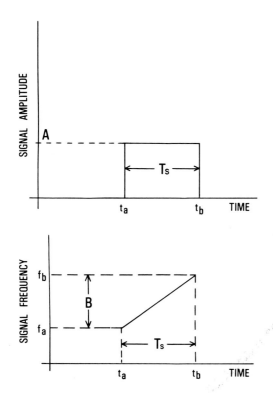

Fig. 2. (a) Time domain envelope of a typical pulse compression radar signal. (b) Frequency versus time characteristic of a typical linear FM pulse compression signal.

PRINCIPLES OF PULSE COMPRESSION RADAR

One of the physical principles that has come to light in the development of radar systems is that the ability to resolve targets closely spaced in range is directly related to the bandwidth of the interrogating signal [3]. That is, greater resolution requires greater bandwidth. This is intuitively obvious in the case of a conventional pulse amplitude modulation radar. If one expects to separate closely spaced targets then the pulse width must be reduced so that returns from individual targets do not overlap. As the targets become more closely spaced in range then the pulses must be shorter in duration which results in wider signal bandwidth. Unfortunately, decreasing the width of the transmitted pulse also reduces its energy which results in decreased total range capability because of reduced signal-to-noise ratio. In the example case of pulsed radar the range resolution is a function of pulse duration but the signal parameter that fundamentally determines range resolution is the bandwidth of the signal. This fact ultimately resulted in the development of PCR systems [4–9]. There are many modulation techniques that will result in broad bandwidths even though the signal has long duration. The term PCR, refers to a class of ranging systems in which time is related to some parameter of the signal modulation. This can be frequency, phase or amplitude modulation. In the most common implementa-

tion, time is linearly related to the instantaneous frequency of the transmitted signal. This is referred to as linear FM PCR or "chirp" radar.

Figure 2 depicts the time and frequency domain characteristics of a typical linear FM pulse compression signal. The waveform of the envelope of the signal, shown in Fig. 2(a) is usually of constant amplitude for the duration of the pulse. This waveform is typical of pulse compression radar systems and may not be the most desirable for biological interrogation. In the biological case, since attenuation through the body increases considerably with frequency, it may be desirable to have the interrogating signal amplitude increase as frequency increases. For example, in muscle tissue the attenuation at 2 GHz is about 4 dB per centimeter while at 4 GHz the attenuation is about 8 dB per centimeter. If the interrogating signal were swept over this frequency range, then there would be a 4 dB amplitude tilt for each centimeter of muscle tissue that the signal propagated through. This amplitude tilt as a function of frequency would result in a reduction of resolution of the ranging system. Also, amplitude modulation may be used to reduce range sidelobes [3] and thus reduce false indications of multiple paths in the target. Figure 2(b) shows the relationship of frequency and time during the pulse. Frequency is varied in a linear fashion over a bandwidth of B Hz in a time of T_s seconds. It should be noted that in biological dielectrics, a linear frequency sweep is probably not the most desirable. Again, using muscle as an example, the velocity of propagation in this tissue varies by about 5 percent from 2 GHz to 4 GHz. A linear sweep is not the optimum interrogating signal for this tissue. This situation is further complicated by the fact that different tissues have different attenuation and velocity of propagation characteristics. This means that the optimum interrogating signal is a function of the tissue type.

A simplified implementation of a PCR is shown in Fig. 3. A signal is transmitted whose frequency and time characteristics are shown in Figs. 2(a) and 2(b). The signal propagates to the targets, where it is reflected and returns to the receiver. It is heterodyned to some intermediate frequency where it is processed by a pulse compression filter. The filter frequency vs time characteristics are shown in Fig. 4. In this illustration, the delay characteristics of the filter are adjusted so that lower frequencies take longer to pass through the filter than higher frequencies. By proper matching of the delay characteristics of the filter to the frequency slope of

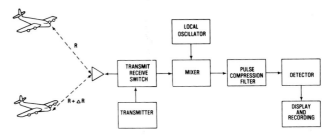

Fig. 3. Simplified block diagram of a typical pulse compression radar.

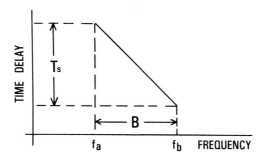

Fig. 4. Pulse compression filter time delay characteristic.

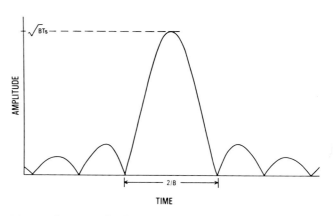

Fig. 5. Compressed pulse at output of pulse compression filter.

the transmitted signal, all of the energy that enters the filter from f_a to f_b will emerge at the same time in a single sharp pulse. This is the origin of the term "pulse compression." Pulse compression filters may assume many forms. They may be ultrasonic delay lines, aluminum or steel strips, surface acoustic wave (SAW) devices, piezoelectric materials, lumped circuits or waveguide near cutoff [1]. Probably the two most important characteristics of pulse compression systems are the bandwidth and time duration of the transmitted signal. The bandwidth of the signal determines the ability to resolve paths of different electrical length [3]. As one can see from Fig. 5, the width of the compressed pulse to its first zero crossings is determined entirely by the bandwidth of the transmitted pulse. The time duration of the signal is a factor in determining the amplitude of the compressed pulse and therefore, partially determines the signal-to-noise ratio. Clearly, wider bandwidth results in better range resolution and longer pulse duration results in higher signal-to-noise ratio. These two factors, bandwidth and pulse duration, are often combined into a product, BT_s, called the pulse compression ratio. In some earlier publications it is called the dispersion factor. The pulse compression ratio can be expressed in several ways:

$$BT_s = \frac{P_o}{P_i} = \frac{T_s}{\lambda} \qquad (1)$$

where

  B = Bandwidth of the transmitted signal (Hz)
  T_s = Duration of the transmitted signal (sec)
  λ = Width of the compressed filter output pulse (sec)
  P_o = Amplitude of the compressed output pulse. Assumes no loss in the filter (watts)
  P_i = Amplitude of the compression filter input pulse (watts)

Another feature of the compressed pulse that should be noted is the ringing that occurs outside the main lobe of the compressed pulse. These lesser peaks are referred to as range sidelobes. With a transmitted signal whose amplitude characteristics are rectangular as shown in Fig. 2(a) the compressed pulse has the form

$$f(t) = BT_s \frac{\sin(\pi Bt)}{\pi Bt} \qquad (2)$$

where t = time (sec). With a uniform amplitude characteristic, the first range sidelobes are only 13.2 bB below the main lobe. Clearly this is undesirable since it will mask nearby signals of interest or the range sidelobes may be mistaken for multipath signals. The range sidelobes can be reduced by a considerable amount by amplitude weighting the received signal [1]. For example, the Taylor weighing function (N = 8) reduces the sidelobes to −40 dB while the Hamming function reduces the level to −42.8 dB. The cost for using weighting functions is that the main lobe is broadened and there is a reduction in signal-to-noise ratio. The main lobe is broadened by 40 and 50 percent and the signal-to-noise ratios are reduced by 1.14 and 1.34 dB respectively for the Taylor and Hamming functions. These penalties must be weighed against the advantages of range sidelobe reduction.

The range resolution of a monostatic PCR is given by:

$$\Delta R = \frac{C}{2B} \qquad (3)$$

where C = velocity of propagation in free space (3×10^{10} cm/sec).

Monostatic radar has the advantage that the signal travels to the target, where it is reflected and returns to the receiver thus traveling over the distance twice. When imaging biological systems, the measurement is usually of time delay and insertion loss through the target thus resembling a bistatic radar system. In the case of forward scatter systems, the signal travels over the propagation path only once, thus the range resolution is degraded by a factor of two:

$$\Delta R = \frac{C}{B} \qquad (4)$$

In the case of biological systems, the velocity of propagation is not the same as in free space. It is reduced because the real part of the complex permittivity is greater than that of free space. For biological systems, the range resolution equation should be modified appropriately.

$$\Delta R = \frac{C}{B\sqrt{\epsilon}} \qquad (5)$$

where
ϵ' = Real part of the complex permittivity of the tissue in which the wave propagates.

Note that in biological systems, the range resolution is much better than for free space for a given signal bandwidth. The velocity of propagation is also effected by the imaginary part of the complex permittivity but that effect is so small that it is generally considered to be negligible.

It is informative to evaluate the resolution equation for parameters that might be encountered in biological systems. Suppose, as an example, images were attempted in muscle tissue where ϵ' is about 50. If one wished to discriminate between paths whose physical lengths were different by 2 cm then, using Eq. (5), and solving for bandwidth:

$$B = \frac{3 \times 10^{10}}{2\sqrt{50}} = 2.12 \times 10^9 \text{ Hz} \qquad (6)$$

A resolution of 2 cm would require a bandwidth of slightly over 2 GHz. An obvious question is to what extent conventional radar technology may be adopted to achieve this resolution. In order to use receivers of reasonable sensitivity the system should operate at a frequency somewhere below 6 GHz. Bandwidths on the order of 2 GHz are never used in radars operating in this frequency range. There are a number of reasons for this. A resolution of 2 cm is not a usual radar requirement, frequency allocations this broad are not available in the frequency bands below 6 GHz and suitable high power transmitting tubes are not available with such wide bandwidth. The most important reason is that pulse compression filters with 2 GHz bandwidth do not exist at this time. A further consideration with regard to imaging biosystems is pulse duration. With the extreme attenuation in biosystems one would like a very long duration pulse so that the energy per pulse would be maximized. However, very long duration pulses increase the image formation time to the point where the system becomes less useful for a medical diagnosis. Therefore, the factor that limits the duration of the pulse is the image formation time. If one is forming an image with 4096 pixels and the pulse duration is 1 millisecond per pixel then the total image formation time (neglecting other delays) is about 4.1 seconds. This would be a reasonable upper limit for a medically useful system. This pulse width is also consistent with the current state-of-the-art in electronically controlled oscillators such as YIG tuned devices. If the bandwidth and pulse width consistent with biological applications are combined to obtain the pulse compression ratio then:

$$BT_s = 2 \times 10^9 \times 1 \times 10^{-3} = 2 \times 10^6 \qquad (7)$$

Pulse compression ratios up to 10^5 are used in radar systems. Therefore, another unique feature of imaging biosystems using pulse compression techniques is the requirement for an extremely large pulse compression ratio compared to radar systems.

The purpose of this introduction to PCR is to acquaint the reader with fundamental limitations in range resolution, techniques that are used in PCR systems and why these techniques cannot be applied directly to imaging biosystems.

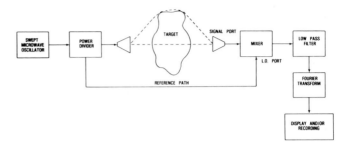

Fig. 6. Simplified block diagram of a microwave time delay spectrometer.

It has been pointed out that there are two characteristics of PCR for biological imaging that are beyond the current state-of-the-art in radar technology. These are bandwidth and pulse compression ratio. However, it is possible to use the same principles and apply a somewhat different, signal processing technique.

MICROWAVE TIME DELAY SPECTROSCOPY (MTDS)

Consider the simplified block diagram in Fig. 6. The swept microwave oscillator generates the linear FM signal just as in a conventional PCR. This signal is divided into two parts. One part of the signal is connected directly to the local oscillator port of a mixer by a cable. The other part is connected to a transmitting antenna which illuminates the target. The transmitted signal travels through and around the target over various paths and is picked up by the receiving antenna which is connected to the signal port of the mixer. The time delay through the reference path is deliberately chosen so that it is not the same as the time delay through the target. This produces a difference in the instantaneous frequencies appearing at the local oscillator and signal ports of the mixer. This is illustrated graphically in Fig. 7. The frequency difference between the reference path signal and the signals transmitted through the target is di-

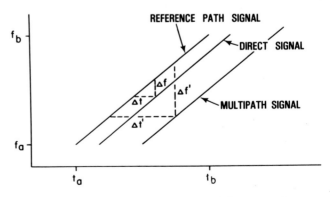

Fig. 7. Time and frequency relationships of the swept microwave reference signal, direct ray path and a multipath signal.

rectly proportional to the difference in time delay between the paths. The mixer detects these differences and produces beat notes whose frequencies are proportional to the differential propagation delay between the target ray paths and the reference signal path. By taking the Fourier transform of the compositie signal at the mixer output and analyzing only that spectral component that corresponds to the direct ray path then it is possible to discriminate against multipath signals.

We can consider the case in which there is only one ray path. The real situation in which there is an infinite number of paths is the summation of an infinite number of individual ray paths. The signal at the reference port of the mixer can be represented as a linear swept signal:

$$f_r(t) = A \cos(\omega_1 t + k_s t^2/2) \tag{8}$$

where

A = Peak amplitude of the signal (volts)
ω_1 = Start frequency of the sweep (rad/sec)
ω_2 = Stop frequency of the sweep (rad/sec)
k_s = System constant = $(\omega_2 - \omega_1)/T_s$

The signal which propagates through the target and arrives at the signal port of the mixer with differential delay of T seconds is a delayed version of the signal at the reference port:

$$f_s(t) = \alpha f_r(t - T) = \alpha A \cos[\omega_1(t - T) + (k_s/2)(t - T)^2] \tag{9}$$

The signal at the mixer output is:

$$f_0(t) = \alpha \beta A^2 f_r(t) f_s(t) \tag{10}$$

If this product is expanded and the high frequency terms (ω_1, ω_2, $\omega_1 + \omega_2$ etc.) and filtered out then the signal at the output of the low pass filter due to a single ray path is:

$$E = \frac{\alpha \beta A^2}{2} \cos[k_s T t - \omega_1 t - k_s T^2][U(t - T) - U(t - T_s)] \tag{11}$$

where β = system gain factor (determined experimentally)

$$U(x) = 0 \text{ for } x < 0, 1 \text{ for } x \geq 0$$

This expression shows that for any given path, there is a spectral component whose frequency, $k_s T$, is proportional to the time delay difference between the reference path and the path through the target.

It is of interest to analyze the range resolution of this approach since it uses a different signal processing technique. Due to the fact that the interrogating signal has a finite duration of T_s seconds, the Fourier transform of the signal will have the form $\sin(X)/X$. Consider the situation where the spectra due to two separate ray paths overlap by one half the width to the first zeroes of their spectra. This situation is shown in Fig. 8. Under this assumption, the main lobes of the spectra overlap at approximately the −4 dB points. The separation between the main lobes of the spectra is:

$$S = \frac{B}{T_s} |T_1 - T_2| = \frac{B\delta}{T_s} \tag{12}$$

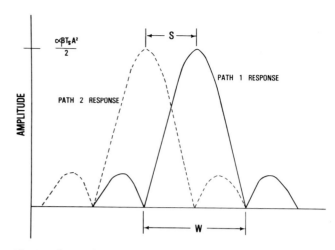

Fig. 8.   Composite spectrum resulting from two discrete ray paths with equal attenuation.

where

S = Separation between the peak frequencies (Hz)
T_1 = Differential time delay between the reference path and path number 1 (sec)
T_2 = Differential time delay between the reference path and path number 2 (sec)
$\delta = |T_1 - T_2|$

The total width of each of the spectra is:

$$W = \frac{2}{T_s} \tag{13}$$

where W = total width of the spectra to the first nulls (Hz). Given that the peaks are separated by half the width to the first nulls then:

$$\frac{S}{W} = 0.5 \tag{14}$$

Dividing Eq. (12) by Eq. (13):

$$\frac{S}{W} = \frac{B\delta}{2} = 0.5 \tag{15}$$

Therefore, the resolution of the system in terms of time delay is:

$$\delta = \frac{1}{B} \tag{16}$$

Another way to specify time delay is in terms of range and velocity of propagation:

$$\delta = \frac{\Delta R \sqrt{\epsilon'}}{C} \tag{17}$$

where ΔR = range resolution (cm). Substituting Eq. (16) into Eq. (17) and rearranging terms:

$$\Delta R = \frac{C}{B \sqrt{\epsilon'}} \tag{18}$$

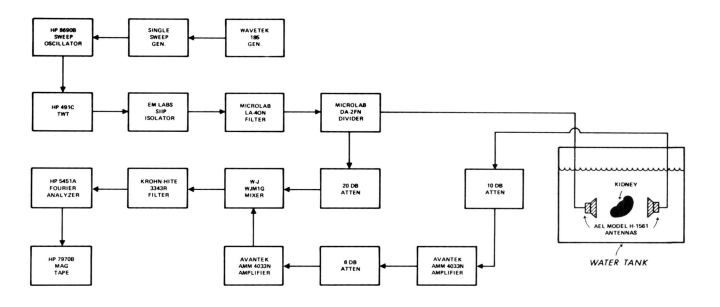

DATA ACQUISITION SYSTEM

Fig. 9.   Detailed block diagram of a microwave time delay spectrometer.

This is exactly the expression given for the system that processed the swept signal in the time domain using pulse compression filters. Therefore, nothing has been sacrificed in terms of range resolution by adopting this approach which is more suitable for use in biosystems.

METHODS

Data Acquisition

A detailed block diagram of the MTDS system is shown in Fig. 9. The swept microwave signal was produced by a Hewlett Packard (HP) 8690B Sweep Oscillator with a HP 8699B R.F. Plug-in Unit. The frequency sweep was generated by driving the external sweep input on the HP 8690B withu a sawtooth waveform. This single sweep sawtooth waveform generator was triggered by a Wavetek Model 185 Function Generator at a rate of about one pulse per second. This trigger function was asynchronous with the scan of the electromechanical scanner but the Fourier analyzer was programmed to ignore data inputs until the scanner had reached the desired position relative to the kidney. The swept microwave signal was amplified to a level of 1 Watt with an HP 491C traveling wave tube (TWT) amplifier in order to help overcome the losses in the water and kidney. The TWT was followed by an EM Laboratories S11P isolator to protect the TWT from excessive reflected power. The signal was then filtered using a Microlab LA-40N Low Pass Filter to remove harmonics. The signal was then divided into two equal parts using a Microlab DA-2FN Power

Divider. One part of the signal was connected to an American Electronics Laboratories (AEL) Model H-1561 transmitting antenna. The other part was adjusted in magnitude by a 20 dB attenuator and connected to the local oscillator port of a Watkins Johnson (WJ) WJM1G mixer. The swept microwave signal propagated through the kidney and was received by an identical AEL H-1561 antenna. The receiving antenna was followed by a 10 dB attenuator whose purpose was to reduce multiple reflections between the antenna and the input of the Avantek AMM 4033N low noise amplifier. It was found that the system was highly vulnerable to reflections between components in the system. For example, there was obviously a multiple reflection between the receiving antenna and the first R.F. amplifier and there was also a reflection between the first and second R.F. amplifiers. These reflections manifested themselves as spectral lines in the Fourier transform and could be eliminated by appropriate use of attenuators. It was also apparent that there was a reflection from the transmitting antenna which went back through the power divider and entered the mixer. Therefore, one of the necessary parameters of components in this system is that they have very low VSMR. In most cases, the state-of-the-art in components such as broadband amplifiers, mixers, and antennas is not adequate and necessitated using attenuators to reduce reflections. After amplification by the Avantek amplifiers, the signal was connected to the mixer. The output of the mixer was filtered in a Krohn-Hite 3343R filter which was set to bandpass the signal from 500 Hz to 2000 Hz. The low frequency was set to remove a rather large D.C. component, some low frequency noise (noise up to 100 or 200 Hz) and power line induced interference. The high frequency was set at 2000 Hz to act as an antialiasing filter prior to the digital Fourier transform. The signal was digitized by the Fourier analyzer and recorded on magnetic tape. The Fourier transform and subsequent processing took place off line after the scan was finished. This was done to reduce the overall scan time. It took nearly 5 hours to scan the kidney because of the slowness of the electromechanical scanner.

The computer in the Fourier analyzer not only digitized the signal, it also controlled the electromechanical scanner. The scanner consisted of a cubical water tank 0.914 meters on a side, vertical and horizontal translation axes, independent computer controlled drive motors for each axis, independent position transducers with computer readouts for each axis and mounting frames upon which to connect the antennas and specimen to be scanned. This was constructed especially for this project by the Middlestadt Machine Co., Baltimore, MD. The Heidenhein Pos-Econ 1 position transducers had a total range of about 200 mm and are accurate to ±0.002 mm. This accuracy was not totally usable since the drive motors minimum increment was 0.003 mm. The position transducers had +8421 binary coded decimal (BCD) outputs that were interfaced to an HP 2100S minicomputer using HP 12604B Data Source Interface cards. The interface required construction of an intermediate buffer that provided higher voltage levels, buffered the data to drive longer lines and provided special signals such as flag and control logic that was required by the 12604B but was available from the Pos-Econ 1.

The horizontal and vertical axes were driven by a Superior Electronic M112-FJ25 synchronous stepper motor through a gear reduction box and worm gear. The motors were interfaced to the HP 2100S minicomputer through a Slosyn Translator Model BTR103RT and a 16 bit duplex register in the HP 2100S. This arrangement allowed the minicomputer to control the horizontal (X) and vertical (Y) coordinates of the positioner by sending pulses to the stepping motors under software control. In his manner, the increments between pixels was set to 1.4 ± 0.01 mm. There were 64 samples per line and 64 lines in the raster scan for a total of 4096 pixels in the image. Although higher positioning accuracy was available, it was not used because it considerably dilated the data acquisition time and it was apparently not necessary.

The cubical tank contained the water into which the antennas and specimen were immersed. The advantages of water immersion are discussed in another publication [10]. The water was continuously filtered through an ion exchange column and iodinated as a bacteriostatic and bacteriocidal measure.

Signal Processing

The signal that was recorded on magnetic tape was the digitized version of the composite waveform at the output of the Krohn-Hite bandpass filter. This contained data for all the possible ray paths. During the system setup, the frequency of the beat note corresponding to the direct ray path was

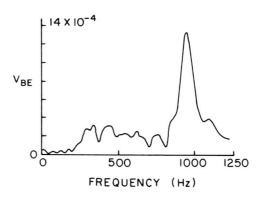

Fig. 10. Typical time delay spectrum. Large peak near 1000 Hz corresponds to direct ray path.

identified. This was accomplished by holding the antennas stationary and repeatedly transmitting the swept microwave signal, digitizing the composite signal at the Krohn-Hite filter output, calculating the Fourier transform and displaying the power spectrum. A typical spectrum is shown in Fig. 10. Note that this spectrum is quite complicated corresponding to many propagation paths. In the absence of the kidney specimen, the direct ray path is represented by the largest peak in the power spectrum. This can be verified by occluding the direct ray path with a small piece of metal and observing which spectral peak disappears. The images were formed by taking the Fourier transform of the composite signal at each spatial position on the kidney and measuring the amplitude of the spectral component corresponding to the direct ray path. The data are, therefore, a hybrid time-amplitude transmission magnitude. This number was then converted to a gray scale or a color as will be described in the section on image processing.

Image Processing

The primary data consisted of an array of 4096 hybrid time-amplitude measurements of transmitted microwave energy. The basic image processing steps subsequently applied may be grouped into two types: amplitude domain and spatial frequency domain. The first two processing steps are in the amplitude domain. The original data was 64 by 64 and 15 bits plus sign deep. It was truncated to 7 bits plus sign and interpolated from 64 by 64 to 256 to 256. The interpolation was accomplished by a cubic spline which performed a discrete convolution with an approximate kernal. The kernal was 4 by 4 and symmetric. It was truncated to include only the first two nulls. The interpolated data were further processed in the amplitude domain to map the amplitudes into a scale which suited the dynamic range of the video monitor. The process maps the data from the original scale onto a new scale by means of a piecewise linear function with a single hinge point. Subsequently, the image was processed by a 3 by 3 Laplacian to enhance gradients. Contour plots at this stage of processing were also performed.

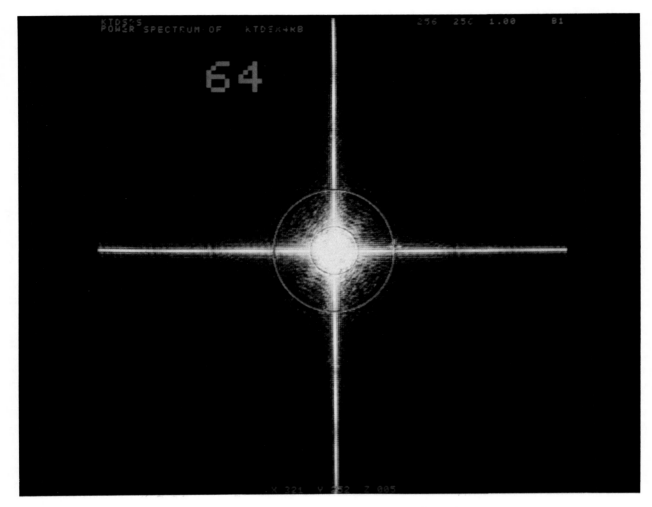

Fig. 11.   A two dimensional MTDS image power spectral density display with magnitude overlay.

Spatial frequency processing consisted of a two-dimensional bandpass filter applied after interpolation. The bandpass filter transfer function magnitude is shown overlaid on the image power density as shown in Fig. 11. Note that the origin of the power density is centered. Positive and negative spatial frequencies in elevation are shown on the ordinate whereas positive and negative frequencies in azimuth are on the abcissa. The outer circle is the −3 dB point for high spatial frequencies and the inner circle is the −3 dB point for low spatial frequencies. The two-dimensional bandpass filter is multiplied with the two-dimensional Fourier transform of the interpolated (and zero padded) image after which a two-dimensional inverse transform recreates the image with rejection of low and high spatial frequencies. In the present case, spatial frequencies at the Nyquist rate are normalized to 64. The high spatial frequency "cutoff" was 64 and the low spatial frequency "cutoff" was 20. The image was finally rescaled by the piecewise linear mapping function and pseudocolored.

Image Interpretation

Inasmuch as the prime motivation for microwave imagery of biosystems is enhanced relevance to physiology and pathophysiology (cf. Larsen and Jacobi in this volume), image interpretation is keyed to regional specialization in the kidney. The basic anatomy and its functional correlations are discussed in the companion paper. The same general features of renal imagery are present in the MTDS and scattering parameter methods with the exception that multipath artifact is reduced in the MTDS method.

The MTDS kidney image after amplitude domain processing and gradient enhancement is shown in Fig. 12. The absence of multipath in the region outside of the capsule is evident. The capsule is well demarcated. The cortex corticis is well visualized along most of the perimeter of the organ. The juxtamedullary cortex and corticomedullary junction is best visualized above the cursor. The outer medullary stripe appears best in this location also. The inner medullary

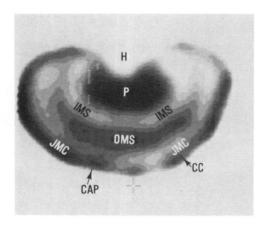

Fig. 12. An MTDS image of isolated canine kidney prior to bandpass spatial frequency filtering. The key for this and subsequent images is as follows: H, hilus; P, pelvis; IMS, inner medullary stripe; OMS, outer medullary stripe; CC, cortex corticis; CAP, capsule: JMC, juxtamedullary cortex; F, renal sinus fat; R, support rod; A, artifact.

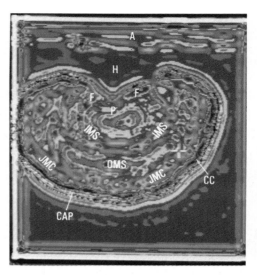

Fig. 14. The MTDS image after two dimensional bandpass filtering and pseudocolor processing (see Fig. 12 legend for key).

stripe appears around the perimeter of the pelvis. The data variations are displayed with greater dynamic range in the contour plot of Fig. 13. Now some suggestion of cortical lobulations is evident as variations within the pelvis that were previously lost in the monochrome display due to disparity between the dynamic range of the data as compared to the display.

The bandpass filtered and pseudocolored MTDS image is shown in Fig. 14. The additional processing steps enhanced the corticomedullary junction, the apparent lobulations of cortex and variations within the pelvis. Note also that the predominant "ringing" outside of the capsule in the pseudocolored and bandpass filtered copolarized scattering pa-

rameter image (cf. Fig. 15) is virtually absent in the similarly processed MTDS image.

Thus it would appear that the cortex is distinguishable from cortex corticis which implies that the relative paucity of glomeruli in the cortex corticis is represented in the spatial distribution of complex permittivity at microwave frequencies. Similarly, medullary structures are distinguishable which implies that water and electrolyte functions of the medulla are also represented in the MTDS image. The pelvis represents greatest insertion loss as would be expected.

Verification of these interpretations must await pathophysiological correlation studies. The phased array antenna

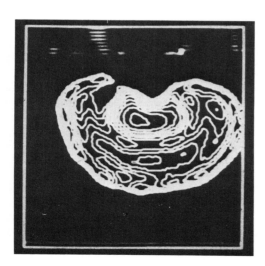

Fig. 13. The same data as in Fig. 12 is displayed in contour format.

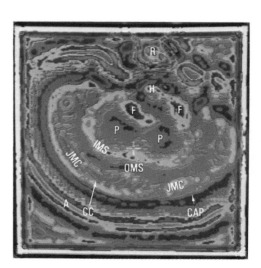

Fig. 15. Copolarized magnitude of S_{21} image shown for comparison with the MTDS images of the preceding three figures. Note the ringing artifact labeled A which is missing in the MTDS images.

described elsewhere in this volume would be one approach to achieve reduced data collection times such that pharmacologic means to effect glomerular filtration and tabular function may be employed.

CONCLUSIONS

MTDS imagery is an effective technique for reduction of multipath in forward scatter microwave imaging systems. The general features of regional specialization in the canine kidney are consistent with image interpretation based on established renal anatomy. It would appear that regions of filtration are distinguishable from regions where the ultrafiltrate is concentrated. Further, the two medullary zones are distinguishable which may represent increasing osmolarity of medullary interstitial tissue from the inner stripe as opposed to the outer stripe. Clearly, the collecting cistern or pelvis is distinguishable from the other regions of the kidney where urine processing takes place.

Many promising areas for future research exist in the MTDS imagery. Prime among these is time resolution improvement and use of combined amplitude/frequency modulation to better suit the frequency dependent attenuation of biological dielectrics.

References

1. M. I. Skolnik, *Introduction to Radar Systems*, McGraw Hill, 1980.
2. J. C. Lin, *Microwave Auditory Effects and Applications*, Charles C Thomas, Springfield, Ill., 1978.
3. J. R. Klauder, A. C. Price, S. Darlington, and W. J. Albersheim, "The Theory and Design of Chirp Radars," *Bell Syst. J.*, Vol. XXXIX, No. 4, pp. 745–809, July 1960.
4. British Patent 604,429, "Improvements in and Relating to System Operating by Means of Wave Trains," July 5, 1948.
5. U.S. Patent 2,624,876, "Object Detection System," R. H. Dicke, Jan 6, 1953.
6. U.S. Patent 2,678,997, "Pulse Transmission," S. Darlington, May 18, 1954.
7. C. E. Cook, "Pulse Compression—Key to More Efficient Radar Transmission," *Proc. I.R.E.*, 48, No. 3, pp. 310–316, Mar. 1960.
8. H. O. Ramp and E. R. Wingrove, "Principles of Pulse Compression," *I.R.E. Trans.*, MIL-5, No. 2, pp. 109–116, Apr. 1961.
9. D. K. Barton, *Radars*, Vol. 3, Pulse Compression, Artech House, Dedham, Mass., 1978.
10. J. H. Jacobi, L. E. Larsen, and C. Hast, "Water Immersed Microwave Antennas and Their Applications to Microwave Interrogation of Biological Targets," *IEEE Trans. Microwave Tech.*, MTT-27, pp. 70–78, Jan. 1979.

A Water-Immersed Microwave Phased Array System for Interrogation of Biological Targets

S. J. Foti,* R. P. Flam,* J. F. Aubin,* L. E. Larsen,† J. H. Jacobi†

The design and theoretical performance of a non-invasive-water-immersed microwave antenna system for interrogation of biological targets is described. Interrogation is achieved by the transmission of S-Band microwave energy through the target. Water-immersion is used to provide improved resolution, multipath rejection, "surface-matching" of the target, and an anechoic environment.

The system comprises a 151 element transmit array antenna and a 127 element receive phased array antenna which are computer controlled. The transmit array may be operated in an "instantaneous" illumination mode covering a 15cm diameter cross section or in a "sequential" illumination mode producing a time sequence of "pencil-beams" each of which is 2 cm in diameter. The receive phased array, when successively placed in five axial positions, synthesized a 635 element "volumetric" phased array capable of producing three-dimensionally focused samples of the perturbed electromagnetic wave distribution.

The overall system performance is characterized by the ability to interrogate biological targets as large as 15 cm in cross section with a data collection time as little as 1 minute. The biological target is characterized by sampling the transmitted field. Total data collection time is governed by three factors: a) the sampling volume; b) the illumination mode of the transmit array, and c) the number of frequencies of measurement. The total dynamic range for this system is 166 dB. The total data collection time for the maximum sampling volume is less than 3.5 minutes if the transmit array's "instantaneous" illumination mode is utilized at a single frequency. This time must be multiplied by the number of "pencil-beams" desired if the transmit array's "sequential" illumination mode is selected and also by the number of frequencies of measurement. Upon collection of the complex field amplitudes, the data is processed off-line in software for three-dimensional focusing in order to obtain the sampled distribution of the electromagnetic wave as perturbed by the organ under study. Based upon the conservative estimate for resolution of 0.8×3 dB beamwidth, the predicted resolution of the subject system is 7.2 mm (5.6 mm), 4.8 mm (4.3 mm), and 16 mm (12.8 mm) in the vertical, horizontal and axial directions respectively, at 3 GHz (at 4 GHz). The horizontal resolution predictions may be unrealistic as the predicted values are smaller than one-half wave length in the medium (i.e., 5.7 mm at 3 GHz and 4.4 mm at 4 GHz). Axial resolution could be improved by increasing the physical size of the array. However, larger size arrays would not be consistent with the noise floor in the present system configuration. Calibration steps compensate for array distortion and complex permittivity of the coupling medium.

1. INTRODUCTION

Previous chapters (Larsen and Jacobi; Jacobi and Larsen) have illustrated the fact that transmission imagery is possible at microwave frequencies in the upper half of the radar S-band for targets of biological interest. This imagery has many potential applications, in both dosimetry and medical diagnosis. These are derived from the low photon energy which allows efficient interaction at the level of weak intra and intermolecular forces and low flux density which prevents significant perturbation of the biological dielectrics under study. Such a system produced images that depend upon the frequency(ies) of operation, polarization and the spatial distribution of complex permittivity within the biosystem under study.

The use of water coupled elements is an essential requirement for successful imagery at microwave frequencies in dosimetric and diagnostic applications as described in Jacobi et al. [1]. In summary, the pertinent benefits of water coupling include: (a) reduction in aperture area by a factor of approximately 77 while retaining the manageable trans-

* Flam & Russell, Inc., Horsham, Pa. † Walter Reed Army Institute of Research, Dept. of Microwave Research, Washington, D.C.

mission losses of the radar S-band, (b) the prevention of mutual coupling in arrays of elements, (c) the production of anechoicity and the attenuation of many multipath components, (d) improved match to biosystems which are either well hydrated or hydratable (as in the case of the stratum corneum of the skin), and (e) the provision of a means for coupling to allow the use of non-contacting antennas.

The previous uses of water-coupled elements in forward-scatter systems were seriously limited by the use of an electromechanical scanner and requisite high accuracy spatial positioning for useful data collection. Typical data acquisition times were 4.5 hours for a single raster scan on a 64 × 64 matrix. This feature precludes many imagery applications of dosimetric and/or diagnostic interest. For example, in the case of dosimetry, circulatory responses to a thermalizing dose of radiofrequency radiation could be studied by the use of a difference mode, where images collected before and after a brief exposure would be compared to extract the desired circulatory response on the basis of permittivity changes consequent to the irradiation. An excessively long data collection interval would degrade performance in this application as a result of thermal diffusion and/or further circulatory changes. Similarly, in the case of diagnostic studies, many dynamic physiological tests would be degraded by an excessively long data collection interval. For example, in the context of isolated kidney studies, pathophysiological correlates before and after administration of agents such as tubular poisons or effectors of the glomerular filtration rate would require data acquisition times of no more than a few minutes for meaningful interpretation. Furthermore, the multiple projection views required for three dimensional reconstruction would impose a totally unacceptable duration for data collection. Although it is true that the 64 × 64 raster collects data for 64 slices, there still exists the need for about 180 positions in angle to accomplish the reconstruction. Also, the 64 samples in each projection are too sparse to achieve adequate resolution for targets greater than 8–10 cm in major dimension.

A major motivation, therefore, in the development of the system was operation with water immersion to accomplish the same wavelength contraction and manageable losses of the existing water loaded element, but in a two dimensional array to permit electronic beam steering in elevation and azimuth. The same data collected in 4.5 hours with the scanned single transmit-receive pair may now be collected in about 1 minute. In addition, near field focusing in azimuth and elevation becomes possible within the shorter data collection interval. Of course, focusing in azimuth and elevation would also be possible with a single pair of scanned transmit-receive elements, but the long data collection time makes this unattractive as a routine method. Furthermore, very near field operation (minimum distance of a few mm) with the single pair of elements would preclude focusing to the extent that the nearby dielectric may alter the impedance of the elements and multiple reflections between the element and target may corrupt the data. Near field focusing is better applied at somewhat larger distances (in the order of a few cm) between the target and the receiving antenna with the use of a large aperture, low f number system as

discussed by Guo et al. elsewhere in this volume.

Focusing functions may be applied off-line once the complex field amplitudes transmitted through the target are sampled. This results in great flexibility in the choice of weighting functions, and the application of complex (time-consuming) data analysis techniques. Clearly, sampling does not require an array, but speed of operation is much enhanced with the use of an array (by a factor of about 300 in this case). Furthermore, a single pair of transmit-receive elements cannot accomplish effective focusing in range when scanned only over a planar surface in the near field. Better performance in near field focusing for range is accomplished by the use of a volumetric (three-dimensional) array, which can be realized with a two dimensional array, electromechanically scanned in the axial direction. The chief advantage of this method is the ability to sample a defined region of the transmitted field (in the "shadow" of the target) with a three dimensional volume element rather than a series of elements with resolution in only two dimensions. The array is focused in three dimensions in its near field where the focused functions are applied to a homogeneous coupling medium (water) rather than within the target of interest. The application of focusing functions within the target is interently faulted by the need for prior knowledge of the spatial distribution of the complex propagation constants within the target. Alternatively, some form of an internal marker could be used for adaptive focusing within simplified media, but the variations of permittivity in actual biological targets would require either multiple or moving markers. In any case, the use of markers would violate the original premise of a non-invasive method.

Another relevant issue is the question of how often the transmitted field must be sampled. This question, of course, depends upon the scene and frequency of operation. A complete development of the generalization of Nyquist sampling to a two dimensional (Fresnel) process is found in Lohmann. The related topic of aliasing is discussed elsewhere in this volume by Farhat. In the present case, if the frequency of operation is set at 4 GHz, the wavelength in water at 25°C is about 8.8 mm. The maximum required sampling rate on simple application of the (one-dimensional) Nyquist principle is, therefore, 4.4 mm. This dimension is also the order of target sizes that would offer maximum scattering cross-section. However, any attempt to reconstruct the complex field amplitudes in such a way as to form a two or three dimensional image requires a greater sampling density. Also sampling at half wave length intervals implies that phase data is useful only at π resolution. Greater phase accuracy may possibly be translated to a Rayleigh criterion at a higher frequency of operation as discussed by Farhat elsewhere in this volume. While further analysis is required in this area, results obtained to date with the mechanically scanned system indicate that a sampling rate of about 6 samples per wavelength (1.5 mm) results in good image quality.

2. CHARACTERISTICS OF MEDIA

Before proceeding with the design of an antenna system in the unusual medium of water, an understanding of wave propagation in such a medium must be established. The two differences in the electrical properties of water in contrast to those of air are: the significantly higher permittivity and the existence of a finite conductivity. The former causes wavelength contraction and the later allows conduction currents to flow thereby dissipating significant energy in the medium. The use of Maxwell's equations for a dielectric conducting medium allows derivation of the wave equation for the vector electric field in the time domain (vector quantities are denoted by a bar above the symbol)

$$\nabla^2\overline{E} - \mu_0\epsilon\frac{\partial^2\overline{E}}{\partial t^2} - \mu_0\sigma\frac{\partial\overline{E}}{\partial t} = 0$$

Where: μ_0 = permeability of free space
$\epsilon = \epsilon_r'\epsilon_0$
ϵ_0 = permittivity of free space
ϵ_r' = relative permittivity of water
σ = conductivity of water

If we assume sinusoidal time variations, then the electric field $\overline{E}(\overline{r},t)$ may be expressed in terms of a complex phasor $\overline{e}(\overline{r},\omega)$ by the relation:

$$\overline{E}(\overline{r},t) = \text{Re}\{\overline{e}(\overline{r},\omega)e^{j\omega t}\} \quad (2)$$

Where (f) is the frequency, (ω) is the angular frequency ($2\pi f$) and $\overline{r}$ is the radius vector. Using this relation, the time domain wave equation is transformed into the frequency domain resulting the vector Helmholtz equation:

$$\nabla^2\overline{e}(\overline{r},\omega) - \gamma^2\overline{e}(\overline{r},\omega) = 0 \quad (3)$$

Where the propagation constant is given by:

$$\gamma = \sqrt{j\omega\mu_0(\sigma + j\omega\epsilon)} = \alpha + j\beta \quad (4)$$

α being the attenuation factor and β is the wave number or phase factor. The general solution of this equation is:

$$\overline{e}(\overline{r},\omega) = \frac{Ae^{-\gamma r}}{4\pi r}\hat{r} + \frac{Be^{+\gamma r}}{4\pi r}\hat{r} \quad (5)$$

Where (A and B) are complex constants, (r) is the magnitude of the radius vector in a spherical coordinate system and ($\hat{r}$) is the unit radius vector. The first term in this solution represents an outgoing sperical wave eminating at the source of energy while the second term represents an incoming spherical wave. Clearly, if one is dealing with an unbounded region (or an anechoicly bounded region) then the constant B is zero by virtue of the Sommerfield readiation condition at infinity. As a convenience, the real parameters of relative permittivity and conductivity may be replaced by a single complex relative permittivity without changing the mathematical form of the solution, i.e., let:

$$\epsilon_r = \epsilon_r' - j\epsilon_r'' \quad (6)$$

where:

$$\epsilon_r'' = \epsilon_r'\tan\delta \quad (7)$$

$$\tan\delta = \text{loss tangent of medium} = \epsilon_r''/\epsilon_r' \quad (8)$$

It is then straightforward to derive

$$\alpha = 2\pi f\sqrt{\mu_0\epsilon_0\epsilon_r'[\sqrt{1 + \tan^2\delta} - 1]} \quad (9)$$

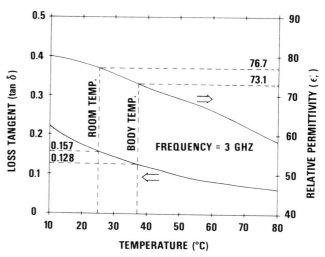

Fig. 1. Temperature dependence of electrical parameters of water.

$$\beta = 2\pi f\sqrt{\mu_0\epsilon_0\epsilon_r'[\sqrt{1 + \tan^2\delta} + 1]} \quad (10)$$

consider the wave traveling in the +r direction

$$\frac{e^{-\gamma r}}{4\pi r} = \frac{e^{-\alpha r}\cdot e^{-j\beta r}}{4\pi r} \quad (11)$$

($e^{-\alpha r}$) represents the decaying envelope of the wave and ($e^{-j\beta r}$) represents the sinusoidal nature of the wave whose phase is (βr). The total loss encountered by a wave over a distance r consists of dissipation loss (L_{DISS}) due to conduction currents being excited in the water and diffusion loss (L_{DIFF}) due to the spherical spreading of the energy. They are given by

$$L_{DISS} = 20\log_{10}e^{\alpha r} \quad (12)$$

or

$$L_{DISS} = 8.686\cdot\alpha\cdot r \text{ (dB)} \quad (13)$$

and from the radar range equation,

$$L_{DIFF} = 10\log_{10}(2\beta r)^2 \quad (14)$$

hence,

$$L_{DIFF} = 20\log_{10}(\beta r) - 29.14 \text{ (dB)} \quad (15)$$

and

$$L_{TOTAL} = 8.686\cdot\alpha\cdot r + 20\log_{10}(\beta r) - 29.14 \text{ (dB)} \quad (16)$$

Figure 1 depicts the temperature dependence of ϵ_r' and $\tan\delta$ for water at a frequency of 3 GHz. It is interesting to note that $\tan\delta$ decreases and hence the dissipation loss also decreases as the temperature is increased over the range shown. This is true because increased temperature shifts the frequency of the peak in $\epsilon''f(\omega)$ (relaxation frequency) further to the right with respect to the radar S band (2 to 4 GHz). The two particular temperatures of greatest interest for the subject system are room temperature (25°) and body temperature (37°C). Using the propagation parameters at 25°C and 37°, the loss curves for a spherical wave traveling in water are presented in Figs. 2(a and b). Not only is the total loss extremely high for relatively short distances, but the rate

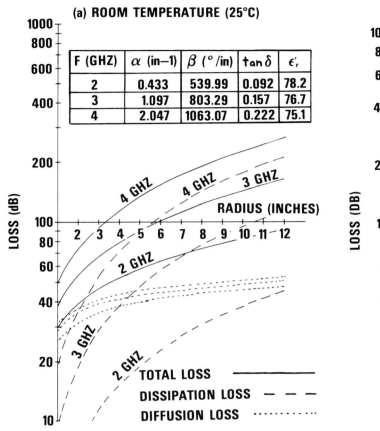

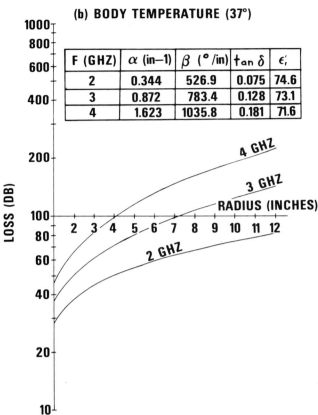

(a) ROOM TEMPERATURE (25°C)

F (GHZ)	α (in−1)	β (°/in)	$\tan \delta$	ϵ'_r
2	0.433	539.99	0.092	78.2
3	1.097	803.29	0.157	76.7
4	2.047	1063.07	0.222	75.1

(b) BODY TEMPERATURE (37°)

F (GHZ)	α (in−1)	β (°/in)	$\tan \delta$	ϵ'_r
2	0.344	526.9	0.075	74.6
3	0.872	783.4	0.128	73.1
4	1.623	1035.8	0.181	71.6

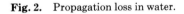

Fig. 2. Propagation loss in water.

of increase of loss vs. distance is much higher in water than that which occurs in air because of the dominant dissipation term in Eq. (16). Hence, for a given source transmitter power level and a receiver sensitivity, the distance at which a signal can be successfully detected will be limited. In particular, in the subject system a 20 watt (+43 dBm) Travelling Wave Tube Amplifier (TWTA) transmitter is used in conjunction with a receiver whose sensitivity is −123 dBm, resulting in a maximum tolerable loss of 166 dB for a 0 dB signal-to-noise ratio at the receiver input. At body temperature the loss equations predict the maximum separation distances between isotropic transmit and receive antennas in distilled water at 92.96 cm, 38.10 cm, and 21.08 cm for 2, 3, and 4 GHz, respectively. These distances may be increased if the particular antennas exhibit gain or in the likely situation that the biological target exhibits lower transmission loss than water. These numbers are reasonably considered worst case examples for homogeneous dielectrics because biological dielectrics generally have significantly less than 100% bulk water content (see the data compiled by Lin and presented elsewhere in this volume). Measured insertion losses (which include reflection and transmission losses) at 2.7 GHz in a human thorax (right side) are ca. 90 dB as reported by Yamaura [2]. The range of attenuation factors predicted at 3 GHz are ca. 60–65 Np/M for various locations on the human

torso [2]. Given the 166 dB maximum tolerable loss of the present system, this would imply a range of ca. 30 cm at 3 GHz in a thoracic application where severe impedance discontinuities are encountered.

As another general comment concerning wave propagation in water, owing to the high wave number (β) tabulated in Figs. 2(a) and (b) (essentially the rate of change of phase vs. distance), it is clear that small changes in antenna locations will cause substantial changes in the phase of the S-band energy received. This fact is an important consideration in the design of a phased array operating in water since an accurate knowledge of each array element location within less than one-tenth millimeter will be necessary in order to determine required element phase delays for coherent signal addition at a given focus point unless calibration procedures are employed to compensate the measured complex field amplitude for array "distortion." Methods for self calibration of the array are discussed by Guo *et al.* elsewhere in this volume.

3. CHARACTERISTICS OF INDIVIDUAL ANTENNA ARRAY ELEMENT

Based on the significantly different wave propagation characteristics in water compared to those in air, great care must be exercised in the design of an underwater antenna. Because of the relative mathematical simplicity of the equations describing the radiation characteristics of antennas in air, asymptotic solutions have been developed and have been successfully used in efficiently predicting radiation characteristics with reasonable accuracy. However, before using any radiation equations approximately valid in air, one must examine the approximations and determine whether or not they are still good approximations in water. In particular, as pointed out in Reference 3, the usual simple techniques applied for determining radiation characteristics for antennas in air no longer apply for antennas in conducting media. Consequently, the normal concepts of gain and antenna pattern are misleading and erroneous when the antenna is in a conducting medium. The pattern, for example, becomes highly dependent upon the choice of the origin of the coordinate system. This is demonstrated pictorially in Fig. 3 where directivity is apparently introduced in an isotropic radiator because of the large dissipation loss incurred even for a small antenna displacement from the center of rotation for pattern measurement. This fact adds to the difficulty in a water immersed phased array design because each element in the array will exhibit a different element radiation pattern.

To analyze array performance, we first derive a mathematical model for the individual array element in the form of an equation which predicts the electric field strength (and direction) at any observation point defined by a radius vector drawn from the element to the observation point. Phased array fields are calculated by superposition of the individual array element fields at each observation point of interest.

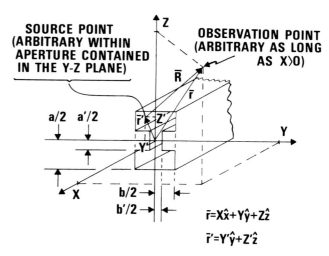

Fig. 4. Geometry and orientation of double ridged waveguide aperture in a rectangular coordinate system.

The concept of an antenna pattern is used only for a measurement verification of the radiation characteristics of an individual element predicted by our mathematical model. One must take great care to insure that the center of rotation in the pattern measurement is located at the center of the radiating aperture of the antenna element.

The basic specifications for the water-immersed antenna array element in the subject system are: octave bandwidth, linear polarization, broad beamwidth (~70°) and low VSWR. Small element cross sectional size to permit close packing in the array is required for the suppression of spurious spatial responses (grating lobes) in the phased array design. Element gain is of little importance as a few dB change in gain has little effect on operating range in the highly dissipative medium. The type of radiating element chosen to achieve these requirements is a double-ridged waveguide aperture. In particular, a WRD-180-C24 waveguide was utilized because the recommended operating frequency range of this standard double-ridged waveguide when air filled is 18 to 43 GHz. Hence, when filled with water ($\epsilon_r' \sim 77$) the operating frequency range scales down by $\sqrt{77}$ (due to wavelength contraction) to the desired 2 to 4 GHz. In addition to broad bandwidth, ridged waveguide has a smaller cross section than that of a standard rectangular waveguide operating in the same frequency range. Further, the use of water filled waveguide will facilitate impedance matching as will be discussed later in this section.

The details of the mathematical formulation for the near and far fields radiated by a double-ridged waveguide aperture in a conducting medium depicted in Fig. 4 are presented in Appendix A. This formulation for the electric field in terms of rectangular coordinates is readily usable for array field calculations when the radiators are located at arbitrary points and are tilted in any direction. The x,y,z coordinates of the center of the aperture and the orientation of the axis of each radiator must be known. A simple calculation will suffice to derive the resultant x,y,z coordinates of an obser-

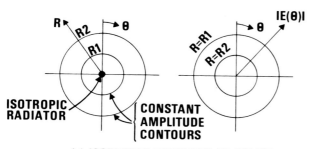

(a) ISOTROPIC RADIATOR AT ORIGIN

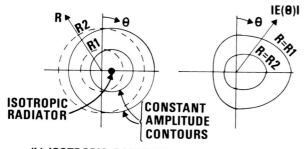

(b) ISOTROPIC RADIATOR DISPLACED FROM ORIGIN

Fig. 3. Effect of coordinate system origin on radiation patterns in water.

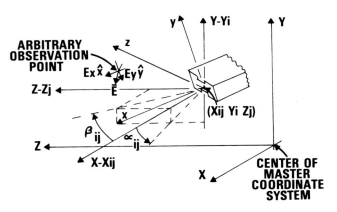

Fig. 5. Transformation of the electric field vector for translation in 3-dimensions and rotation in 2-planes.

Coordinate Transformation

$x = (X - X_{ij}) \cos\alpha_{ij} \cos\beta_{ij}$
$y = (Y - Y_i) \cos\alpha_{ij}$
$z = (Z - Z_j) \cos\beta_{ij}$

Electric Field Vector Transformation:

$E_x(x,y,z)\hat{x} = E_x ((X - X_{ij}) \cos\alpha_{ij} \cos\beta_{ij}, (Y - Y_i) \cos\alpha_{ij},$
$(Z - Z_j) \cos\beta_{ij}) \{\cos\alpha_{ij} \cos\beta_{ij}\hat{X} - \sin\alpha_{ij}\hat{Y} + \cos\alpha_{ij} \sin\beta_{ij} \hat{Z}\}$

$E_y(x,y,z)\hat{y} = E_y ((X - X_{ij}) \cos\alpha_{ij} \cos\beta_{ij}, (Y - Y_i) \cos\alpha_{ij},$
$(Z - Z_j) \cos\beta_{ij}) \{\sin\alpha_{ij} \cos\beta_{ij} \hat{X} + \cos\alpha_{ij} \hat{Y} + \sin\alpha_{ij} \sin\beta_{ij} \hat{Z}\}$

vation point with respect to each radiator. Therefore, there is no need to compute or store amplitude or phase patterns at multiple frequencies and perform complicated interpolations in order to compute the contribution of each radiator at each point of interest. The electric field is calculated only at one point in space for each radiator at each focus point. This is very advantageous for array calculations in a conducting medium since it avoids the pitfalls of radiation pattern and gain ambiguities which can occur when the radiator is displaced and/or tilted with respect to the master coordinate system center. The final transformation equations linking the electric fields expressed in each individual element's coordinate system with the master coordinate system which will be used for array calculations are given in Fig. 5.

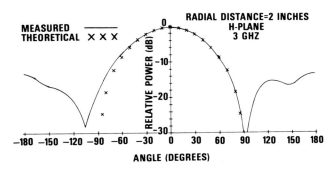

Fig. 6. Measured and theoretical H-plane radiation patterns of double ridged waveguide array element.

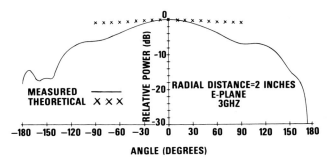

Fig. 7. Measured and theoretical E-plane radiation patterns of double ridged waveguide array element.

This mathematical model for the individual array element has been experimentally verified. Figure 6 shows H-plane patterns theoretically predicted and experimentally measured, both at a 2 inch radial distance (recall no far field approximations have been made and the patterns have a radial distance dependence on the observation point for small radial distances), and show excellent agreement over ±90° of the pattern. No comparison beyond ±90° is possible since the mathematical model is not valid there. However, in this application, pattern levels beyond 90° are of no interest. Figure 7 shows E-plane pattern comparisons and, although the agreement between theory and experiment degrades as we approach ±90°, the model is sufficiently accurate for the phased array design. The disagreement near ±90° arises from one of the inherent approximations of the aperture field technique presented in Appendix A. Namely,

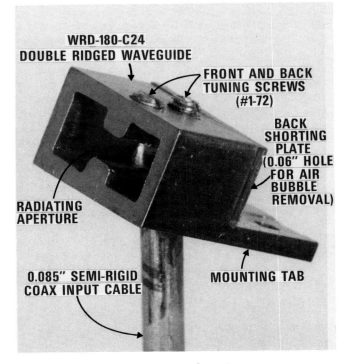

Fig. 8. Underwater array antenna element.

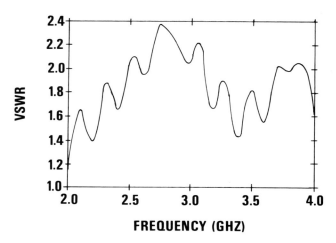

Fig. 9.   VSWR of underwater array antenna element.

that the E-field must approach zero in the plane of the aperture but outside the aperture. The small errors incurred in the phased array design due to this disagreement will be shown to be compensated during system calibration.

Figure 8 is a photograph of the individual array element. The double-ridged waveguide is fed by a miniature semi-rigid coaxial cable. Efficient coupling between the transverse electric mode in the waveguide and the TEM mode in the coax is achieved by the appropriate positioning of a back shorting plate and allowing the center conductor of the coax to extend through the ridge gap and protrude into a center hole in a tuning screw in the opposite wall. When the waveguide becomes water-filled, rather than closed at the aperture with a dielectric window, two benefits result: First, an abrupt change in relative dielectric constant at the aperture is avoided. This prevents a large reflection which would be difficult to "tune out" over an octave band. Second, when water comes in contact with the excitation probe (coax center conductor), the conduction currents yield a more diffusely located distribution of current (which would have otherwise been constrained to flow only in the probe and waveguide walls (within the radiator. In this way, the bandwidth of the transition between the coax and the waveguide is increased. The VSWR vs. frequency plot for this element is shown in Fig. (9).

A practical consideration of operational importance is that air bubbles must not be allowed to exist within the waveguide since these would result in dielectric discontinuities causing high VSWR's and phase changes. Hence, a small hole has been placed in the back shorting plate for air bubble removal. This hole is "cut-off" at the frequencies of interest and therefore does not result in r.f. leakage or detuning.

4.   SYSTEM DESCRIPTION

The system is designed to allow interferences concerning the spatial distribution of the internal dielectric and conducting properties of biological targets in a noninvasive manner. The spatial distribution of complex permittivity can be related

to the perturbation of electromagnetic waves after transmission through the biological target by means of profile inversion as described in companion papers by Boener and by Slaney et al. elsewhere in this volume. The details involved are beyond the scope of this paper; however, it suffices to state that an accurate sampling of the perturbed electromagnetic wave distribution with and without the target at all points in a cylindrical interrogation volume in the "shadow" or exit point of the target will provide the necessary information for the reconstruction to be applied. The measurement must be performed with three-dimensional resolution, in such a way that the measurement system does not alter the wave to be sampled.

A previously developed system [7] utilizing a single transmitting antenna element and a single receiving antenna element provided sampling with two dimensional resolution by means of mechanically scanning the transmit and receive element pair over a plane "filling" a 64 × 64 array (4096) of measurement positions. This system, although slow (measurement time ~4.5 hours), has produced excellent microwave images as shown in companion papers by Larsen and Jacobi and by Jacobi and Larsen. Hence, the present system is a second generation design with the added goals of three-dimensional resolution and increased speed.

Figure (10) depicts the general system block diagram. An H.P. 1000-45 computer system (which includes a 50 Mega-Byte disk drive and magnetic tape drive along with a printer, keyboard, and CRT) controls other system components by means of the H.P. Multiprogrammer which utilizes the HP-IB Interface Bus. An S-Band signal from a frequency synthesizer is amplified by a TWTA to a 20-watt level. This 20-watt signal then enters a transmit array control unit which distributes power to the transmit array for illumination of the target over its cross section uniformly or in a series of time-sequenced "spots." The details of the transmit array design and operation will be presented in Section VI. The electromagnetic wave is then transmitted through the target to the receive phased array which samples the wave distribution in a three-dimensional grid of points within a cylin-

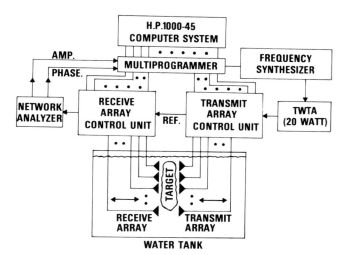

Fig. 10.   System block diagram.

drical volume in the "shadow" of the target. The complex field amplitudes at each of the samples are then computer processed, using an algorithm which will be described in Section V, to generate a three-dimensionally focused spot which can be "scanned" throughout the sample volume. The resultant data is then stored on magnetic tape for use as an input to an off-line image processing system.

5. RECEIVE PHASED ARRAY DESIGN

Based on the system performance requirements discussed in Section 4, the specifications invoked on the receive phased array design are as follows: a) Perform 3-D focused sampling of the transmitted electromagnetic wave; b) provide a sample volume of ca. 6 mm in each direction; c) scan the sample volume in the "shadow" of the target over a cylinder 10 cm length, and 15 cm dia; d) provide vertical polarization; e) suppress unwanted polarization components (horizontal and axial) ≥ 20 dB; f) operate over a frequency range of 2–4 GHz; g) suppress spurious spatial response ≥ 20 dB.

The last specification requires that when the array is focused on a particular point, its output must be basically caused by the presence of energy only at the desired focus point. It must not respond to energy coming from any other point. Phased arrays which scan in any direction off boresight (i.e., two-dimensionally focused at infinity) are subject to the possibility of being simultaneously scanned to other directions. These unwanted, extraneous beams (or lobes) other than the main beam are referred to as grating lobes. Grating lobes will occur for planar arrays whose interelement spacing is too large for the angle of scan desired. The design of a three-dimensionally focused array must suppress these effects which are identified as "spurious spatial responses." It will be shown that these unwanted responses will occur more easily in concave surfaced arrays than in planar surface arrays.

The equations governing the existence (or nonexistence) of grating lobes are summarized in the following. Figure 11(a) depicts an array geometry consisting of N equally spaced isotropic radiators on a line. For simplicity, we will assume lossless media for demonstration purposes. The desired direction of the main beam is denoted by θ_m. If each radiator is excited with an amplitude and phase of $a_i e^{j\rho_i}$ than far from the array the radiation pattern ($|\overline{E}(\theta)|$) is given by

$$|\overline{E}(\theta)| = |a_1 e^{j\psi_1'} + a_2 e^{j\psi_2'} + a_3 e^{j\psi_3'} + \ldots + a_n e^{j\psi_N'}| \quad (17)$$

Where:

$$\left. \begin{array}{l} \psi_1' = \rho_1 \\ \psi_2' = \dfrac{2\pi d}{\lambda} \sin\theta + \rho_2 \\ \psi_3' = \dfrac{4\pi d}{\lambda} \sin\theta + \rho_3 \\ \vdots \\ \psi_N' = \dfrac{(N-1)2\pi d}{\lambda} \sin\theta + \rho_N \end{array} \right\} \quad (18)$$

if

$$\left. \begin{array}{l} \rho_1 = 0 \\ \rho_2 = \rho \\ \rho_3 = 2\rho \\ \vdots \\ \rho_N = (N-1)\rho \end{array} \right\} \quad (19)$$

and

$$a_1 = a_2 = a_3 = \ldots = a_N = a \quad (20)$$

then

$$\left. \begin{array}{l} \psi_1' = 0 \\ \psi_2' = \dfrac{2\pi d}{\lambda} \sin\theta + \rho = \psi' \\ \psi_3' = 2\psi' \\ \vdots \\ \psi_N' = (N-1)\psi' \end{array} \right\} \quad (21)$$

and

$$|\overline{E}(\theta)| = a|1 + e^{j\psi'} + e^{j2\psi'} + \ldots + e^{j(N-1)\psi'}|. \quad (22)$$

Now, in order to scan the beam in the θ_m direction, all the complex terms in the above equation must add coherently. Hence, $\psi = 0$ at $\theta = \theta_m$ (main beam) or

$$0 = \frac{2\pi d}{\lambda} \sin\theta_m + \rho \quad (23)$$

yielding

$$\rho = \frac{-2\pi d}{\lambda} \sin\theta_m. \quad (24)$$

This is the value of ρ, the phase difference between elements, to produce a main beam at θ_m, the desired scan angle. Using this value of ρ, we find, however, that

$$\psi' = \frac{2\pi d}{\lambda} \sin\theta - \frac{2\pi d}{\lambda} \sin\theta_m \quad (25)$$

can equal zero or a multiple of 2π at some other angles (θ_{Gn}) given by

$$\theta_{Gn} = \arcsin\left\{\frac{\lambda}{2\pi d}\left[\frac{2\pi d}{\lambda}\sin\theta_m \pm 2\pi n\right]\right\} \quad (26)$$

or

$$\theta_{Gn} = \arcsin\left\{\sin\theta_m \pm \frac{\lambda n}{d}\right\} \quad (27)$$

Clearly, this equation for the direction of the nth grating lobe (θ_{Gn}) yields imaginary values when

$$\left|\pm\frac{n\lambda}{d} + \sin\theta_m\right| > 1 \quad (28)$$

The first grating lobe (n = 1) will first appear at end fire ($-90°$) when d/λ is increased so as to cause

$$-\frac{\lambda}{d} + \sin\theta_m = -1 \quad (29)$$

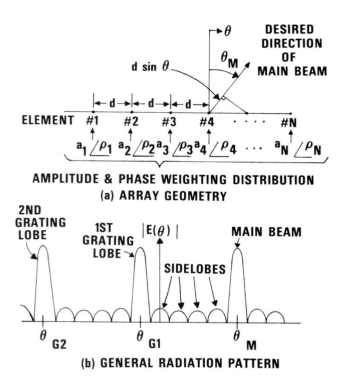

AMPLITUDE & PHASE WEIGHTING DISTRIBUTION
(a) ARRAY GEOMETRY

(b) GENERAL RADIATION PATTERN

Fig. 11. Grating lobe effects in phased arrays.

or

$$d = \frac{\lambda}{1 + \sin\theta_m} \qquad (30)$$

For example, at $\theta_m = 60°$ the spacing (d) required for the first grating lobe to enter real space or the so-called visible region is d = 0.5359λ. This is the spacing for the peak of the first grating lobe to occur; however, if d is slightly less than 0.5359λ the grating lobe will still be partially present. Figure 11(b) shows the general radiation pattern of an N element array. If d $\ll$ λ/1 + $\sin\theta_m$ then only the main beam will be present. As d is increased the first grating lobe will start to appear at −90°. When d = λ/1 + $\sin\theta_m$ it will be equal in amplitude with the main beam but still pointing at −90°. Further increase in d will cause the first grating lobe to "slide-in" toward the main beam while the main beam direction remains invariant. Even further increase in d will cause the second grating lobe to appear, etc.

Before discussing the design of the array configuration itself in terms of its shape, inter-element spacing and element tilt angles, consideration must be given to the array feed network which will combine the signals received by each individual array element in such a way as to produce the desired three dimensionally focused response at its output. Figure 12 depicts alternative feed network focusing techniques. Alternative (a) employs focusing at r.f. This requires an octave band r.f. phase shifter for every element in the array as well as an r.f. combining network capable of coherently combining a number of signals equal to the number of array elements. The operation basically requires a calcula-

tion of the propagation delays incurred between the particular focus point and each of the array elements. The phase shifters must then be controlled to compensate for these propagation delays and thus cause all signals entering the combiner to be in phase. This would be a prohibitively expensive network to realize. Furthermore, it will be shown later in this section that to obtain the desired performance, the coupling coefficients in each arm of this r.f. combiner would have to be electronically controllable, further increasing the cost. In addition, since the phase shift values and coupling coefficients in the combiner network must be different for each focus point, sampling over the entire interrogation volume will require large amounts of real-time computation that, in turn, result in long measurement times. Technique (b) is basically the same as (a) with the exception that the r.f. signals are first down-converted to a convenient i.f. frequency and then the focusing is accomplished at i.f. This was conceived as a possible means of cost reduction; but the cost would still be excessive. The last technique, (c), is that of "software" focusing. This achieves a focused response in the following manner: The signals received on all array elements are sequentially routed into a dual channel receiver (a reference signal is supplied from the transmitter section) by means of an r.f. switch matrix. The analog video amplitude and phase outputs are then passed through A/D converters and stored in computer memory. The target may be removed at this point in the procedure. Then, for each focus

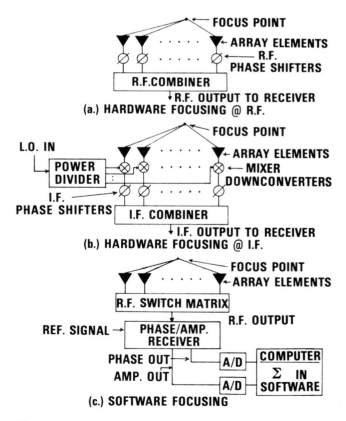

Fig. 12. Alternative focusing techniques.

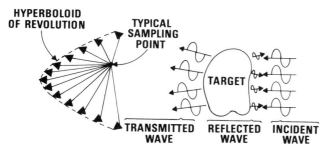

Fig. 13.   Concave hyperboloidal phased array geometry.

point, the same propagation delays mentioned above are computed. This performs the phase compensation and desired amplitude weighting (coupling coefficient control) in software. The details of this algorithm will be presented later in this section. This technique is much less expensive, faster, and offers freedom in selection of the amplitude weighting distribution which will be shown to be a key requirement. In addition, the use of measurements with and without the target present facilitates calibration and correction of errors caused by hardware imperfections and changes in the propagation parameters of the water which may result from temperature changes and/or impurities. The details of the calibration and propagation parameter evaluation techniques are discussed later in this section.

Now that the software focusing technique for processing the array element signals has been established, the design procedure must concentrate on the array configuration geometry. Since, not only vertical and horizontal resolution but also axial or range resolution is required, an untuitive candidate array configuration is that of a concave surface distribution of elements as shown in Fig. 13. The particular concave surface shown is a hyperboloid of revolution. The projection of the surface of this array onto a plane perpendicular to the axis of revolution yields a square lattice distribution of elements contained within a circle. At the relative position of the particular sampling point shown, it would appear that phase compensation of the relative propagation delays to each of the elements should yield a response with three-dimensional resolution. A phased array analysis program was written for this configuration utilizing the mathematical model for the individual array element along with the transformation equations for translation and element tilting developed in Section 3. One of the representative typical hyperboloidal concave arrays analyzed consisted of 113 elements with a 1.5cm square lattice distribution in an 18cm diameter, 12.7 cm in axial depth. An amplitude weighting distribution equal to the *inverse* of the corresponding signal strengths was used so as to compensate for relative propagation losses to each element. The amplitude weighting distribution used in software focusing corresponds to the coupling factors between the input ports and output port of a power combiner network which would be utilized in "real-time" hardware focusing. This yielded the best performance of the various amplitude weighting distributions considered. The response of this array at 3 GHz for a

typical focus point 5 cm from the array vertex is shown in Fig. 14(a,b,c) in the vertical (y), horizontal (z), and axial (x) directions respectively. The solid curves represent the vertically polarized (desired) response and the dashed curves represent the axially polarized (unwanted) response. This computer modeling predicts satisfactory vertical and horizontal resolution but also several undesirable performance characteristics, they are (in order of severity):

1. Horizontal Spurious Spatial Response Suppression: 0.8 dB (located ±3.2 cm from focus point).
2. Vertical Spurious Spatial Response Suppression: 3.0 dB (located ±3.2 cm from focus point).
3. Axially polarized Response Suppression: 4.9 dB
4. Axial 10 dB width: 5.0 cm; 3 dB width: 1.8 cm; 1 dB width: 0.8 cm.
5. Vertical 10 dB width: 1.9 cm; 3 dB width: 1.1 cm; 1 dB width: 0.7 cm.
6. Horizontal 10 dB width: 1.5 cm; 3 dB width: 1.0 cm; 1 dB width: 0.6 cm.

Clearly, this is an unacceptable design. The spurious spatial response locations can be further removed from the desired focus point by reducing inter-element spacing. However, the reduction required would cause overlapping which is physically unrealizable. The axially polarized re-

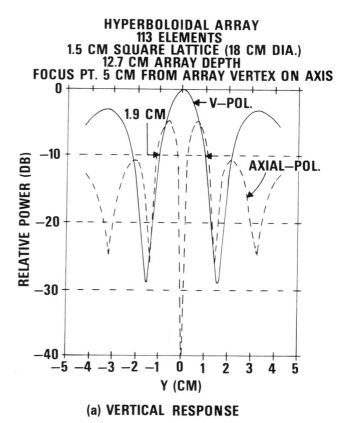

(a) VERTICAL RESPONSE

Fig. 14.   Three-dimensionally focused response of concave hyperboloidal phased array (computer generated).

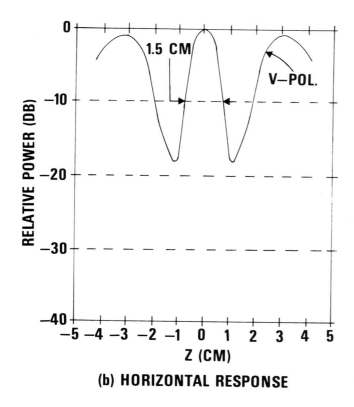

(b) HORIZONTAL RESPONSE

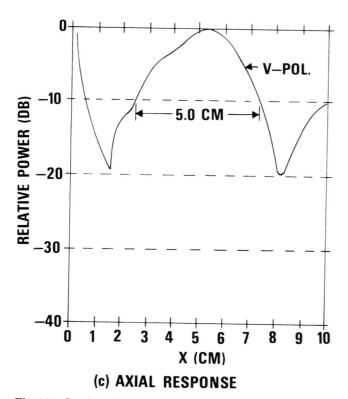

(c) AXIAL RESPONSE

Fig. 14. *Continued*

sponse which results from steep vertical ray angles, can be reduced to an acceptable level by flattening the array (i.e., reducing its axial depth) and locating the array further from the focus point. However, this would further degrade the already poor axial resolution and, thereby, only facilitate two-dimensional resolution.

These shortcomings are overcome by the use of a Synthesized Volumetric Phased Array, shown in Fig. 15, made possible by software focusing. The array is actually a planar array that is successively repositioned to various axial locations. Thus, a volumetric array is synthesized. The effective number of elements is equal to the number of elements in the real planar array times the number of axial locations. Clearly, such an array could never operate instantaneously in real time since the elements in the rear axial locations could not "see-through" the forward elements. Axial movement of a single aperture to synthesize "end-fire" gain has been reported in the literature [8], [9]. However, this does not yield axial focusing, but only improves resolution transverse to the axial direction. In order to focus at a point (not at infinity) in the axial direction, off-axis elements are also required.

The previously described phased array analysis computer program was modified to facilitate analysis of the synthesized volumetric phased array. The particular configuration analyzed consisted of 127 real elements on a plane distributed in a hexogonal lattice with 1.5cm inter-element spacing located within a 18cm circular diameter. This planar array was successively located at 5 axial positions with 1cm separation, thus synthesizing a 635 element volumetric array. The amplitude weighting distribution utilized was inverse (in that the weighting factor compensated for the propagation loss from the focus point to each element) in the axial and horizontal directions within the array but uniform in the vertical direction (for axial polarization suppression). The performance of this array at 3 GHz for a typical focus point 5 cm in front of the front-most array position on the axis of the cylindrical interrogation volume is shown in Figs. 16(a,b,c) in the vertical (y), horizontal (z), and axial (x) directions. The performance of this array is clearly far better than that of the concave hyperboloidal array. Examination of the responses yields the following performance characteristics shown below.

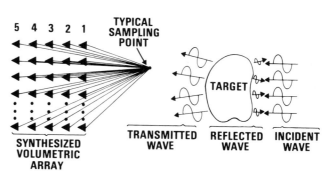

Fig. 15. Synthesized volumetric phased array geometry.

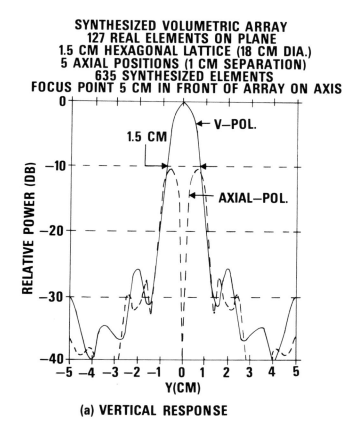

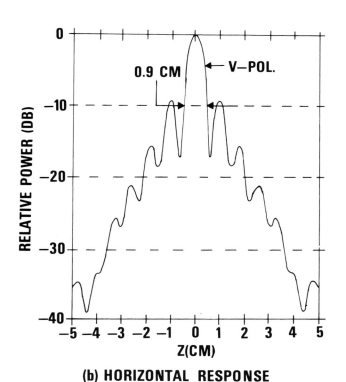

Fig. 16. Three-dimensionally focused response of synthesized volumetric phased array (computer generated).

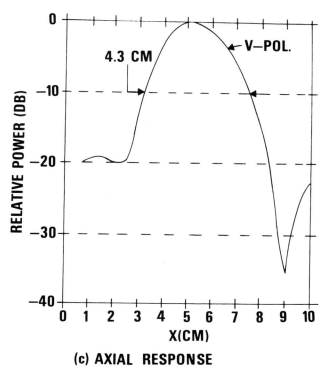

(c) AXIAL RESPONSE

Fig. 16. *Continued*

1. Vertical 10 dB width: 1.5 cm; 3 dB width: 0.9 cm; 1 dB width: 0.6 cm.
2. Horizontal 10 dB width: 0.9 cm; 3 dB width: 0.6 cm; 1 dB width: 0.35 cm.
3. Axial 10 dB width: 4.3 cm; 3 dB width: 2 cm; 1 dB width: 1.2 cm.
4. Spurious Spatial Response Suppression: 20 dB typ., 16 dB min.
5. Cross and Axial Polarization Suppression: 20 dB typ., 10 dB min.

This performance exceeds design goals for vertical and horizontal resolution and meets all other goals with the one exception of axial resolution. However, the axial resolution obtained is judged to be adequate for this application. This can be further improved by using a larger diameter array, however this would significantly impact cost by increasing the number of elements. Furthermore, it will be shown in Section 7 that the sensitivity of the receiver utilized in this system limits the array diameter because the outermost elements would receive signals below the noise floor, especially in the rear-most axial positions. Subsequent system improvement will require both the implementation of a more sensitive receiver and a larger diameter array.

The length of the cylindrical interrogation volume is limited only by the number of axial positions at which samples are taken and the dynamic range of the system. The desired 10cm long interrogation volume requires 15 axial positions, at 1cm intervals. This is true since 5 axial positions are required for focusing at points within a single plane.

Fig. 17. Underwater receive phased array (left side of photo).

A photograph of the receive array is shown in Fig. 17. The receive array is the left most array shown. The array to the right is the transmit array (as seen from its rear) which will be described in Section 6. The 127 semi-rigid cable runs are shown to be dressed vertically to a bulkhead plate where additional interface semi-rigid cabling is employed for connection to the receive array control unit shown in Fig. 18. The control unit contains a series of electromechanical r.f. switches which route each of the element signals, in turn, to the input of a low noise amplifier (3.5dB noise figure, 30dB gain) prior to the "Test In" port of the harmonic converter of the network analyzer. In the case of those elements of the array where greater loss is encountered, another identical amplifier is added in cascade by means of two signal pole double throw r.f. switches. In this way signals are maintained within the operating range of the harmonic converter (−10 to −78 dBm). On the basis of the 3.5dB noise figure of the amplifier, amplifier gain of 30 dB, the noise figure of the network analyzer (56 dB) and its effective i.f. bandwidth (10 KHz), the sensitivity of this receiver is calculated to be −123 dBm. Without the low noise pre-amplifier, the sensitivity is −78 dBm. This −123dBm sensitivity has been experimentally verified. The r.f. switch matrix is computer controlled by means of relay cards within the multiprogrammer. The instantaneous switch positions are visually displayed on the front panel of the control unit shown in Fig. 18. In the case shown in Fig. 18, the 93rd array element is routed into the receiver and the second pre-amplifier is not connected at the moment shown. The time to sequentially route the 127 element signals into the receiver and store their amplitude and phase has been experimentally verified to be 8 seconds.

The software focusing equations will now be presented. In the case of the 127 element planar array, an individual element will be identified by the index n. The five required axial positions of the array for focusing on a given plane within the cylindrical interrogation volume are denoted by

the index m. The calibration factors for each of the n real elements are specified by $A_n e^{j\psi_n}$. These factors are the measured relative losses and phase delays encountered by element signals in being routed through the receive array control unit. The signal received by the nth element when the array is in the mth axial position without the target present is denoted by $a_{nm} e^{j\phi_{nm}}$. The signal received with the target present is denoted by $a_{nm}' e^{j\phi'_{nm}}$. The focus point coordinates are specified by (x_k, y_k, z_k). When the focus point coordinates and the coordinates of each array element in the mth axial position are known, the propagation transfer function between the nth element and the kth focus point can be computed using the equations presented in Section 2. This is denoted by $t_{nmk} e^{j\xi_{nmk}}$. The weighting factor distribution is then given by $W_{nmk} e^{-j\xi_{nmk}}$. Note that the phase of the weighting factor is the conjugate of the phase of the propagation transfer function, thus, coherent addition in assured. The amplitude of the weighting factor (W_{nmk}) is given by $1/t_{nmk}$ in the axial and horizontal directions through the volumetric array. This is unity in the vertical direction through the array. Furthermore, W_{nmk} is truncated to zero if $(a_{nm} < a_{nm}^{max}/T)$ or $(a_{nm}' < a_{nm}^{max'}/T)$. T is a variable threshold level. Typical values range from −20 to −40 dB. This insures that only signals with sufficient signal-to-noise ratios are combined. The focused response without a target (for reference) is

$$M_k e^{j\eta_k} = \sum_{m=m_1}^{m_1+4} \sum_{n=1}^{127} W_{nmk} e^{-j\xi_{nmk}} \frac{a_{nm}}{A_n} e^{j(\phi_{nm}-\psi_n)} \quad (31)$$

The focused response with a target is

$$M_k' e^{j\eta'_k} = \sum_{m=m_1}^{m_1+4} \sum_{n=1}^{127} W_{nmk} e^{-j\xi_{nmk}} \frac{a_{nm}'}{A_n} e^{j(\phi_{nm}'-\psi_n)} \quad (32)$$

The differential focused response which represents the perturbation of the electromagnetic wave distribution by the target and hence contains the information necessary to collect the profiles needed for inversion and reconstruction

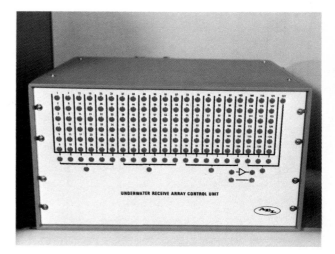

Fig. 18. Underwater receive array control unit.

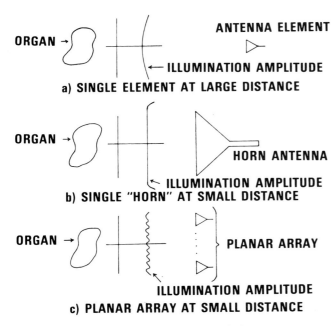

Fig. 19. Alternative target illumination techniques.

is

$$\Delta M_k e^{j\Delta\eta k} = M_k' e^{j\eta k'} - M_k e^{i\eta k} \qquad (33)$$

6. TRANSMIT ARRAY

The function of the transmitting antenna system is to provide a uniform distribution of S-Band energy over selectable target cross-sections ranging of up to 15 cm in diameter. The required polarization of the system is vertical. Maximum "ripple" in the illumination distribution is to be ±1dB. Cross polarized and axially polarized components are to be suppressed greater than 20 dB relative to the vertically polarized component. The minimum separation distance from the target is 2.54 cm (closer separation would perturb the near fields of the radiator thus possibly detuning it).

Several alternative target illumination techniques were considered in the design process. These are depicted in Fig. 19 and include: a) a single waveguide radiator at a large separation distance, b) a large aperture horn radiator at a small separation distance, and c) a planar array of waveguide radiators at a small separation distance.

The use of a single waveguide radiator is the simplest solution; however, in order to illuminate the target area uniformly, the radiator must be placed at a large distance. Alternatively, a closer single element could be used; but mechanical scanning would then be required for each focal point of the receive array and data acquisition time would suffer. Larger separation to a single element is of little consequence in air; but, in the medium of water, separation distance from the target is critical insofar as the increased loss places a formidable burden on the detection capability of the system. This is illustrated in Fig. 20 wherein a waveguide radiator

placed 2.54 cm from the target provides illumination that decreased 3 dB from the peak at a radial distance from target (illumination) center (in the cross sectional plane) of 1.2 cm at 3 GHz. A radiator placed 25 cm from the target decreases 3 dB from the peak at a radial distance of 6.25 cm. However, the additional loss incurred in the latter case is approximately 80 dB, which cannot be tolerated. Thus, the single waveguide radiator at either large or small range is not a viable alternative for this application.

The second alternative considered, a large aperture horn fixed at a small separation distance, has the ability to generate a uniform illumination over the E-plane direction of its aperture and a cosine illumination over its H-plane provided it is *not* filled with water which would taper the illumination due to its loss. In order to match the horn to the medium of water, it would be necessary to implement multiple-step impedance transformers within the horn so as to match it over the octave bandwidth. Although this may be possible, the resultant design would *not* have the capability of variable illumination width which will be shown to be advantageous.

The alternative selected, a planar array of waveguide radiators fixed at a small separation distance, not only has the ability to uniformly illuminate the target, but it also has the advantage of selectable illumination for small target cross-sections. This is accomplished by the use of only a few elements of the array in a tight hexagonal lattice which is sequentially switched over the transmit array area. This is beneficial to the extent that refractive errors are minimized. Furthermore, the scanned transmit/receive pair offers improved spatial resolution over either scanned receiver or scanned transmitter systems. That is, in holographic systems, the maximum spatial frequency for scanned transmit/receive systems is twice as high as that when only one member is scanned. Thus, the target is sequentially illuminated over small areas of the organ. The responses of the system to each of the spot illuminations is then summed. The

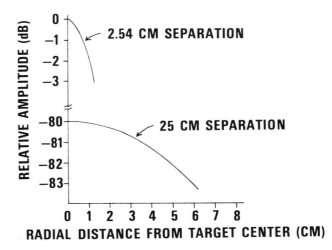

Fig. 20. Single element illumination at 2.5 and 25.0 cm separation (computer generated).

Fig. 21. Transmit array.

spot illumination mode of operation will be discussed in more detail later in this chapter.

The element chosen for use in the transmit array was the same WRD 180-C24 as used in the receive array. The inter-element spacing is minimized to provide the smoothest possible illumination. The performance of several types of lattice distributions were simulated by a computer program. These included rectangular, circular, and hexagonal structures. Both the rectangular and circular arrangements were found to have severe pattern anomalies and illumination ripple. Thus, these lattice structures were not chosen. The hexagonal lattice structure proved to be far superior in performance, with extremely low ripple over the target cross-section and adequate axial polarization suppression. There is some pattern roll-off around the outer circumference of the target area; however, this may be compensated for by making the cross-sectional area of the array larger than that of the target. Thus, the roll-off region is moved

outside of the required cross-section when instantaneous rather than sequential illumination is desired. The transmit array is comprised of a hexagonal lattice structure of 151 elements packed as densely as possible. The center-to-center element spacing is 1.15 cm. The array elements are shown mounted in the supporting plate in Fig. 21. As mentioned previously, the array has the capability of operating in both an instantaneous illumination mode, where all elements are excited simultaneously, or a sequential spot illumination mode, where small clusters of elements are excited sequentially to minimize refractive error. This is effected through computer control of r.f. switches by means of relay cards within the Multiprogrammer. The r.f. switches are located at each of the outputs of the 151-way power divider network within the transmit array control unit shown in Fig. 22. Each of the power divider outputs is either routed to its associated array element or into an internal termination. The r.f. semi-rigid coaxial cables between the transmit array control unit and the array elements must be phase-matched to ±10° since the phase coherency of the elements is critical for low ripple performance. Phase errors less than ±10° will cause

(a.) TRANSMIT ARRAY-INSTANTANEOUS ILLUMINATION MODE
SEPARATION FROM TARGET = 2.54 CM   FREQUENCY = 3 GHz

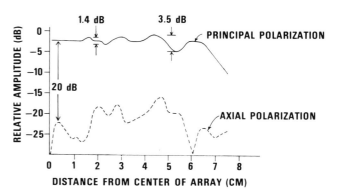

(b.) TRANSMIT ARRAY-SEQUENTIAL ILLUMINATION MODE (7 ELEMENTS)
SEPARATION FROM TARGET = 2.54 CM
FREQUENCY = 3 GHz

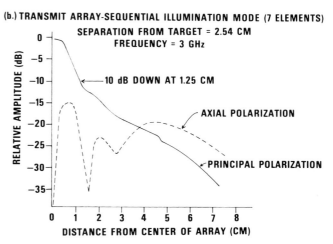

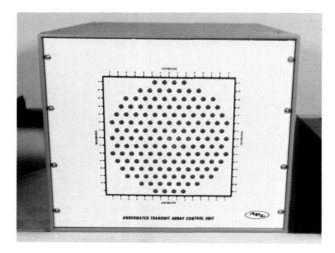

Fig. 22. Transmit array control unit.

Fig. 23. Typical instantaneous and spot illumination modes of transmit array (computer generated).

only small deviations (~1 dB) from the predicted illumination. However, these small deviations may be compensated by differential measurements with and without the target present. The control unit front panel indicator lights show which elements are excited. This is illustrated in Fig. 22 where a typical spot illumination comprising of a 7 element hexagonal cluster is being excited. The seven elements represented a hexagon and one central element.

Computer predicted performance of the array in both the instantaneous mode and a typical spot illumination mode is shown in Figs. 23(a,b). The instantaneous illumination pattern at 3 GHz has a typical ripple of 1.5 dB peak-to-peak. Maximum ripple of 3.5 dB is the edge of the target area. Steep pattern roll-off occurs at the outer edge of the cross-sectional area. Typical axial polarization response is 20 dB below the principal vertical polarization. In the case of sequential spot-illumination by a series of 7 element clusters the 10 dB width is 2.5 cm with rapid pattern roll-off after that point. In order to sequentially illuminate the maximum target area, 109 spot illuminations are required. The number of spot illuminations required to cover a specific target area is initiated in the system software by operator specification of the target diameter.

7. MEASUREMENT PROCEDURE

Before an actual measurement can be correctly executed, two procedures must be carried out. They are: (a) the calibration of the receive phased array its associated control unit and (b) the measurement of the propagation parameters of the water. These must be accomplished at closely spaced frequency intervals in order to accomodate small deviations of the actual frequency of operation from the exact frequencies of array calibration and coupling medium constitutive parameter measurements. In this way, no significant errors will be introduced by data interpolation between frequencies.

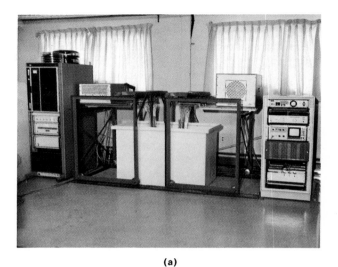

(a)

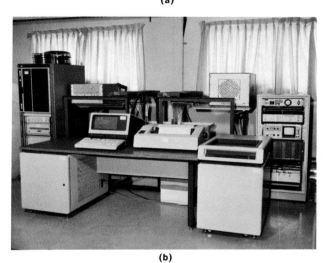

(b)

Fig. 25. Overall system implementation.

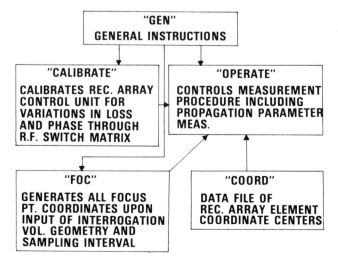

Fig. 24. System computer program architecture for biological target characterization.

The receive array and its associated control unit calibration is necessary because of the loss and phase delay differences between the aperture of each array element and the receiver input (r.f. switch matrix output). These differences are caused by slight nonuniformity in fabrication and tuning of each array element, cable length variations, and r.f. switch differences within the control unit. In actual system operation, the measured complex field amplitudes are divided by the calibration amplitudes and phases are reduced by the calibration phases; thus, array/receiver interface errors are compensated.

The second preparatory procedure required is propagation parameter measurement. Once again the central element of the transmit array and the central element of the receive array are placed in geometrical alignment. The received signal is measured and stored at several axial distances. Given the differential axial displacement, the relative amplitude ratio and the phase difference of the two measurements, it is straight forward to calculate the propagation parameters α & β at each measurement frequency by use of the propagation equations presented in Section 2.

The system software architecture for biological target characterization is presented in Fig. 24. At each interrogation frequency and at each transmit illumination mode, the receive array is positioned to fifteen axial locations. At each position, each of the array element signals are measured and stored. This must be done first without a target for reference and then with each successive target to be interrogated. Each interrogation condition (combination of frequency and transmit illumination mode) will require approximately 3.5 minutes in order to collect the data necessary for synthetic scanning of the entire cylindrical interrogation volume. After this data is collected, the target can be removed and the focused responses can then be computed off-line using the software focusing algorithm described in Section 5. Any number of sampling points within the interrogation volume may be specified (within computer memory limits). The number of samples will not affect data collection time (as it would have if hardware focusing were employed), but only data processing time.

Figure 25(a,b) is a photograph of the complete system. The transmit and receive arrays along with their associated control units are temporarily mounted on rolling frames which facilitate relative axial movement between the arrays. These frames are a temporary feature of the development. The arrays along with their control units will be remounted onto the computer controlled mechanical scanner illustrated in Chapter X.

8. SYSTEM PERFORMANCE SUMMARY

In summary, the system presented herein can be described by the following performance characteristics; 1) Cylindrical interrogation volume, 10 cm in length and 15 cm in diameter; 2) frequency range, 2–4 GHz; 3) illumination modes, instantaneous and sequential "Pencil-Beams"; 4) illumination ripple, 1.4 dB typical, 3.5 dB maximum; 5) polarization vertical; 6) cross and axial polarization suppression, 20 dB typical, 10 dB minimum at 3 GHz; 7) spurious spatial response suppression; 20 dB typical, 16 dB minimum at 3 GHz; 8) predicted 3-D Focusing Resolution,

  a) 7.2 mm at 3 GHz and 5.6 mm at 4 GHz in the vertical direction,

  b) 5.7 mm at 3 GHz and 4.3 mm and at 4 GHz in the horizontal direction,

  3) 16 mm at 3 GHz and 12.8 mm in the axial direction

The data collection time can be as little as one minute if a single frequency and single illumination condition is utilized when interrogation is only required over a plane. If the entire cylindrical interrogation volume is to be interrogated, then the data collection time increases to 3.5 minutes. This speed may be further reduced by the use of continuous rather than incremental scanning in range and by the use of faster RF switches such as PIN diode switches.

9. ACKNOWLEDGMENT

The development of this system was performed at American Electronic Laboratories, Lansdale, Pa., under the sponsorship of the U.S. Army Medical Research and Development Command (contract No. DAMD 17-78-C-8060).

APPENDIX A

Mathematical Solution for the Near and Far Fields of a Double Ridged Waveguide Aperture Radiator in a Conducting Medium

A solution is desired for the electric field at an arbitrary observation point in the semi-infinite region in front of the radiating aperture. We shall be interested in the amplitude, phase and direction of the electric field. The solution which follows utilizes the aperture field technique and hence is subject to only two approximations. First we assume that the field in the double ridged waveguide aperture is known; and second we assume that the field in the remainder of the plane containing the aperture is zero. This second approximation forces the electric field to zero at $\pm 90°$ from the axis normal to the aperture and although this is not exactly valid the field is negligible there. These are the only assumptions made and the derivation which relates the radiated field to the assumed aperture field in the plane is mathematically rigorous and uses no far field approximations. The true test as to whether these two approximations lead to negligible error is to experimentally verify the predicted results.

Referring to Fig. 4 the aperture field is approximately given by [4]:

$$\overline{E}_a(y,z) = \begin{cases} -\hat{y}\, E_0 \cos\left(\dfrac{2\pi z}{\lambda_c'}\right) > |z| \le \dfrac{a'}{2} \\[2em] -\hat{y}\, \dfrac{b'}{b}\, E_0\, \dfrac{\cos\left(\dfrac{\pi a'}{\lambda_c'}\right)}{\sin\left(\dfrac{\pi(a-a')}{\lambda_c'}\right]\right)}\, \sin\left[\dfrac{2\pi}{\lambda_c'}\left(\dfrac{a}{2}-|z|\right)\right],\ \dfrac{a'}{2} \le |z| \le \dfrac{a}{2} \\[2em] 0,\ |y| > \dfrac{b'}{2}\ \text{and}\ |z| \ge \dfrac{a'}{2} \\[1.5em] 0,\ |y| > \dfrac{b}{2}\ \text{and}\ |z| \ge \dfrac{a}{2} \end{cases}$$

where:

$\hat{y} \equiv$ unit vector in y-direction
$\lambda_c' = \sqrt{\epsilon_r'}\lambda_c$
ϵ_r' = relative dielectric constant of the medium
λ_c = wavelength in free space at the ridged guide cutoff frequency.

Transforming this aperture electric field into an equivalent magnetic surface current density via Love's equivalence principle [5] we have:

$$\overline{J}_{ms}(y,z) = -2\hat{x}X\overline{E}_a = -2\hat{x}X(-|\overline{E}_a|\hat{y})$$

or

$$\overline{J}_{ms}(y,z) = +2\hat{z}|\overline{E}_a(y,z)|$$

where $\hat{x}$ and $\hat{z}$ are the unit vectors in the x and z directions. Now, the Green's function in the conducting medium [6] (i.e., the potential at point r due to a delta function *magnetic* source current at r') is given by:

$$G_m(\overline{r}|\overline{r}') = \frac{\epsilon}{4\pi}\frac{e^{-\gamma|\overline{r}-\overline{r}'|}}{|\overline{r}-\overline{r}'|}$$

where:

$$\gamma = \alpha + j\beta$$

Note: $G_m(\overline{r}|\overline{r}')$ was derived from the following:

$$\nabla \times \overline{E} = -j\omega\mu_0\overline{H} - \hat{z}\,\delta(\overline{r}-\overline{r}')$$

$$\nabla \times \overline{H} = \gamma\overline{E}$$

$$\nabla \cdot \overline{B} = 0$$

$$\nabla \cdot \overline{D} = 0$$

$$\overline{D} = -\nabla \times \hat{z}G_m(\overline{r}|\overline{r}')$$

where $\delta(\overline{r}-\overline{r}')$ is a delta function from which $G_m(\overline{r}|\overline{r}')$ satisfies:

$$(\nabla^2 - \gamma^2)\,G_m(\overline{r}|r') = -\mu_0\delta\,(\overline{r}-\overline{r}')$$

Then, the electric vector potential is given by:

$$\overline{F} = \iint_s \overline{J}_{ms}\,(\overline{r}')\,G_m\,(\overline{r}|\overline{r}')\,ds'$$

and the electric field is by definition:

$$\overline{E} = -\frac{1}{\epsilon}\nabla \times \overline{F}$$

carrying out the above procedure:

$$|\overline{r}-\overline{r}'| = \sqrt{x^2 + (y-y')^2 + (z-z')^2}$$

let:

$$R\,(x,y,z,y',z') = |\overline{r}-\overline{r}'|$$

then,

$$F_z(x,y,z) = \frac{\epsilon}{2\pi}\Bigg\{\int_{-b/2}^{b/2}\int_{-a/2}^{-a'/2}|\overline{E}_a(y',z')|\,\frac{e^{-\gamma R}}{R}\,dz'dy'$$

$$+ \int_{-b'/2}^{b'/2}\int_{-a'/2}^{a'/2}|\overline{E}_a(y',z')|\,\frac{e^{-\gamma R}}{R}\,dz'dy'$$

$$+ \int_{-b/2}^{b/2}\int_{a'/2}^{a/2}|\overline{E}_a(y',z')|\,\frac{e^{-\gamma R}}{R}\,dz'dy'\Bigg\}$$

now,

$$\nabla \times F_z(x,y,z)\hat{z} = \begin{vmatrix} \hat{x} & \hat{y} & \hat{z} \\ \dfrac{\partial}{\partial x} & \dfrac{\partial}{\partial y} & \dfrac{\partial}{\partial z} \\ 0 & 0 & F_z(x,y,z) \end{vmatrix}$$

or,

$$\nabla \times F_z(x,y,z)\,\hat{z} = \hat{x}\frac{\partial F_z(x,y,z)}{\partial y} - \hat{y}\frac{\partial F_z(x,y,z)}{\partial x}$$

then:

$$\overline{E}(x,y,z) = E_x(x,y,z)\,\hat{x} + E_y(x,y,z)\,\hat{y}$$

where:

$$E_x(x,y,z) = \frac{1}{2\pi}\Bigg\{\int_{-b/2}^{b/2}\int_{-a/2}^{-a'/2}|\overline{E}_a(y',z')|\,(y-y')\frac{e^{-\gamma R}}{R^2}$$

$$\cdot\left(\gamma+\frac{1}{R}\right)dz'dy' + \int_{-b'/2}^{b'/2}\int_{-a'/2}^{a'/2}|\overline{E}_a(y',z')|\,(y-y')$$

$$\cdot\frac{e^{-\gamma R}}{R^2}\left(\gamma+\frac{1}{R}\right)dz'dy' + \int_{-b/2}^{b/2}\int_{a'/2}^{a/2}|\overline{E}_a(y',z')|$$

$$\cdot(y-y')\frac{e^{-\gamma R}}{R^2}\left(\gamma+\frac{1}{R}\right)dz'dy'\Bigg\}$$

and

$$E_y(x,y,z) = \frac{-1}{2\pi}\Bigg\{\int_{-b/2}^{b/2}\int_{-a/2}^{a'/2}|\overline{E}_a(y',z')|$$

$$\cdot\frac{xe^{-\gamma R}}{R^2}\left(\gamma+\frac{1}{R}\right)dz'dy'$$

$$+\int_{-b'/2}^{b'/2}\int_{-a'/2}^{a'/2}|\overline{E}_a(y',z')|\,\frac{xe^{-\gamma R}}{R^2}\left(\gamma+\frac{1}{R}\right)dz',dy'$$

$$+\int_{-b/2}^{b/2}\int_{-a/2}^{a/2}|\overline{E}_a(y',z')|\,\frac{xe^{-\gamma R}}{R^2}\left(\gamma+\frac{1}{R}\right)dz'y'\Bigg\}$$

Computing real and imaginary components, first for E_x,

$$\text{Re}\{E_x(x,y,z)\} = \frac{1}{2\pi}\Bigg\{\int_{-b/2}^{b/2}\int_{-a/2}^{a'/2}|\overline{E}_a(y',z')|\,(y-y')\frac{e^{-\alpha R}}{R^2}$$

$$\cdot\left[\left(\alpha+\frac{1}{R}\right)\cos(\beta R) + \beta\sin(\beta R)\right]dz'dy'$$

$$+\int_{-b'/2}^{b'/2}\int_{-a'/2}^{a'/2}|\overline{E}_a(y',z')|\,(y-y')\frac{e^{-\alpha R}}{R^2}$$

$$\cdot\left[\left(\alpha+\frac{1}{R}\right)\cos(\beta R) + \beta\sin(\beta R)\right]dz'dy'$$

$$+\int_{-b/2}^{b/2}\int_{a'/2}^{a/2}|\overline{E}_a(y'z')|\,(y-y')\frac{e^{-\alpha R}}{R^2}$$

$$\cdot\left[\left(\alpha+\frac{1}{R}\right)\cos(\beta R) + \beta\sin(\beta R)\right]dz'dy'\Bigg\}$$

$$\text{Im}\{E_x(x,y,z)\} = \frac{1}{2\pi}\left\{ \int_{-b/2}^{b/2}\int_{-a/2}^{-a'/2} |\overline{E}_a(y',z')| \frac{(y-y')e^{-\alpha R}}{R^2} \right.$$

$$\cdot \left[\beta\cos(\beta R) - \left(\alpha + \frac{1}{R}\right)\sin(\beta R) \right] dz'dy'$$

$$+ \int_{-b'/2}^{b'/2}\int_{-a'/2}^{a'/2} |\overline{E}_a(y',z')| \frac{(y-y')e^{-\alpha R}}{R^2}$$

$$\cdot \left[\beta\cos(\beta R) - \left(\alpha + \frac{1}{R}\right)\sin(\beta R) \right] dz'dy'$$

$$+ \int_{-b/2}^{b/2}\int_{a'/2}^{a/2} |\overline{E}_a(y',z')| \frac{(y-y')e^{-\alpha R}}{R^2}$$

$$\left. \cdot \left[\beta\cos(\beta R) - \left(\alpha + \frac{1}{R}\right)\sin(\beta R) \right] dz'dy' \right\}$$

and then for E_y

$$\text{Re}\{E_y(x,y,z)\} = \frac{-1}{2\pi}\left\{ \int_{-b/2}^{b/2}\int_{-a/2}^{-a'/2} |\overline{E}_a(y',z')| \frac{xe^{-\alpha R}}{R^2} \right.$$

$$\cdot \left[\left(\alpha + \frac{1}{R}\right)\cos(\beta R) + \beta\sin(\beta R) \right] dz'dy'$$

$$+ \int_{-b'/2}^{b'/2}\int_{-a'/2}^{a'/2} |\overline{E}_a(y',z')| \frac{xe^{-\alpha R}}{R^2}$$

$$\cdot \left[\left(\alpha + \frac{1}{R}\right)\cos(\beta R) + \beta\sin(\beta R) \right] dz'dy'$$

$$+ \int_{-b/2}^{b/2}\int_{a'/2}^{a/2} |\overline{E}_a(y',z')| \frac{xe^{-\alpha R}}{R^2}$$

$$\left. \cdot \left[\left(\alpha + \frac{1}{R}\right)\cos(\beta R) + \beta\sin(\beta R) \right] dz'dy' \right\}$$

$$\text{Im}\{E_y(x,y,z)\} = \frac{-1}{2\pi}\left\{ \int_{-b/2}^{b/2}\int_{-a/2}^{-a'/2} |\overline{E}_a(y'z')| \frac{xe^{-\alpha R}}{R^2} \right.$$

$$\cdot \left[\beta\cos(\beta R) - \left(\alpha + \frac{1}{R}\right)\sin(\beta R) \right] dz'dy'$$

$$+ \int_{-b'/2}^{b'/2}\int_{-a'/2}^{a'/2} |\overline{E}_a(y',z')| \frac{xe^{-\alpha R}}{R^2}$$

$$\cdot \left[\beta\cos(\beta R) - \left(\alpha + \frac{1}{R}\right)\sin(\beta R) \right] dz'dy'$$

$$+ \int_{-b/2}^{b/2}\int_{a'/2}^{a/2} |\overline{E}_a(y',z')| \frac{xe^{-\alpha R}}{R^2}$$

$$\left. \cdot \left[\beta\cos(\beta R) - \left(\alpha + \frac{1}{R}\right)\sin(\beta R) \right] dz'dy' \right\}$$

Therefore, for E_x

$$|E_x(x,y,z)| = \sqrt{\text{Re}^2\{E_x(x,y,z)\} + \text{Im}^2\{E_x(x,y,z)\}}$$

$$\arg\{E_x(x,y,z)\} = \arctan\left\{ \frac{\text{Im}\{E_x(x,y,z)\}}{\text{Re}\{E_x(x,y,z)\}} \right\}$$

and for E_y

$$|E_y(x,y,z)| = \sqrt{\text{Re}^2\{E_y(x,y,z)\} + \text{Im}^2\{E_y(x,y,z)\}}$$

$$\arg\{E_y(x,y,z) = \arctan\left\{ \frac{\text{Im}\{E_y(x,y,z)\}}{\text{Re}\{E_y(x,y,z)\}} \right\}$$

The above solution must be evaluated numerically if no far field approximations are to be made. The double integrals involved are easily evaluated using a Simpson's Rule algorithm.

References

1. J. Jacobi, L. Larsen, and C. Hast, "Water-Immersed Microwave Antennas and Their Application to Microwave Interrogation of Biological Targets," *IEEE Trans. Microwave Theory Tech.*, Vol. 27 (1), Jan., 1979.
2. Yamaura, *IEEE Trans. Microwave Theory Tech.*, Vol. 25 (8), pp. 707–710, 1977.
3. R. K. Moore, Effects of a Surrounding Conducting Medium on Antenna Analysis, *IEEE Trans. Antennas Propag.*, May, 1963.
4. S. Cohn, "Properties of Ridged Waveguide," *Proc. IRE*, Aug., 1947.
5. Collin and Zucker, "Antenna Theory Part I," New York, McGraw Hill, 1969, p. 20.
6. M. B. Kraichman, "Hardbook of Electromagnetic Propagation in Conducting Media," Headquarters Naval Material Command, 1970, pp. 1–6.
7. L. Larsen and J. Jacobi, "Microwave Scattering Parameter Imagery of an Isolated Canine Kidney, *Med. Phys.*, Vol. 6 (5), Sept./Oct. 1979.
8. W. E. Kock, "A Holographic (Synthetic Aperture) Method for Increasing Gain of Ground-to-Air Radars," *Proc. IEEE*, pp. 426–427, March, 1971.
9. W. E. Kock, "Synthetic End-Fire Hologram Radar," *Proc. IEEE*, pp. 1858–1859, Nov., 1970.

Recent Developments in Microwave Medical Imagery—Phase and Amplitude Conjugations and the Inverse Scattering Theorem

Theodore C. Guo,* Wendy W. Guo,* and Lawrence E. Larsen†

A theoretical analysis of the local field of microwave lattice radiation source and an inverse scattering theorem are presented. The results are applied to a water-immersed microwave array system for medical imaging. It is shown that, using a technique of phase and amplitude conjugations, a satisfactory three-dimensional focusing for a target located in the neighborhood of the array may be achieved. The focusing resolutions for transverse and longitudinal directions are approximately $\lambda/2$ and λ, respectively, where λ is the wavelength in the dielectric. By increasing the element spacing of the array, the resolutions can be as good as 5.3 mm and 11.7 mm, respectively, at the operating frequency of 3 GHz. Combining the technique of phase and amplitude conjugations with the inverse scattering theorem, the array will be able to provide a three-dimensional imaging system with satisfactory resolution.

1. INTRODUCTION

The use of imagery methods for microwave dosimetric analysis has many advantages. Chief among these are the non-invasive nature of the data collection and the ease with which imagery as a form of data display can be related to the medical traditions of anatomy and pathology. In fact, a considerable literature is being developed on the subject of medical applications of microwave technology in, e.g., the cardiovascular system as described elsewhere in this volume by Lin [1]–[13]. Nevertheless, medical imagery with microwave radiation faces a number of technical and theoretical problems. Historically, the most prominent technical problem has been the dilemma cast by the contradictory requirements of resolution and propagation loss with respect to the choice of the frequency of operation. When imagery is formed by sampling the scattered fields in air, system operation at frequencies below ca. 10 GHz produces unacceptably low spatial resolution whereas higher frequencies of operation impose excessive propagation loss [14]. Simultaneously, the use of air coupling in system design introduces two additional undesirable features: multipath contamination from propagation paths exterior to the target, including lateral or Beverage wave propagation at the air-dielectric interface [15]; and poor power transfer between the target and the antennas used for data collection [16]. These prob-lems are solved by operation of the antennas in a medium of high permittivity and loss tangent comparable to the biologic target under study. The medium of choice is water since it provides a resolution enhancement by wave length contraction of nearly a factor of 9 at 3 GHz [3] while simultaneously the impedance match is improved and reflection is reduced in comparison to air coupling. Water coupled arrays offer all the advantages of water coupled elements plus the enormous increase in data collection speed made possible by electronic scanning and focusing. Additional discussion is presented in a companion paper by Foti *et al* elsewhere in this volume.

In spite of these operational advantages, water coupling introduces its own complications. The worst of these is the fact that the antennas must be placed in close proximity to the subject to reduce coupling losses. Thus, the antennas must operate within their reactive or local zone. This is especially troublesome in the case of array antennas since Fresnel or Fraunhofer diffraction theory cannot be used to effectively focus or stear the array. We address this problem in terms of a new method for array focusing which accommodates not only local-zone operation with 3-dimensional focus control, but also element-to-element variation in network parameters. This method is based on a phase-amplitude conjugation of fields sampled by the array when illuminated by a half-space omnidirectional radiator.

The major theoretical problem in medical application of microwave imagery is recovery of resolution in the direction of propagation for forward scattering (bistatic) based systems. In medical imagery from radiologic disciplines, this is known as the tomography problem; in microwave and

* Department of Physics, The Catholic University of America, Washington, D.C. 20064.
† Department of Microwave Research, Walter Reed Army Institute of Research, Washington, D.C. 20012.

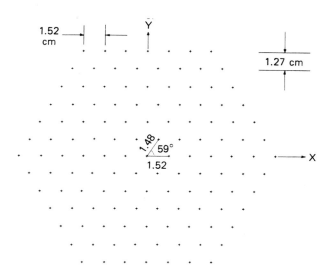

Fig. 1a. Lattice structure of the planar array of 127 elements.

'electromagnetic propagation disciplines, this is the inverse scattering problem. A brief review of the inverse scattering problem and its historical solutions is presented elsewhere in this volume by Boerner and Chan. We express the problem chiefly in terms of axial resolution since usable resolution is obtainable in the transverse plane for isolated organs. Obviously, operation *in vivo* would require solution of the inverse scattering problem in any case. We present a new inverse scattering theorem which may be considered as a generalization of the Lorentz reciprocity theorem to the case of lossy media [19]–[21]. It is applied to forward scattered fields in combination with the phase-amplitude conjugation in a model data collection system which simulates the DART (Dosimetric Analysis by Radiofrequency Tomography) Mark 4 system under development at the Walter Reed Army Institute of Research [2], [3], [6]–[8], [12]–[13].

The next section provides a brief description of the DART system. It is followed by a theoretical analysis of the local field of the receiving array in Section 3. Section 4 introduces the topic of phase-amplitude conjugation and the application to three-dimensional focusing. Section 5 presents the new inverse scattering theorem and its use in combination with the phase-amplitude conjugation method for image reconstruction. The proof of the new inverse scattering theorem is given in Section 6 and concluding remarks appear in Section 7.

2. DESCRIPTION OF THE SYSTEM

The system is composed of two antenna arrays, one for transmission and another for reception, submerged in a cylindrical water container of about 3 feet in diameter and 3 feet high. Both antenna arrays are of hexagonal shape. The elements are placed in a brick-staggered arrangement, corresponding to a planar lattice with one lattice vector at 59° from another and 0.97 times the length (Fig. 1a). The re-

ceiving array is composed of 127 elements whereas the transmitting array contains 151 elements. Each element is a short, water-filled waveguide. The cross section, 4 mm × 7 mm, is that of a degenerate ellipse (see Fig. 1b). The feed structure consists of waveguide-to-coaxial adapter with an insulated end feed which is shorted to the broad wall of the waveguide as shown in Fig. 1c. This element differs from the one described by Foti *et al* (elsewhere in this volume) in that it is more amenable to series production in a monobloc array by numerically controlled milling machines. It is designed for fixed tuning (VSWR $\leq$ 1.5) over a 1GHz band centered on 3GHz.

At an axial distance of 5 cm or farther, the underwater field pattern of each element in the forward direction is similar to that of a dipole. Both transmitting and receiving antennas are mounted in adjustable frames, facing each other for forward scattering imagery. The target is to be placed between the two antennas. The axial distance between the antennas and the target may be adjusted from as close as 5 cm to a distance of about 35 cm. Other engineering details on water coupled antennas for medical microwave imagery may be found in Ref. [3].

In order to compensate for the differences of the distance from each of the elements to the target, a method of phase and amplitude conjugations is used. That is, a factor which includes both phase and amplitude is applied to each element depending on its distance to each focal point [see Section 4, cf. Eq. (25)]. Instead of applying the conjugations to the transmitting array, which would require RF attenuators and phase shifters, the conjugations are applied to the receiving elements in the form of off-line data processing,

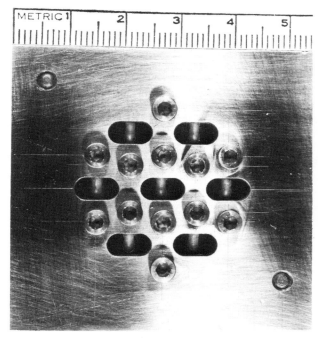

Fig. 1b. Subarray in a stainless steel monobloc.

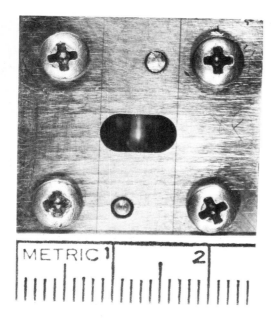

Fig. 1c. Close view of the array element and its feed structure, assembled (left) and disassembled (right).

i.e., by multiplying the received complex field amplitude for each element by a complex factor that corresponds to the conjugation of the phase and amplitude of the scattering parameter S_{21} [17] measured for each element illuminated by an omnidirectional source (this will be described in detail in Section 5). On the other hand, the phasing of the emitting array is designed to produce a near plane-wave. Alternatively, sequentially overlapping subarrays may be energized to provide illumination of selected areas of the target.

In order to describe the application of this technique to a water coupled microwave imaging system, a brief digression into the design of a multiplex receiver is necessary. The receiver consists of 127 open ended waveguide elements with 127 coax lines are routed to 23 6P1T diode switches in a 3-tier reverse corporate power divider network. This network provides the switching to connect each receiver array element to a harmonic converter. Two low noise amplifiers of 20 dB gain are inserted under operator selection prior to the harmonic converter to compensate for path losses through the coupling medium. The local oscillator for the harmonic converter is derived via a directional coupler from a digital synthesizer which serves as the signal source for the transmitting array. The RF port of the harmonic converter is attached to each element of the receiver array via the switch matrix. The IF port returns the down-converted receiver signal to a complex ratiometer which compares another sample of the transmitted signal with the IF signal from the receiver.

The amplitude and phase conjugation, therefore, includes not only element-to-element variations in the array geom-

etry, but also path length and insertion loss variations in the switch matrix enroute to the harmonic converter. The complex ratiometer provides this measurement by estimating S_{21} for each element over the entire RF signal path from the source to the RF port of the harmonic converter.

The source used for the S_{21} measurement should ideally

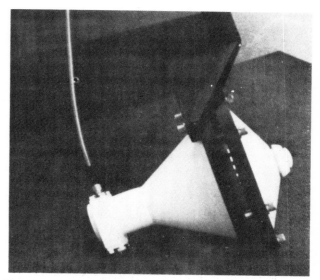

Fig. 1d. A water-immersed lens antenna for receiver calibrations is shown. This derived the measured data for phase and amplitude conjugations to accomplish three-dimensional focusing.

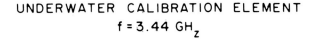

UNDERWATER CALIBRATION ELEMENT
f = 3.44 GH$_z$

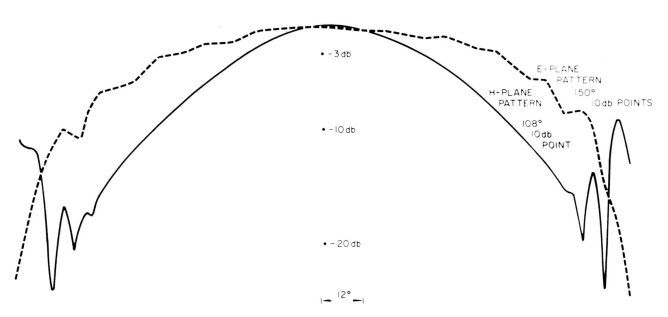

Fig. 1e.   Pattern cuts from the dielectric lens calibration antenna.

consist of a 2π steradian omnidirectional radiator. One realization of such a calibration source is shown in. Fig. 1d. This design is based upon a dielectric lens. It provides a 3-dB beamwidth of 130° in azimuth and in elevation as shown in Fig. 1e. The measured amplitude for each element, inclusive of its path to the harmonic converter, for a given position of the calibration source provides the amplitude taper needed to compensate for path losses through the coupling medium and insertion losses in the switch matrix. That is, the needed amplitude taper is the inverse of the measured amplitude taper for that position of the calibration source. The needed phase taper to insure coherent addition at each position of the calibration source is the conjugate of the measured phase taper. All focal spot positions are provided by translation of the calibration source.

The calibration data set derived as described above is then applied to the S_{21} measurements in the presence of a target for each element for each focal point. In this way, the forward scattered fields are scanned by a sensitive volume element. The receiver array focusing takes place off-line. The array is focused only in the coupling medium, not in the target. The forward scattered fields from the target are differenced from the beam pattern of the illuminator recorded at the same plane in the absence of the target.

3. THE LOCAL FIELD ANALYSIS

In this section, the formula for the electromagnetic field in the neighborhood of a lattice radiation source is derived. The formula will be used to calculate the field pattern of the antenna array and the beam characteristics. The antenna array is treated as a localized distribution of charge and current in a lattice structure. The following three assumptions are made:

Assumption 1: If d_ℓ and d_t represent, respectively, the longitudinal and transverse dimensions of each array element with respect to the direction of the point of observation, and r_n the distance from the nth element to the observation point, it is assumed that, for every element, the following magnitude comparisons are valid:

$$r_n \gg d_\ell \text{ and } d_t, \lambda \gg d_\ell, \lambda \gg d^2{}_t/r_n \qquad (1)$$

where λ is the wave length of the microwave signal in water. Under this assumption, the field due to each radiating element may be approached by the dipole approximation.

Assumption 2: Mutual couplings between the radiating elements are included in the local field formula to the extent that the effect of all coupling is assumed to be identical in every element. In other words, the difference between the peripheral elements and the interior elements with regard to the effect of mutual coupling is assumed to be negligible.

Assumption 3: Mutual coupling between the radiating elements is linear with respect to the phase and amplitude of the power input to the elements. Measured mutual coupling using a Hewlett Packard 8542C Automated Network Analyzer proved this assumption to be true and demonstrated that mutual coupling is less than −30 dB.

It is remarked that Assumption 1 is of a quantitative nature, in the sense that its degree of satisfaction depends on the degree of quantitative precision needed for the field pattern. Although the broad dimension of each array element is about half a wavelength in the coupling medium, it is the actual current distribution that determines the size of the source. Since the dominant mode (TE_{10}) of the electric field in the aperture with respect to boresight angle is known to be sinusoidal, the effective size of the current distribution is shorter than the broad dimension of the guide due to center weighting. Our calibration measurement shows that the field of each individual array element resembles that of a dipole, indicating that this assumption is valid for the system. It will be made clear where this assumption, as well as the other two assumptions, enter into the derivation, so that the percentage error of the derived quantities may be determined.

To derive the local field formula, consider a localized charge density, ρ, and current density, $\vec{J}$, distributed in a space region V. For an antenna in a free space, V indicates the space occupied by the antenna array, as well as its accessories. Without losing any generality, monochromatic time-variation is assumed, so that

$$\rho(\vec{x},t) = \rho(\vec{x})e^{-i\omega t}$$

and

$$\vec{J}(\vec{x},t) = \vec{J}(\vec{x})e^{-i\omega t}. \qquad (2)$$

Accordingly, all other field quantities resulting from ρ and $\vec{J}$ also vary with time monochromatically. Any other time-variation can always be obtained by superposition of monochromatic waves. From Maxwell's equation, the vector potential at any point outside of V is given by, in the Gaussian system of units [18],

$$\vec{A}(\vec{x}) = \int_V \frac{1}{c}\vec{J}(\vec{x}') \frac{\exp(ik|\vec{x} - \vec{x}'|)}{|\vec{x} - \vec{x}'|} \, d\vec{x}', \qquad (3)$$

from which one obtains the magnetic induction $\vec{B}$ and electric field strength $\vec{E}$:

$$\vec{B}(\vec{x}) = \nabla \times \vec{A}(\vec{x}) \qquad (4)$$

and

$$\vec{E}(\vec{x}) = \frac{1}{k} \nabla \times \vec{B}(\vec{x}), \qquad (5)$$

where k is the magnitude of the wave vector in the medium. For water, k is a complex quantity $k_1 + ik_2$ where k_1 is equal to 2π times the inverse of the wave length and k_2 is the inverse of the distance over which the field is attenuated by a factor of e = 2.72 (equivalent to a power loss of 8.7 dB). At an operating frequency of 3 GHz in water, the values of k_1 and k_2 are

$$k_1 = 5.5 \text{ cm}^{-1} \text{ and } k_2 = 0.44 \text{ cm}^{-1}. \qquad (6)$$

The current-charge volume, V, is divided into a number of subvolumes, each denoted by V_n, which represents the space occupied by the nth radiating element. Let $\vec{x}_n$ be the center of V_n and denote by $\vec{J}_n$ and ρ_n, respectively, the current density and the charge density in V_n with respect to its center, then

$$\vec{J}_n(\vec{x}) = \vec{J}(\vec{x} + \vec{x}_n) \text{ and } \rho_n(\vec{x}) = \rho(\vec{x} + \vec{x}_n). \qquad (7)$$

Clearly, Equation 3 may also be written as:

$$\vec{A}(\vec{x}) = \sum_{n=1}^N \int_{V_n} \frac{1}{c}\vec{J}(\vec{x}) \frac{\exp(ik|\vec{x} - \vec{x}'|)}{|\vec{x} - \vec{x}'|} \, d\vec{x}'. \qquad (8)$$

Making the change of variable $\vec{x}' \rightarrow \vec{x}_n + \vec{x}'$, where the new $\vec{x}'$ is a vector from the center of each element to the volume $d\vec{x}'$, which is identical for every n and thus may simply be regarded as a vector in V_1, using Eq. 7, Eq. 8 becomes

$$\vec{A}(\vec{x}) = \sum_{n=1}^N \int_{V_1} \frac{1}{c}\vec{J}_n(\vec{x}') \frac{\exp ik|(\vec{x} - \vec{x}_n) - \vec{x}'|}{|(\vec{x} - \vec{x}_n) - \vec{x}'|} \, d\vec{x}'. \qquad (9)$$

Note that each integral in the right hand side of Eq. 9 is the same as that in Eq. 3, except that $\vec{x}$ is replaced by $\vec{x} - \vec{x}_n$ and the space of integration is over only the center element, V_1, instead of the entire array, V. Therefore, even though the observation point is in the neighborhood of the array, as long as $|\vec{x} - \vec{x}_n| = r_n$ is much greater than the size of each array element, d, one may expand the integrand in Eq. 9 in powers of $\vec{x}'/r_n$ (note that $|\vec{x}'| \lesssim d$).

So, denote by $\vec{x}_\ell'$ and $\vec{x}_t'$, respectively, the longitudinal and transverse components of $\vec{x}'$, i.e., the projections of $\vec{x}'$ in the directions parallel and perpendicular to $\vec{x} - \vec{x}_n$, then

$$\vec{x}_\ell' = \frac{1}{r_n{}^2}[\vec{x}' \cdot (\vec{x} - \vec{x}_n)](\vec{x} - \vec{x}_n) \qquad (10)$$

and

$$\vec{x}_t' = -\frac{1}{r_n{}^2}[\vec{x}' \times (\vec{x} - \vec{x}_n)] \times (\vec{x} - \vec{x}_n). \qquad (11)$$

Define

$$x_\ell' = |\vec{x}_\ell'|, \, x_t' = |\vec{x}_t'|. \qquad (12)$$

Then one has the identity

$$|\vec{x} - \vec{x}_n - \vec{x}'| = [(r_n - x_\ell')^2 + x_t'^2]^{1/2}$$

or

$$\qquad (13)$$

$$|\vec{x} - \vec{x}_n - \vec{x}'| = r_n\left(1 - \frac{x_\ell'}{r_n}\right)\left[1 + \frac{(x_t'/r_n)^2}{(1 - x_\ell'/r_n)^2}\right]^{1/2}.$$

Under Assumption 1, x_t'/r_n and x_ℓ'/r_n are both small quantities. So expand $|\vec{x} - \vec{x}_n - \vec{x}'|$ and other functions of it in powers of x_ℓ'/r_n and x_t'/r_n, then the following series result:

$$k|\vec{x} - \vec{x}_n - \vec{x}'| = kr_n\left\{1 - \frac{x_\ell'}{r_n} + \frac{1}{2}\left(\frac{x_t'}{r_n}\right)^2 - \frac{1}{2}\left(\frac{x_\ell'}{r_n}\right)\left(\frac{x_t'}{r_n}\right)^2\right.$$
$$\left. + \left(\frac{x_t'}{r_n}\right)^2 \cdot O\left[\left(\frac{x_\ell'}{r_n}\right)^2, \left(\frac{x_t'}{r_n}\right)^2\right]\right\} \qquad (14)$$

and

$$\frac{1}{|\vec{x} - \vec{x}_n - \vec{x}'|} = \frac{1}{r_n}\left\{1 + \frac{x_\ell'}{r_n} + \left(\frac{x_\ell'}{r_n}\right)^2 - \frac{1}{2}\left(\frac{x_t'}{r_n}\right)^2 \right.$$
$$\left. + \left(\frac{x_\ell'}{r_n}\right) \cdot O\left[\left(\frac{x_\ell'}{r_n}\right)^2, \left(\frac{x_t'}{r_n}\right)^2\right] + O\left[\left(\frac{x_t'}{r_n}\right)^4\right]\right\}. \quad (15)$$

As to the factor $\exp(ik|\vec{x} - \vec{x}_n - \vec{x}'|)$, its expansion depends not only on the relative magnitude of r_n and d, but also on the magnitudes of kr_n and kd. If the real part of kd is small, which is valid under Assumption 1, then, except for the first term, every term on the right hand side of Eq. 14 is much smaller than 1. The exponential of the series then gives

$$\exp(ik|\vec{x} - \vec{x}_n - \vec{x}'|) = \exp(ikr_n)(1 - ikx_\ell' + \ldots). \quad (16)$$

Combining this expansion with Eq. 15, the integrand of Eq. 9 becomes

$$\frac{1}{c}J_n(\vec{x}')\frac{\exp(ik|\vec{x} - \vec{x}_n - \vec{x}'|)}{|\vec{x} - \vec{x}_n - \vec{x}'|}$$
$$= \frac{1}{c}\frac{\exp(ikr_n)}{r_n}J_n(\vec{x}')\left\{1 + \frac{x_\ell'}{r_n} - ikx_\ell' + \ldots\right\}. \quad (17)$$

Substituting the leading term on the right hand side of Eq. 17 into Eq. 9, an approximation for the vector potential is obtained as

$$\vec{A}(\vec{x}) = \sum_{n=1}^{N}\frac{1}{c}\frac{\exp(ikr_n)}{r_n}\int_{V_1}\vec{J}_n(\vec{x}')d\vec{x}'. \quad (18)$$

It can be shown that the integral in the right hand side is proportional to the total dipole moment, $\vec{p}_n$, of the array element v_n:

$$\vec{p}_n = \frac{i}{ck}\int_{V_1}\vec{J}_n(\vec{x}')d\vec{x}' = \frac{i}{ck}\int_{V_n}J(\vec{x}')d\vec{x}', \quad (19)$$

where the dipole, $\vec{p}_n$, is defined as the moment of the charge distribution of the nth radiating element with respect to its center:

$$\vec{p}_n = \int_{V_1}\vec{x}'\rho(\vec{x}')d\vec{x}'. \quad (20)$$

Therefore, the vector potential may also be written as

$$\vec{A}(\vec{x}) = -ik\sum_{n=1}^{N}\vec{p}_n\frac{\exp(-ik|\vec{x} - \vec{x}_n|)}{|\vec{x} - \vec{x}_n|}, \quad (21)$$

which is the field due to N radiating dipoles.

Assumptions 2 and 3 are now applied to Eq. 21. Noting that all elements have the same geometry, the only factors that could contribute to different values of $\vec{p}_n$ for different elements are the input power and phase and the differences in the current-charge distributions due to mutual coupling. Under Assumption 2, the last factor is assumed to be negligible, and, under Assumption 3, $\vec{p}_n$ must be proportional to the input phase and amplitude factor. Therefore $\vec{p}I_nC_n$ may be substituted for $\vec{p}_n$ in Eq. 21, where $\vec{p}$ is the dipole moment for each radiating element at a certain standard input, I_n is the illumination factor for the nth element, and C_n is a complex factor representing the phase and amplitude conjugations. I_n is used as a controlling factor to modify the mainbeam shape. Equation 21 then becomes

$$\vec{A}(\vec{x}) = -ik\vec{p}\sum_{n=1}^{N}I_nC_n\frac{e^{ik|\vec{x} - \vec{x}_n|}}{|\vec{x} - \vec{x}_n|}. \quad (22)$$

The electric and magnetic fields may be obtained from the above equation by applying Eqs. 4 and 5 on $\vec{A}(\vec{x})$. Again applying Assumption 1, the results are

$$\vec{B}(\vec{x}) = -k^2\vec{p} \times \sum_{n=1}^{N}I_nC_n\frac{\vec{x} - \vec{x}_n}{|\vec{x} - \vec{x}_n|}\frac{\exp(ik|\vec{x} - \vec{x}_n|)}{|\vec{x} - \vec{x}_n|} \quad (23)$$

and

$$\vec{E}(\vec{x}) = -k^2\sum_{n=1}^{N}I_nC_n\left(\vec{p} \times \frac{\vec{x} - \vec{x}_n}{|\vec{x} - \vec{x}_n|}\right)$$
$$\times \frac{\vec{x} - \vec{x}_n}{|\vec{x} - \vec{x}_n|}\frac{\exp(ik|\vec{x} - \vec{x}_n|)}{|\vec{x} - \vec{x}_n|}. \quad (24)$$

The definition of the quantities in Equations 22 through 24 are summarized below:

$\vec{A}$ = the vector potential, in the Gaussian system of units

$\vec{B}$ = the magnetic induction, in the Gaussian system of units

C_n = the complex number representing the phase and amplitude conjugation for the nth radiating element (see next section)

$\vec{E}$ = the electric field intensity, in the Gaussian system of units

I_n = the illumination factor for the nth element; this factor is used to control the beam shape (see next section)

k = the complex number representing the magnitude of the wave vector of the radiation in water; the values of its real and imaginary parts for a 3 GHz radiation are given in Equation 6

n = a subscript denoting the nth radiating element

N = the total number of radiating elements in the array

$\vec{p}$ = the dipole moment of each radiating element at a standard phase and amplitude input (i.e., for I_n and C_n being unity), in the Gaussian system of units

$\vec{x}$ = the vector representing the observation point with respect to the center of the array

$\vec{x}_n$ = the vector representing the center of the nth radiating element with respect to the center of the array.

4. PHASE AND AMPLITUDE CONJUGATIONS AND THREE-DIMENSIONAL FOCUSING

The field patterns presented in Eqs. (22–25) depend on the set of factors $\{I_nC_n\}$. In this section we introduce a formula for assigning the values of C_n to provide a maximum relative field at a desirable focal point. The idea is that, if one wishes to focus the field of the array at a point, say, $\vec{x}_f$, one can maximize the field at that point by applying a phase and amplitude taper which compensates the propagation loss

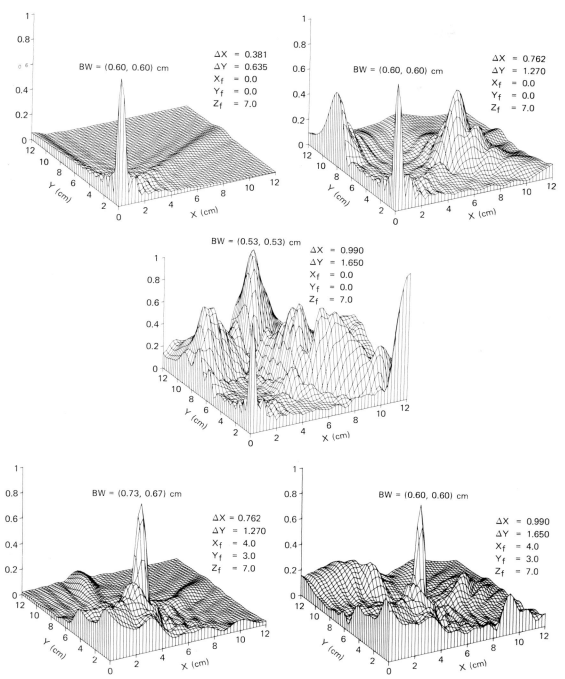

Fig. 2a–2c.   Relative amplitude of the vector field in the surface z = 7 cm with phase and amplitude conjugations focused at $\check{x}_f$ = (0, 0, 7) cm. The element spacings are (a) ΔX = 0.381 cm and ΔY = 0.635 cm, (b) ΔX = 0.762 cm and ΔY = 1.270 cm, (c) ΔX = 0.990 cm and ΔY = 1.650 cm. The 3 dB widths of the main-beam in both x- and y-directions are as indicated.

Figs. 2d–2e.   Relative amplitude of the vector field in the surface z = 7 cm with phase and amplitude conjugations focused at $\check{x}_f$ = (4, 3, 7) cm. The element spacings are (d) ΔX = 0.762 cm and ΔY = 1.270 cm, (e) ΔX = 0.990 cm and ΔY = 1.650 cm. The 3 dB widths of the main-beam in both x- and y-directions are as indicated.

and phase differential from each array element to the focal point. Thus, to focus the main beam at the point $\check{x}_f$, the factor C_n is assigned as

$$C_n = |\check{x}_f - \check{x}_n| \exp(-ik|\check{x}_f - \check{x}_n|). \qquad (25)$$

Noting that k is a complex number, the exponential factor in the above equation includes a phase factor and an amplitude factor to compensate for the absorption by the coupling medium. For this reason, we call C_n the phase-ampli-

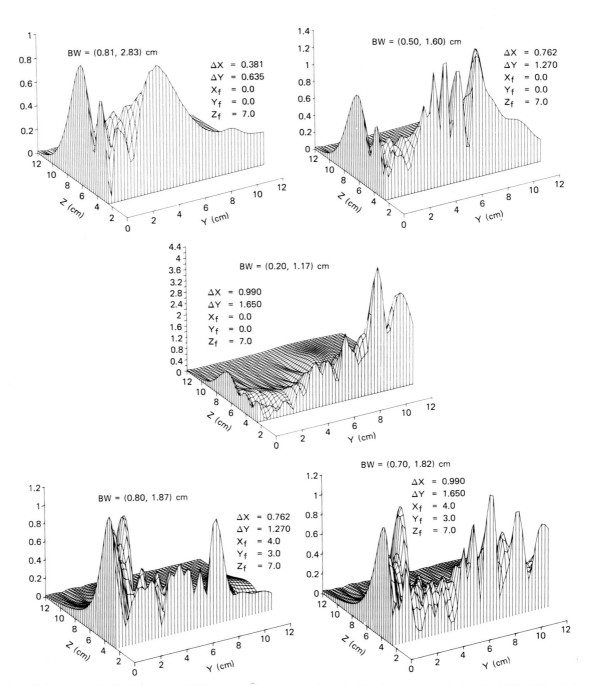

Figs. 3a–3e.   Relative amplitude of the vector field in the yz-plane at x = 0, with focusing and element spacings corresponding to those in Figs. 2a–2e, respectively. The 3 dB widths of the main-beam in both cm y- and z-directions are as indicated.

tude conjugation factor. The factor I_n in Eqs. (22–24), which is called the illumination factor, may be used as additional leverage to provide optimal resolution and minimize the side lobes.

For a planar array of N × M radiating dipole elements, Eqs. 22 and 23 may be written as

$$\vec{A}(\vec{x}) = -ik\vec{p} \sum_{n=1}^{N} \sum_{m=1}^{M} I_{nm} C_{nm} \frac{e^{ik|\vec{x} - \vec{x}_{nm}|}}{|\vec{x} - \vec{x}_{nm}|} \quad (26)$$

and

$$C_{nm} = |\vec{x}_f - \vec{x}_{nm}| \exp(-ik|\vec{x}_f - \vec{x}_{nm}|), \quad (27)$$

As we shall see below, the phase-amplitude conjugation as presented in Eq. (25) does not make the field peak at exactly the point $\vec{x}_f$. This is simply due to the fact that we are in the local field region. For the array structure described in

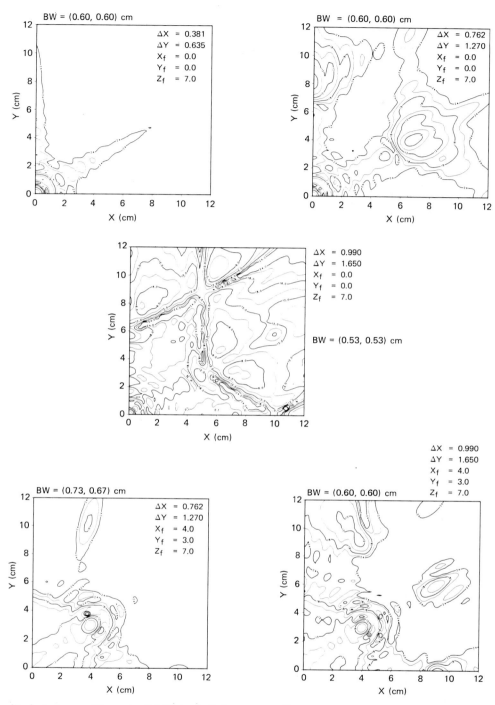

Figs. 4a–4e.   Amplitude contours of the vector field in isodecibels corresponding to the cases indicated in Figs. 2a–2e respectively.

Section 2, the main lobe beamwidth is minimized when I_{nm} is taken to be a uniform illumination.

Now we present some result of the field patterns using the phase and amplitude conjugations. Only the absolute value of vector potential and the corresponding field characteristics are presented here. The structure of the array lattice is illustrated in Fig. 1a, and the equations used are (26), (27)

and (6), with illumination factor $I_{nm} = 1$, and polarization taken to be in y-direction. In all the figures, the plane of the array is taken to be the xy-plane and the z-axis is perpendicular to the array plane and pointing to the forward direction. In the following discussion and in all figures, the phrase "mainbeam" is used in reference to field characteristics of the 3-dimensional focal region. Figures 2(a) through

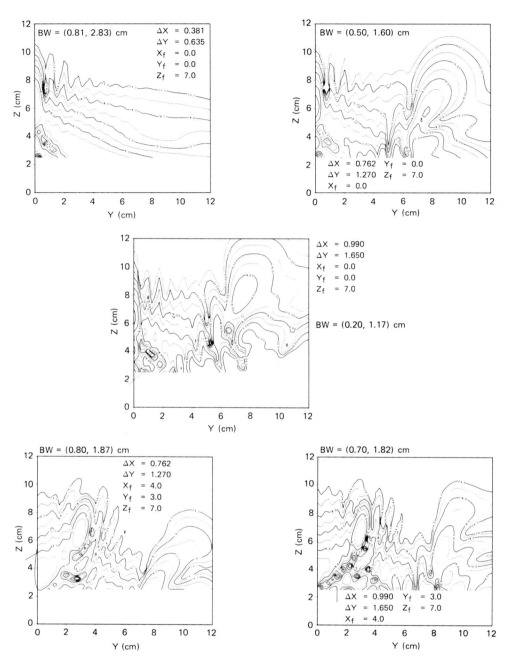

Figs. 5a–5e.   Amplitude contours of the vector field in isodecibels corresponding to the cases indicated in Figs. 3a–3e respectively.

2(e) show various field patterns with different interelement spacings in the transverse plane at an axial distance of 7 cm, which is also the axial distance of the desired focal point. Similarly, Figs. 3(a) through 3(e) are the longitudinal field patterns at different interelement spacings and focal points. The corresponding 3-dB full mainbeam width is given on each figure. Note that Figs. 2(b),2(c) and 3(b),3(c) are for the array with lattice structure shown in Fig. 1a. Figures 2(d)-,2(e) are for arrays that are similar to the one shown in Fig. 1a, except with different lattice spacings. Figures 2(a) and

3(a) are for the same size array as shown in Fig. 1a but with 419 instead of 127 elements. Figures 4 and 5 are the corresponding pictures of Figs. 2 and 3, and plotted in isodecibel field contours. For easy comparison, the beam characteristics of these figures are tabulated in Table I. These data show that larger interelement spacings result in narrower 3-dB beamwidths and more accurate focusing; however, larger interelement spacings also bring grating lobes closer to the mainbeam. This result demonstrates that, with respect to beamwidth reduction, the array size plays a more important

Table 1

ARRAY BEAM CHARACTERISTICS OF THE FIELD PATTERNS AT 3 GHz

FOCAL POINT (x y z) cm	ELEMENT SPACING (dx dy) cm	PEAK AT Z=7 (x y z) cm	PEAK AT X=X$_f$ (x y z) cm	BEAMWIDTH AT Z=7 (dx dy) cm	BEAMWIDTH AT X=X$_f$ (dy dz) cm
0 0 7	.381 .635	0 0 7	0 0 6	.60 .60	.81 2.83
0 0 7	.762 1.27	0 0 7	0 0 6	.60 .60	.50 1.60
0 0 7	.990 1.65	0 0 7	0 0 6.5	.53 .53	.20 1.17
4 3 7	.762 1.27	4 3 7	4 2.75 6.25	.74 .67	.80 1.87
4 3 7	.900 1.65	4 3 7	4 3 6.5	.60 .60	.70 1.82

role than the number of elements. The resolutions of $\frac{1}{2}\lambda$ in transverse direction and 1λ in longitudinal direction can be achieved with the use of phase and amplitude conjugation. Note also that the peaks along $z = 2.5$ cm shown in Figs. 3(a)–3(e) and Figs. 5(a)–5(e) are due to the single element that is closest to each of these peaks. As the distance becomes so close to an individual element, the coherent addition from other elements is negligible in comparison. As long as the target is not closer than 3 cm from the array, these peaks will not cause any problem for actual applications in microwave imagery.

5. THE INVERSE SCATTERING THEOREM AND ITS APPLICATION TO IMAGE RECONSTRUCTION

The objective of all inverse scattering problems is to reconstruct the target from the scattered field. From Maxwell's electromagnetic theory, if one knows the field everywhere in space, the polarization charge-current distribution of the scattering source can be derived completely. However, in practice, one can only measure the scattered field at a limited number of points in space which are often confined in a small region. The question is then how much information on the scattering target one can infer based on a limited knowledge of the scattered field. Here we present a theorem [8], which is indeed a generalization of the Lorentz reciprocity theorem [19]–[21], and show that it can provide a good facility in answering the above question.

In this section we shall state the theorem and describe its application to microwave biological imaging, and leave the details of the proof to the next section. Consider a dielectric target immerged in a homogeneous and dissipative medium of dielectric constant ϵ_m and dielectric susceptibility χ_m (both are complex numbers). Let $\chi(\vec{x})$ describe the dielectric

susceptibility of the entire system, including the homogeneous medium and the target, so that $\chi - \chi_m$ is null outside the target. It is assumed that there is no free charge or current distribution (including ionic charge and current) in the target or the medium, and that both have the same homogeneous magnetic permeability μ_m; however, we remark that the theorem may be generalized to include free charge-current and magnetization. Let the target be illuminated by a plane wave $\vec{E}_{inc}$ of frequency $\omega/2\pi$, which induces electric polarization $\vec{P}$ in the target and produces a scattered field $\vec{E}_{scatt}$. Let $\vec{J}_w$ be some weighing function and $\vec{A}_w$ be the corresponding field derivable from $\vec{J}_w$ in the same way as the vector potential is derivable from a current density. More precisely,

$$(\nabla^2 + k_m^2)\vec{A}_w = -\frac{4\pi\mu_m}{c}\vec{J}_w, \qquad (28)$$

or its reverse equivalence,

$$\vec{A}_w(\vec{x}) = \frac{\mu_m}{c}\int\frac{e^{ik_m|\vec{x}-\vec{x}'|}}{|\vec{x}-\vec{x}'|}\vec{J}_w(\vec{x}')d\vec{x}'. \qquad (29)$$

Then the inverse scattering theorem may be stated as below:

$$-\frac{1}{c}\int\int\int\vec{A}_w(\vec{x})\cdot\left[\frac{c^2}{\mu_m\epsilon_m}\nabla\rho_s + \frac{\partial}{\partial t}\vec{J}_s\right]d\vec{x} = \int\int\int\vec{J}_w\cdot\vec{E}_{scatt}d\vec{x} \qquad (30)$$

where k_m, defined as $\sqrt{\mu_m\epsilon_m}\omega/c$, is the complex wave number in the homogeneous medium, and ρ_s and $\vec{J}_s$ represent the equivalent charge density and its time derivative, the current density, due to the polarization of the target in excess of the homogeneous background polarization:

$$\rho_s = -\nabla\cdot\frac{\chi-\chi_m}{\chi}\vec{P}, \quad \vec{J}_s = \frac{\chi-\chi_m}{\chi}\frac{\partial\vec{P}}{\partial t} \qquad (31)$$

As we shall see in the next section, ρ_s and $\vec{J}_s$ are indeed the source of the scattered field.

This theorem is more general than the Lorentz reciprocity theorem [19]–[21] since $\vec{J}_w$, as well as the associated vector field $\vec{A}_w$, is only a weighing function. They need not be physical quantities; for example, $\vec{J}_w$ needs not satisfy the equation of continuity and $\vec{A}_w$ needs not satisfy the Lorentz gauge condition (or any alternative gauge). The proof of this theorem is based on the fact that the medium is dissipative and therefore a Hilbert space may be defined in which all the differential operators involved may have their Hermitian conjugations defined. The great facility of this theorem in applications to image reconstuction lies on the flexibility of choosing the weighing function $\vec{J}_w$. If the scattered field $\vec{E}_{scatt}$ is measured at a set of spatial points $\{\vec{x}_n\}$, which are the lattice points of our receiving array, and if, say, only the y-polarization is measured, then the weighing function $\vec{J}_w(\vec{x})$ may be chosen to be only in the y-direction and to be a discrete distribution over the points $\{\vec{x}_n\}$. It then follows from Eq. (29) that the resulting weighing vector potential $\vec{A}_w(\vec{x})$ will also be in the y-direction. One may further adjust the phases and amplitudes of $\vec{J}_w(\vec{x}_n)$ to optimize the weighing potential $\vec{A}_w(\vec{x})$ at any desired point of the target. It was shown in the previous section that applying the phase and amplitude conjugations to a transmitting array could provide a 3-dimensionally focused radiation in the neighborhood of the array. For imagery application with an incident plane wave, this technique may be applied to the receiving array to "focus" the weighing potential. Thus, by setting $\vec{J}_w(\vec{x}_n)$ to be the phase and amplitude conjugation factor, one may make $\vec{A}_w(\vec{x})$ negligibly small inside the target organ, except for a sharp peak near any desired point, say $\vec{x}_f$, which we shall call the focal point. Note that, since ρ_s vanishes outside the target region and so the integration in the left hand side of Eq. (30) is limited to only the target region, $\vec{A}_w(\vec{x})$ may assume any value, however large, outside the target region. Then the right hand side of Eq. (30) is obtainable from the measured data, whereas the integration on the left hand side is dominated by the integrand at the focal point.

It is emphasized that the "focusing" described above does not involve active focusing of the transmitted microwave energy on any point of the target. Rather, it is simply a mathematical management of the measured scattering data so that the retrieved information of the dielectric property may be "focused" on a desired point inside the target. In this sense, it may be considered as a focusing of the receiving array. Since one microwave exposure of the target will provide N samples of the complex scattered field, where N is the number of receiving elements, it is then theoretically possible to retrieve N estimates of the spatial distribution of the complex dielectric properties of the target by optimal management of the measured data. The inverse scattering theorem described above is to provide a basis for such purpose. We shall discuss in Section 7 the maximum limit of retrievable information from a single exposure of microwave radiation.

To express this imagery application more clearly, we set $\vec{J}_w(\vec{x}) = \sum_n \delta(\vec{x} - \vec{x}_n)\hat{y}J_n$ and substitute Eq. (31) for ρ_s and $\vec{J}_s$

in Eq. (30), then replace $\partial^2\vec{P}/\partial t^2$ by $-\omega^2\vec{P}$ and perform partial integration on the first term on the left hand side. Noting that $\vec{A}_w$ vanishes at infinity, one then gets

$$\frac{c}{\epsilon_m\mu_m} \int\int\int \left[-(\nabla \cdot \vec{A}_w)\left(\nabla \cdot \frac{\chi - \chi_m}{\chi}\vec{P}\right) + k_m^2\vec{A}_w \cdot \frac{\chi - \chi_m}{\chi}\vec{P} \right]d\vec{x} = \sum_n J_nE_{scatt}(\vec{x}_n) \quad (32)$$

With the discrete distribution of $J_w(\vec{x})$, Eq. (29) reduces to

$$\vec{A}_w(x) = \sum_n \frac{\hat{y}J_n}{|\vec{x} - \vec{x}_n|}\exp(ik_m|\vec{x} - \vec{x}_n|) \quad (33)$$

which is similar to the vector potential produced by a set of dipoles as given by Eq. (22). To obtain the dielectric information at any given point $\vec{x}_f$ inside the target from the measured scattered field at the points $\{\vec{x}_n\}$, one sets J_n in the right hand side of Eq. (32) to be the phase-amplitude conjugation factor:

$$J_n(\vec{x}_f) = (c/\mu_m)|\vec{x}_n - \vec{x}_f| \cdot \exp(-ik|\vec{x}_n - \vec{x}_f|) \quad (34)$$

Based on the phase-amplitude conjugation discussed in the previous section, the resulting $A_w(\vec{x})$ has a sharp peak at $\vec{x}_f$ and is otherwise negligible in the target region. Owing to the factor $\chi - \chi_m$ which vanishes outside the target region, all sidelobes of $A_w(\vec{x})$, however large, which lie outside the target region will not contribute to the integral on the left hand side of Eq. (32). The main contribution of the integral is the dielectric characteristics of the target at the point $\vec{x}_f$, which is then equal to the weighted version of the measured values $\sum_n J_nE_{scatt}(\vec{x}_n)$. For this reason we refer to the point $\vec{x}_f$ as a focal point since, by Eq. (32), the sum of the products of the conjugation factors and the measured scattered fields, $\sum_n J_nE_{scatt}(\vec{x}_n)$, "focuses" the result to the dielectric polarization at the point $\vec{x}_f$ inside the target. By scanning the vector $\vec{x}_f$ in Eq. (34) through the target region, one then reconstructs an image of the target. The response peak in $\vec{A}_w(\vec{x})$ at the "focal point" determines the spatial resolution, and the flat response of $\vec{A}_w(\vec{x})$ within the target elsewhere contributes to analytical interpretation of the imagery [7]. The size of the target is limited by the locations of the grating lobes, since these grating lobes will not contribute to the integral on the left hand side of Eq. (32) as long as they are outside of the target region. Another method to reduce the grating lobe interference is to limit the illumination region so that the local field is negligible in the grating lobe region. For a larger target, this will require dividing the target into smaller illumination regions.

6. PROOF OF THE INVERSE SCATTERING THEOREM

Our first approach in proving the theorem is to separate the target from the surrounding homogeneous medium so that the scattered field may be considered as due to a localized charge-current source. We shall consider the entire system to be a superposition of a distribution of dielectric susceptibility $\chi(\vec{x}) - \chi_m$ and a homogeneous medium of dielectric susceptibility χ_m; thus $\chi - \chi_m$ may be considered as the excess susceptibility of the system over the homogeneous background. Since the medium is homogeneous, the propagation of the scattered field is then governed by a set of Maxwell equations similar to that in free space except that the free space permitivity and permeability are replaced by that of the homogeneous medium. Using the notations and assumptions that were described at the beginning of the previous section, the Maxwell's equations for the total field $(\vec{E}, \vec{B})$ may be written as

$$\nabla \cdot (1 + 4\pi\chi)\vec{E} = 0$$
$$\frac{1}{\mu_m} \nabla \times \vec{B} - \frac{(1 + 4\pi\chi)}{c} \frac{\partial \vec{E}}{\partial t} = 0 \qquad (35)$$
$$\nabla \cdot \vec{B} = 0$$
$$\nabla \times \vec{E} + \frac{1}{c} \frac{\partial \vec{B}}{\partial t} = 0$$

In order to separate the source of the scattered field from the homogeneous background, we transport all quantities involving excess dielectric susceptibility to the right hand sides of the above equations, so that the left hand sides resemble that of the Maxwell equations in free space. The results are, with the definition of ρ_s and $\vec{J}_s$ given by Eq. (31):

$$\epsilon_m \nabla \cdot \vec{E} = 4\pi\rho_s$$
$$\frac{1}{\mu_m} \nabla \times \vec{B} - \frac{\epsilon_m}{c} \frac{\partial \vec{E}}{\partial t} = \frac{4\pi}{c} \vec{J}_s \qquad (36)$$
$$\nabla \cdot \vec{B} = 0$$
$$\nabla \times \vec{E} + \frac{1}{c} \frac{\partial \vec{B}}{\partial t} = 0$$

where $\epsilon_m = 1 + 4\pi\chi_m$ is the dielectric permitivity of the homogeneous medium, and $\vec{P} = \chi\vec{E}$ is the electric polarization in the target. Replacing the total field $(\vec{E}, \vec{B})$ in Eq. (36) by the sum of the incident wave and the scattered field, since the incident wave satisfies the homogeneous (viz. sourceless) version of Eq. (36), one finds that the scattered field satisfies exactly Eq. (36). Denote the scattered field by a subscript s, one then has

$$\epsilon_m \nabla \cdot \vec{E}_s = 4\pi\rho_s$$
$$\frac{1}{\mu_m} \nabla \times \vec{B}_s - \frac{\epsilon_m}{c} \frac{\partial \vec{E}_s}{\partial t} = \frac{4\pi}{c} \vec{J}_s \qquad (37)$$
$$\nabla \cdot \vec{B}_s = 0$$
$$\nabla \times \vec{E}_s + \frac{1}{c} \frac{\partial \vec{B}_s}{\partial t} = 0$$

Thus the scattered field alone may be considered as that produced by the localized charge-current density $(\rho_s, \vec{J}_s)$.

Due to the dissipation of the medium, the electromagnetic field vanishes exponentially as $|\vec{x}| \to \infty$. Thus it is possible to define a Hilbert space in which the electromagnetic fields and charge densities are vectors. We define the scalar product between any pair of vectors (f, g) as the integral over the entire space:

$$\int\int\int f^*(\vec{x})g(\vec{x})d\vec{x},$$

which exists owing to the fact that f and g diminish exponentially as $|\vec{x}| \to \infty$. With this definition, all differential operators involved in the Maxwell equations as well as all derivative equations may have their Hermitian conjugations defined. It is based on this fact that the inverse scattering theorem can be proved. For mathematical simplicity, we shall prove the theorem using the Hilbert space notations. Noting that the Hilbert space product combined with the hermiticity of linear operators is equivalent to integration by parts involving differential operators, the proof may also be made equivalently in conventional differential equation form.

We shall now derive the inverse scattering theorem from Eq. (37). First note that the third and fourth equations of Eq. (37) show that the scattered electric field and magnetic induction are derivable from a pair of scalar and vector potentials $(\phi_s, \vec{A}_s)$, which shall be called the scattered potentials:

$$\vec{B}_s = \nabla \times \vec{A}_s$$
$$\vec{E}_s = -\nabla\phi_s - \frac{1}{c} \partial\vec{A}_s/\partial t \qquad (38)$$

The relationships of Eq. (38) still leave another degree of freedom on the choice of the potentials. We shall use the conventional choice of Lorentz gauge condition

$$\nabla \cdot \vec{A}_s + \frac{\epsilon_m\mu_m}{c} \frac{\partial\phi_s}{\partial t} = 0. \qquad (39)$$

Then, from the first two equations of Eq. (37), the scattered potentials satisfy the wave equation with ρ_s and $\vec{J}_s$ respectively as the source:

$$(\nabla^2 + k_m{}^2)\phi_s = -4\pi\rho_s/\epsilon_m$$
$$(\nabla^2 + k_m{}^2)\vec{A}_s = -\frac{4\pi}{c} \mu_m\vec{J}_s \qquad (40)$$

where $k_m{}^2 = (\omega^2/c^2)\mu_m\epsilon_m$ and we have replaced $\partial/\partial t$ by $-i\omega$.

To facilitate our proof of the theorem, we shall denote by $|A_s\rangle$ and $|J_s\rangle$ respectively the 4-dimensional potential and the 4-dimensional current density:

$$|A_s\rangle = \begin{pmatrix} \phi_s \\ \vec{A}_s \end{pmatrix}, \quad |J_s\rangle = \begin{pmatrix} \dfrac{c}{\epsilon_m}\rho_s \\ \mu_m\vec{J}_s \end{pmatrix}.$$

Also denote by K_m the 4-dimensional Helmholtz operator associated with k_m:

$$K_m = (\nabla^2 + k_m{}^2)\begin{pmatrix} 1 & 0 \\ 0 & \vec{1} \end{pmatrix} \qquad (41)$$

Then Eq. (40) may be expressed as

$$K_m|A_s\rangle = -\frac{4\pi}{c}|J_s\rangle \qquad (42)$$

To express the relationship between the electric field and the vector potential, we define the 4-dimensional E-vector,

$$|E_s\rangle = \begin{pmatrix} 0 \\ \vec{E} \end{pmatrix} \qquad (43)$$

and the 4-dimensional S-operator and its hermitian conjugate,

$$S_m = \begin{pmatrix} \dfrac{ic}{\omega}k_m^2 & -\nabla \\ -\nabla & \dfrac{i\omega}{c}\bar{1} \end{pmatrix}, \; S_m^\dagger = -\begin{pmatrix} \dfrac{ic}{\omega}(k_m^2)^* & -\nabla \\ -\nabla & \dfrac{i\omega}{c}\bar{1} \end{pmatrix} \qquad (44)$$

Then the second equation of Eq. (38) may be expressed as

$$|E_s\rangle = S_m|A_s\rangle \qquad (45)$$

Note that the above equation actually represents two relationships: the scalar component is equivalent to the gauge condition Eq. (39), and the vector condition gives the electric field in terms of the potentials. Since the operators K_m and S_m commute with each other, operating both sides of Eq. (42) by S_m yields the wave equation for $|E_s\rangle$:

$$K_m|E_s\rangle = -\frac{4\pi}{c}S_m|J_s\rangle \equiv -\frac{4\pi}{c}|F_s\rangle \qquad (46)$$

which shows that $S_m|J_s\rangle$ may be considered as the source of $|E_s\rangle$ as much as the current density $|J_s\rangle$ is the source of the potential $|A_s\rangle$, therefore we denote it by $|F_s\rangle$. By the equation of continuity $\nabla\cdot\vec{J}_s - i\omega\rho_s = 0$, we then have

$$|F_s\rangle \equiv S_m|J_s\rangle = \begin{pmatrix} 0 \\ -\dfrac{c}{\epsilon_m}\nabla\rho_s - \dfrac{\mu_m}{c}\dfrac{\partial\vec{J}_s}{\partial t} \end{pmatrix} \qquad (47)$$

So far we have rewritten all electrodynamic equations for a dissipative medium in their Hilbert space representations. To complete our description of the Hilbert space, we define, for any vector $|F\rangle$, a complex conjugate vector $|\overline{F}\rangle$ which corresponds to the complex conjugate in the $\tilde{x}$-representation,

$$\langle\tilde{x}|\overline{F}\rangle = (\langle\tilde{x}|F\rangle)^*$$

Also, given any pair of 4-dimensional vectors $|F\rangle = (f, \vec{F})$ and $|G\rangle = (g, \vec{G})$, their Hilbert space scalar product is defined as

$$\langle F|G\rangle = \iiint (f^*g + \vec{F}^*\cdot\vec{G})d\tilde{x}$$

With these Hilbert space representations, the proof of the inverse scattering theorem becomes rather straightforward. Let $|J_w\rangle$ be any 4-dimensional vector, and $|A_w\rangle$ be the 4-dimensional vector derivable from $|J_w\rangle$ in the same way as $|A_s\rangle$ is derivable from $|J_s\rangle$. That is,

$$K_m|A_w\rangle = -\frac{4\pi}{c}|J_w\rangle \qquad (48)$$

$|J_w\rangle$ and $|A_w\rangle$ are weighing vectors which satisfy only the wave equations, but not the equation of continuity nor the gauge condition. So they are not physical quantities. For instance, the weighing current $|J_w\rangle$ may be a point function with only the vector component, viz., a delta-function current in space without charge density. From Eq. (41), the complex conjugate of K_m is equal to its hermitian conjugate, therefore the complex conjugate of Eq. (48) gives $(K_m)^\dagger|\overline{A}_w\rangle = -4\pi/c|\overline{J}_w\rangle$, where $\dagger$ denotes the hermitian conjugation. The hermitian conjugate of this equation then gives $\langle\overline{A}_w|K_m = \langle\overline{J}_w|(-4\pi/c)$. Taking the scalar product of both sides with $|E_s\rangle$ and utilizing Eq. (46), it then gives

$$-\frac{4\pi}{c}\langle\overline{J}_w|E_s\rangle = \langle\overline{A}_w|K_m|E_s\rangle = -\frac{4\pi}{c}\langle\overline{A}_w|F_s\rangle$$

Therefore

$$\langle\overline{A}_w|F_s\rangle = \langle\overline{J}_w|E_s\rangle \qquad (49)$$

which is the Hilbert space form of the inverse scattering theorem. Take $|J_w\rangle$ to be $(0, \vec{J}_w(\tilde{x}))$, so that the scalar component of $|A_w\rangle$ also vanishes, then, with the aid of Eq. (47), the above equation gives the inverse scattering theorem stated in the previous section, viz., Eq. (30).

One may wish to express Eq. (49) in a more symmetric form, such as one involves $|J\rangle$ and $|E\rangle$ in both sides of the equality. To do this, we first replace $|F_s\rangle$ in the left hand side of Eq. (49) by $S_m|J_s\rangle$, as defined in Eq. (47), then Eq. (49) becomes

$$\langle\overline{A}_w|S_m|J_s\rangle = \langle\overline{J}_w|E_s\rangle \qquad (50)$$

If we also define a weighing electric field $|E_w\rangle$ in the same way as a physical electric field is related to the vector potential through Eq. (45), then $|E_w\rangle = S_m|A_w\rangle$. Taken its complex conjugation followed by hermitian conjugation, this relationship gives $\langle\overline{A}_w|(S_m)^{*\dagger} = \langle\overline{E}_w|$. From Eq. (44) one sees that $(S_m)^{*\dagger}$ differs from S_m by having opposite signs in all off-diagonal elements while being equal to S_m in all diagonal elements, therefore the left hand side of Eq. (50) is not equal to $\langle\overline{E}_w|J_s\rangle$, as what would have been expected from a symmetric expression. Therefore we define another field $|\tilde{E}_w\rangle = (S_m)^{*\dagger}|A_w\rangle$, so that its hermitian conjugation followed by complex conjugation is $\langle\tilde{E}_w| = \langle\overline{A}_w|S_m$. Then Eq. (50) may also be stated in the following form:

$$\langle\tilde{E}_w|J_s\rangle = \langle\overline{J}_w|E_s\rangle \qquad (51)$$

It is remarked that, while $|A_w\rangle$ may be considered as the potential due to the source current density $|J_w\rangle$, neither $|\tilde{E}_w\rangle$ nor $|E_w\rangle$ defined above has a parallel analogy. From Eq. (43), a parallel analogy would require that the scalar components of $|\tilde{E}_w\rangle$ and $|E_w\rangle$ vanish. But, since $|A_w\rangle$ may not satisfy the Lorentz gauge condition, the scalar components of $S_m|A_w\rangle$ and $(S_m)^{*\dagger}|A_w\rangle$ do not vanish.

Equation (51) appears to be similar to the Lorentz reciprocity theorem as expressed by Carson [19]–[21]. The difference here is that $|J_w\rangle$ and $|\tilde{E}_w\rangle$ are only weighing functions and they need not be physical quantities. Therefore this theorem may be considered as a generalization of the reciprocity theorem. The theorem is valid only if a Hilbert

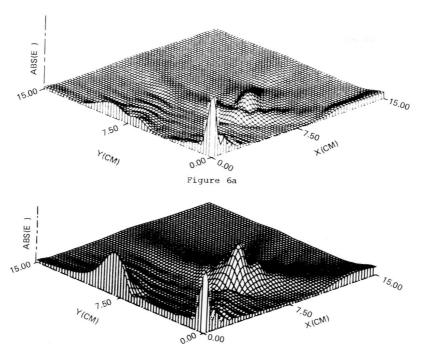

Figs. 6a–6b. Comparison of the electric fields in the z = 7 cm surface generated by a planar array and a volumetric array, with the phase and amplitude conjugation focusing on the axis at 7 cm from the center of the array. (a) For a system of 5 parallel arrays spaced at 1 cm from each other, the 3 dB main-beam width and the mainbeam to first sidelobe ratio in the (x, y) directions are respectively, (0.56, 0.82)cm and (24, 18)dB. (b) For a planar array, the 3 dB main-beam width and the mainbeam to first sidelobe ratio in the (x, y) directions are respectively, (0.64, 0.82)cm and (10, 16)dB.

space may be defined in which the fields, the sources, and the weighing functions are Hilbert space vectors, which is the case when the propagation medium is dissipative.

7. REMARKS AND CONCLUSIONS

Three main topics have been presented in this article: (1) an instrumental and hardware description of a microwave array system being developed for medical application, (2) the phase and amplitude conjugation technique which may be used to actively focus a transmitting microwave array or, with the help of the inverse scattering theorem, to passively focus a receiving array; in either case, it may achieve a 3-dimensional focusing in the neighborhood of the array, and (3) an inverse scattering theorem on retrieving information of a scattering target from limited data of the scattered field, which, if applied in conjunction with the phase and amplitude conjugations, may provide an optimal 3-dimensional imaging from one sampling of the scattering field measured by a receiving array.

On the quality of 3-dimensional focusing using the phase and amplitude conjugations, we have presented several results graphically. Summarizing these results, the following conclusions may be drawn:

1. Using the method of phase and amplitude conjugations, one may achieve a satisfactory degree of 3-dimensional fo-

cusing in the neighborhood of radiation sources in a lattice structure. There is a slight shift of the peak point of the field from the intended focal point as defined in the phase-amplitude conjugation factor. The shift, as outlined in Table I, generally points toward the center of the array. This should not pose any problem for practical applications since it can be calibrated.

2. Upon applying the phase and amplitude conjugations, the field patterns and the beam characteristics in the transverse direction appear to have similar dependency upon the lattice structure and the array size as that of a Fraunhofer field. Thus, the transverse beamwidth becomes narrower as the element spacing increases, at the expense of more grating lobes. Interestingly, this behavior also applies to the longitudinal beamwidth. Therefore, for a smaller target, it is possible to improve the resolution further by increasing the element spacing, as long as the target does not extend to the region covered by grating lobes. Alternatively, the regions covered by the grating lobes may be excluded from illumination by active control on the transmitting array.

3. Along the longitudinal direction, the field patterns and the beam characteristics using the phase and amplitude conjugations differ considerably from that of a Fraunhofer field. A Fraunhofer field is invariant in the longitudinal direction except for the inverse-square dependence, whereas a local field under phase and amplitude conjugations has a diffraction structure in the longitudinal direction, as well as

in the transverse direction. Therefore some degree of focusing of the local field along the longitudinal direction may be achieved using the phase and amplitude conjugation technique. It must be remarked that, as the axial distance increases, the sensitivity of longitudinal focusing to the phase-amplitude conjugation decreases. Our analysis of the field pattern for focusing at 30 cm axial distance indicates that the conjugation factor is totally overcome by the exponential attenuation. However, the transverse focusing remains good even at this axial distance.

4. If all array elements may be represented by parallel dipoles, then clearly the vector potential everywhere must be polarized in the dipole direction. However, in the local region, the electric field will still have strong polarization dependency. This polarization dependency is reduced as the axial distance increases, and, at an axial distance of 30 cm and beyond, the electric field is highly polarized in the direction of the source dipoles [13].

The 3-dimensional resolution and the allowable target volume may further be optimized by varying the lattice structure and the element spacing of the receiving array. However, based on information theory, there is a theoretical limit on the ratio of the target volume and the resolution. If the total number of the array elements is N, then each sampling of the scattered field provides an information equivalent to 2N real numbers, where the factor 2 accounts for the measurements of both phase and amplitude. The value 2N is the maximum information one may expect from the image reconstruction. The theoretical limit of the 3-dimensional resolution from each sampling of the scattered field may then be described by the equation:

$$\frac{\text{volume of the target}}{\text{volume of the focal region}} \lesssim 2N. \qquad (52)$$

Therefore, if the target is smaller, the resolution may be further improved without increasing the number of elements. For example, for the enlarged interelement spacings corresponding to Figs. 2c and 3c, the resolution is improved to about half-wavelength transversely and one-wavelength longitudinally.

It is possible to acquire more target information, and thereby improve the three-dimensional resolution, by making multiple views from different angles with respect to the direction of the incident wave as suggested by the models of diffraction tomography. Multiple views at different ranges from the target for a fixed transmitter position are also possible; however, analysis of such a volumetric synthesis due to the superposition of parallel planar arrays discloses that such an approach may be of limited value, because much of the information contained in parallel samples of the scattered field are redundant. Figure 6(a) shows the field pattern of five parallel arrays separated at 1 cm from one another. Comparing to the field pattern of a single array as shown in Fig. 6(b), the five-array system provides slightly narrower main-beam width and smaller grating lobes. However, the difference may not repay the minimum of five times the data acquisition and data processing time.

ACKNOWLEDGMENT: This work was supported in part by the Walter Reed Army Institute of Research through the U.S. Army Medical R&D Command under the U.S. Naval Sea Systems Command Contract N00024-83-C-5301 to the Johns Hopkins University Applied Physics Laboratory.

References

1. A. S. Pressman, *Electromagnetic Fields and Life*, Plenum Press, New York, 1970.
2. L. E. Larsen and J. H. Jacobi, "Microwave scattering parameter imagery of an isolated canine kidney," *Med. Phys.*, Vol. 6, pp. 394–402, 1979.
3. J. H. Jacobi, L. E. Larsen, and C. T. Hast, "Water-immersed microwave antennas and their application to microwave interrogation of biological targets," *IEEE Trans. Microwave Theory Tech.*, Vol. MTT-27, pp. 70–78, 1979.
4. J. H. Bolomey, A. Izadnegahdar, L. Jofre, Ch. Pichot, G. Peronnet, and M. Solaimani, "Microwave diffraction tomography for biomedical applications," *IEEE Trans. Microwave Theory Tech.*, Vol. MTT-30, pp. 1998–2000, 1982.
5. N. H. Farhat, D. L. Jaggard, T. H. Chu, D. B. Ge, and S. Mankoff, presented at the 3rd Annual Benjamin Franklin Symposium on Advances in Antennas and Microwave Technology, Philadelphia, Pennsylvania, 1983.
6. L. E. Larsen and J. H. Jacobi, "Methods of microwave imagery for diagnostic application," *Diagnostic Imaging in Medicine*, NATO Advanced Science Institute Series E, No. 61, C. R. Reba, ed., Nijhoff Publishers, The Hague, pp. 68–123, 1983.
7. T. C. Guo, W. W. Guo, and L. E. Larsen, "Comment on 'Microwave diffraction tomography for biomedical applications'," *IEEE Trans. Microwave Theory Tech.*, Vol. 32, p. 473, 1984.
8. T. C. Guo, W. W. Guo, and L. E. Larsen, Proc. IEEE 8th International Conference on Infrared and Millimeter Waves, Miami Beach, Florida, December 12–17, 1983.
9. M. Melek and A. P. Anderson, "Theoretical studies of localized tumour heating using focused microwave arrays," *IEE Proc.*, Vol. 127, Pt. F, No. 4, pp. 319–321, 1980.
10. J. Mendecki, E. Friedenthal, and C. Botstein, "Microwave-induced hyperthermia in cancer treatment: apparatus and preliminary results," *Int. J. Radiat. Oncol. Biol. Phys.*, Vol. 4, No. 11, pp. 1095–1103, 1978.
11. D. A. Christensen and C. H. Durney, "Hyperthermia production for cancer therapy: a review of fundamentals and methods," *J. of Microwave Power*, Vol. 16, No. 2, June, 1981.
12. T. C. Guo, W. W. Guo, and L. E. Larsen, "A local field study of a water-immersed microwave antenna array for medical imagery and therapy," *IEEE Trans. Microwave Theory Tech.*, Vol. MTT-32, pp. 844–854, 1984.
13. L. E. Larsen, J. H. Jacobi, W. W. Guo, T. C. Guo, and A. C. Kak, "Microwave imaging systems for medical diagnostic applications," Proceedings—Sixth Annual Conference of IEEE Engineering in Medicine and Biology Society—Frontiers Eng. Computing in Health Care, pp. 532–539, ed. John L. Semmlow and Walter Welkowitz, Los Angeles, CA., September, 1984.

14. I. Yamaura, "Measurements of 1.8–2.7 GHz microwave attenuation in the human torso," *IEEE Trans. Microwave Theory Tech.*, Vol. MTT-25, pp. 707–710, 1977.

15. J. H. Jacobi and L. E. Larsen, "Microwave time delay spectroscopic imagery of isolated canine kidney," *Med. Phys.*, Vol. 7, No. 1, pp. 1–7, 1980.

16. H. P. Schwan, "Radiation biology, medical applications, and radiation hazards," *Microwave Power Engineering*, Vol. 2, E. C. Okress, ed., Academic Press, New York, pp. 215–232, 1968.

17. K. Kurokawa, "Electromagnetic wave and waveguides with wall impedance," *IEEE Trans. Microwave Theory Tech.*, Vol. MTT-13, pp. 314–320, 1962.

18. J. Jackson, *Classical Electrodynamics*, John Wiley and Sons, New York, 1962, Ch. 9.

19. J. R. Carson, "Reciprocal theorems in radio communication," *Proc. IRE*, Vol. 17, pp. 952–956, 1929.

20. J. H. Richmond, "A reaction theorem and its application to antenna impedance calculations," *IRE Trans. Antennas Propag.*, Vol. AP-9, pp. 515–520, 1961.

21. L. D. Landau and E. M. Lifshitz, *Electrodynamics of Continuous Media*, Pergamon Press, Oxford, 1960.

Microwave Imaging with First Order Diffraction Tomography

Malcolm Slaney,* Mani Azimi,** Avinash C. Kak,*** and
Lawrence E. Larsen****

Tomographic imaging with microwave radiation is discussed from the perspective of relating the Fourier transformation of projection views (both bistatic and monostatic) to samples of the two-dimensional Fourier transformation of the scattering object. The limitations of the first order Born and Rytov approximations in scalar diffraction tomography are explored. The role of a complex index of refraction for the coupling medium and/or target is emphasized.

1. INTRODUCTION

During the past ten years the medical community has increasingly called on X-Ray computerized tomography (CT) to help make its diagnostic images. With this increased interest has also come an awareness of the dangers of using ionizing radiation and this, for example, has made X-Ray CT unsuitable for use in mass screening for cancer detection in the female breast. As a result, in recent years much attention has been given to imaging with alternative forms of energy such as low-level microwaves, ultrasound and NMR (nuclear-magnetic-resonance). Ultrasonic B-scan imaging has already found widespread clinical applications; however it lacks the quantitative aspects of ultrasonic computed tomography, which in turn can only be applied to soft tissue structures such as the female breast.

A necessary attribute of any form of radiation used for biological imaging is that it be possible to differentiate between different tissues on the basis of local propagation parameters. It has already been demonstrated by Larsen and Jacobi in a companion paper elsewhere in this volume and in reference [1] that this condition is satisfied by microwave radiation with the relative dielectric constant and the electric loss factor in the 1–10 GHz range. When used for tomography, a distinct feature of microwaves is that they allow one to reconstruct cross-sectional images of the molecular properties of the object. The dielectric properties of the water molecule dominate the interaction of microwaves and

biological systems [2], [3] and thus by interrogating the object with microwaves it is possible to image, for example, the state of hydration of an object.

The past interest in microwave imagery has focused primarily on either the holographic, or the pulse-echo modes. In the holographic mode, most attention has focussed on conducting targets in air, such as that described elsewhere in this volume by Farhat. There are exceptions as represented by the work of Yue et al. [4] wherein low-dielectric-constant slabs embedded in earth were imaged. The approach of Yue et al. is not applicable to the cross-sectional imaging of complicated three-dimensional objects, because of the underlying assumptions made regarding the availability of a priori information about the "propagators" in a volume cell of the object. Another example of microwave imaging with holography is the work of Gregoris and Izuka [5], wherein conductors and planar dielectric voids were holographically imaged inside flat dielectric layers. A reflection from the air-dielectric interface provided the reference beam. Again this work is not particularly relevant for microwave imaging of biosystems since many important biological constituents are dielectrics dominated by water. When used in the pulse-echo mode, microwaves again possess limited usefulness due to the requirement that the object be in the far field of the transmit/receive aperture, although the video pulse technique, described by Kim & Webster elsewhere in the volume mitigate this objection to some extent.

Tomography represents an attractive alternative to both holography and pulse-echo for cross-sectional (or three-dimensional) reconstruction of geometrically complicated biosystems, but there is a fundamental difference between tomographic imaging with x-rays and microwaves. X-rays, being non-diffracting, travel in straight lines, and therefore, the transmission data measures the line integral of some object parameter along straight lines. This makes it possible to apply the Fourier-slice theorem [6], which says that the

---

* Malcolm Slaney was with the School of Electrical Engineering, Purdue University, West Lafayette, IN 47907. He is now with Schlumberger Palo Alto Research, 3340 Hillview Avenue, Palo Alto, CA 94304.

** Mani Azimi was with the School of Electrical Engineering, Purdue University, West Lafayette, IN 47907. He is now with Department of Electrical Engineering, Michigan State University, East Lansing, MI 48824.

*** Avinash C. Kak is with the School of Electrical Engineering, Purdue University, West Lafayette, IN 47907.

**** Lawrence E. Larsen is with Microwaves Department, Walter Reed Army Institute of Research, Washington, DC 20012.

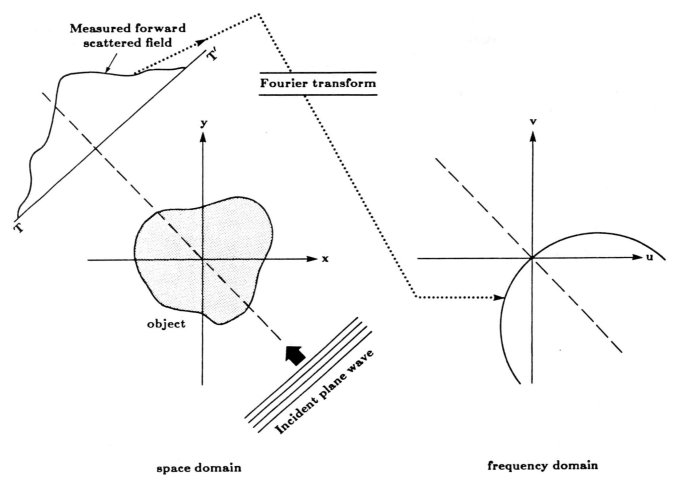

Fig. 1.   The Fourier Diffraction Theorem.

Fourier transform of a projection is equal to a slice of the two-dimensional Fourier transform of the object.

On the other hand, when microwaves are used for tomographic imaging, the energy often does not propagate along straight lines. When the object inhomogeneities are large compared to a wavelength, energy propagation is characterized by refraction and multipath effects. Moderate amounts of ray bending induced by refraction can be taken into account by combining algebraic reconstruction algorithms [7] with digital ray tracing and ray linking algorithms [8].

When the object inhomogeneities become comparable in size to a wavelength, it is not even appropriate to talk about propagation along lines or rays, and energy transmission must be discussed in terms of wavefronts and fields scattered by the inhomogeneities. Polarization or vector fields must also be considered in this circumstance as discussed in a review article on inverse scattering by Boerner and Chan elsewhere in this volume. A tutorial presentation of polarization description and polarimetric imaging is in a companion paper by Larsen & Jacobi elsewhere in this volume. When consideration is limited to scalar fields, it has been

shown [9], [10], [11], [12] that with certain approximations a Fourier-slice like theorem can be formulated. In [13] this theorem was called the Fourier Diffraction Projection Theorem. It may simply be stated as follows:

> When an object is illuminated with a plane wave as shown in Fig. 1, the Fourier transform of the forward scattered fields measured on a line perpendicular to the direction of propagation of the wave (line TT in Fig. 1) gives the values of the 2-D Fourier transform of the object along a circular arc as shown in the figure.

In Section 2 we will review the proof of this theorem. In our review, we will show how the derivation of the theorem points to a FFT-based implementation of higher order Born and Rytov algorithms, which are currently under development by us and other researchers. *The Fourier Diffraction Projection Theorem is valid only when the inhomogeneities in the object are weakly scattering.*

According to the Fourier Diffraction Projection Theorem, by illuminating an object from many different directions and measuring the diffracted data, one can in principle fill up the Fourier space with the samples of the Fourier transform of the object over an ensemble of circular arcs and then reconstruct the object by Fourier inversion.

The above theorem forms the basis of first order diffraction tomography. The work of Mueller et al. [11] was initially responsible for focusing the attention of many researchers on this approach to cross-sectional and three-dimensional imaging, although from a purely scientific standpoint the technique owes its origins to the now classic paper by Wolf [12], and a subsequent article by Iwata and Nagata [14].

This chapter will review the theory, implementation and some of the mathematical limitations of diffraction tomography with microwaves. As will be shown in the review of Section 2, the algorithms for diffraction tomography are derived from the classical wave equation. The wave equation is a non-linear differential equation that relates an object to the surrounding fields. To estimate a cross-sectional iamge of an object, it is necessary to find a linear solution to the wave equation and then to invert this relation between the object and the scattered field. The necessary approximations for this purpose limit the range of objects that can be successfully imaged to those that do not severely change the incident field or have a small refractive index gradient compared to the surrounding media.

In Section 3 we will look at several different methods to collect the scattered data and then invert it to find an estimate of the object. To generate a good estimate of the object it is necessary to combine the information from a number of different fields and this can be done with several different approaches. Then a simple algorithm based on the Fourier Diffraction Projection Theorem can be used to invert the scattered data.

Finally in Section 4 we will show the effects of these approximations by calculating the scattered fields for computer simulated objects using a number of different approaches. For cylindrical objects with a single refractive index it is possible to use the boundary conditions to solve for the exact scattered field. These simulations will establish the first order Born approximation to be valid for objects where the product of the change in refractive index and the diameter is less than 0.35λ and the first order Rytov approximation for changes in the refractive index of less than a few percent, with essentially no constraint on the object size. The scattered fields from objects that consist of more than one cylindrical object will then be calculated using Twersky's multiple scattering theory. These simulations will show that even when each component of the object satisfies the Born approximation the multiple scattering can degrade the reconstruction. Finally simulations will show that when attenuation is included in the model the high frequency information about the object is lost.

2. THE FOURIER DIFFRACTION THEOREM

Diffraction tomography is based on a linear solution to the wave equation. The wave equation relates an object and the scattered field and by linearizing it we can find an estimate of a cross section of the object based on the scattered field. The approximations used in the linearization process are crucial to the success of diffraction tomography and we will be careful to highlight the assumptions.

2.1   The Wave Equation

In a homogeneous medium, electromagnetic waves, $\psi(\vec{r})$, satisfy a homogeneous wave equation of the form

$$(\nabla^2 + k_0{}^2)\psi(\vec{r}) = 0, \tag{1}$$

where the wave number, k_0, represents the spatial frequency of the plane wave and is a function of the wavelength, λ, or $k_0 = 2\pi/\lambda$. It is easy to verify that a solution to Eq. (1) is given by a plane wave

$$\psi(\vec{r}) = e^{j\vec{k}_0 \cdot \vec{r}} \tag{2}$$

where $\vec{k}_0 = (k_x, k_y)$ is the wave vector of the wave and satisfies the relation $|\vec{k}_0| = k_0$. For imaging, an inhomogeneous medium is of interest, so the more general form of the wave equation is written as

$$(\nabla^2 + k^2(\vec{r}))\psi(\vec{r}) = 0. \tag{3}$$

For electromagnetic fields, if the effects of polarization are ignored, $k(\vec{r})$ can be considered to be a scalar function representing the refractive index of the medium. We then write

$$k(\vec{r}) = k_0 n(\vec{r}) = k_0[1 + n_\delta(\vec{r})], \tag{4}$$

where k_0 now represents the average wavenumber of the media, and $n(\vec{r})$ is the refractive index as given by

$$n(\vec{r}) = \sqrt{\frac{\mu(\vec{r})\epsilon(\vec{r})}{\mu_0 \epsilon_0}} . \tag{5}$$

The parameter $n_\delta(\vec{r})$ represents the deviation from the average of the refractive index. In general it will be assumed that the object of interest has finite support so $n_\delta(\vec{r})$ is zero outside the object. Here we have used μ and ϵ to represent the magnetic permeability and dielectric constant and the subscript zero to indicate their average values.

If the second order terms in n_δ (i.e., $n_\delta \ll 1$) are ignored we find

$$(\nabla^2 + k_0{}^2)\psi(\vec{r}) = -2k_0{}^2 n_\delta(\vec{r})\psi(\vec{r}) = -\psi(\vec{r})O(\vec{r}), \tag{6}$$

where $O(\vec{r}) = 2k_0{}^2 n_\delta(\vec{r})$ is usually called the object function.

Note that Eq. (6) is a *scalar* wave propagation equation. Its use implies that there is no depolarization as the electromagnetic wave propagates through the medium. It is known [15] that the depolarization effects can be ignored only if the wavelength is much smaller than the correlation size of the inhomogeneities in the object. If this condition is not satisfied, then strictly speaking the following vector wave propagation equation must be used:

$$\nabla^2 \vec{E}(r) + k_0{}^2 n^2 \vec{E}(\vec{r}) - 2\nabla\left[\frac{\nabla_n}{n} \cdot \vec{E}\right] = 0, \tag{7}$$

where $\vec{E}$ is the electric field vector. A vector theory for diffraction tomography based on this equation has yet to be developed.

In addition $\psi_0(\vec{r})$, the incident field, may be defined as

$$(\nabla^2 + k_0{}^2)\psi_0(\vec{r}) = 0. \qquad (8)$$

Thus $\psi_0(\vec{r})$ represents the source field or the field present without any object inhomogeneities. The total field then is expressed as the sum of the incident field and the scattered field

$$\psi(\vec{r}) = \psi_0(\vec{r}) + \psi_s(\vec{r}), \qquad (9)$$

with ψ_s satisfying the wave equation

$$(\nabla^2 + k_0{}^2)\psi_s(\vec{r}) = -\psi(\vec{r})O(\vec{r}) \qquad (10)$$

which is obtained by substituting Eqs. (8) and (9) in Eq. (6). This form of the wave equation will be used in the work to follow.

The scalar Helmholtz equation (10) cannot be solved for $\psi_s(\vec{r})$ directly but a solution can be written in terms of a Green's function [16]. The Green's function, which is a solution of the differential equation

$$(\nabla^2 + k_0{}^2)G(\vec{r}|\vec{r}') = -\delta(\vec{r} - \vec{r}'), \qquad (11)$$

is written in 3-space as

$$G(\vec{r}|\vec{r}') = \frac{e^{jk_0R}}{4\pi R}, \qquad (12)$$

with

$$R = |\vec{r} - \vec{r}'|. \qquad (13)$$

In two dimensions, the solution of (11) is written in terms of a zero-order Hankel function of the first kind, and can be expressed as

$$G(\vec{r}|\vec{r}') = \frac{j}{4} H_0{}^{(1)}(k_0R). \qquad (14)$$

In both cases the Green's function, $G(\vec{r}|\vec{r}')$, is only a function of the difference $\vec{r} - \vec{r}'$ so the argument of the Green's function will often be represented as simply $G(\vec{r} - \vec{r}')$. Because the object function in Eq. (11) represents a point inhomogeneity, the Green's function can be considered to represent the field resulting from a single point scatterer.

Since Eq. (11) represents the radiation from a two-dimensional impulse source, the total radiation from all sources on the right hand side of (10) must be given by the following superposition:

$$\psi_s(\vec{r}) = \int G(\vec{r} - \vec{r}')O(\vec{r}')\psi(\vec{r}')d\vec{r}'. \qquad (15)$$

In general, it is impossible to solve Eq. (15) for the scattered field, so approximations must be made. Two types of approximations will be considered: the Born and the Rytov.

2.2 The Born Approximation

The Born approximation is the simpler of the two approaches. Consider the total field, $\psi(\vec{r})$, expressed as the sum of the incident field, $\psi_0(\vec{r})$, and a small perturbation, $\psi_s(\vec{r})$, as in Eq. (9). The integral of Eq. (15) is now written as

$$\psi_s(\vec{r}) = \int G(\vec{r} - \vec{r}')O(\vec{r}')\psi_0(\vec{r}')dr' \\ + \int G(\vec{r} - \vec{r}')O(\vec{r}')\psi_s(\vec{r}')dr'. \qquad (16)$$

If the scattered field, $\psi_s(\vec{r})$, is small compared to $\psi_0(\vec{r})$, the effects of the second integral can be ignored to arrive at the approximation

$$\psi_s(\vec{r}) = \int G(\vec{r} - \vec{r}')O(\vec{r}')\psi_0(\vec{r}')dr'. \qquad (17)$$

This constitutes the first-order Born approximation. For a moment, let's denote the scattered fields obtained in this manner by $\psi_s{}^{(1)}(\vec{r})$. If one wished to compute $\psi_s{}^{(2)}(\vec{r})$ which represents the second order approximation to the scattered fields, that could be accomplished by substituting $\psi_0 + \psi_s{}^{(1)}$ for ψ_0 in the right hand side of Eq. (17), yielding

$$\psi_s{}^{(2)}(\vec{r}) = \int G(\vec{r} - \vec{r}')O(\vec{r}')[\psi_0(\vec{r}') + \psi_s{}^{(1)}(\vec{r}')]dr'. \qquad (18)$$

In general, we may write

$$\psi_s{}^{(i+1)}(\vec{r}) = \int G(\vec{r} - \vec{r}')O(\vec{r}')[\psi_0(\vec{r}') + \psi_s{}^{(i)}(\vec{r}')]dr' \qquad (19)$$

for the higher $(i + 1)$'th approximation to the scattered fields in terms of the i'th solution. Since the science of reconstructing objects with higher order approximations is not fully developed, this particular point will not be pursued any further and the first order scattered fields will be represented by ψ_s (i.e., without the superscript).

Note again that the first-order Born approximation is valid only when the magnitude of the scattered field,

$$\psi_s(\vec{r}) = \psi(\vec{r}) - \psi_0(\vec{r}), \qquad (20)$$

is smaller than that of the incident field, ψ_0. If the object is a cylinder of constant refractive index it is possible to express this condition as a function of the size of the object (radius $= a$) and the refractive index. Let the incident wave, $\psi_0(\vec{r})$, be a plane wave propagating in the direction of the unit vector, $\vec{k}_0$. For a large object, the field inside the object will not be given by

$$\psi(\vec{r}) = \psi_{\text{object}}(\vec{r}) \neq Ae^{j\vec{k}_0 \cdot \vec{r}}, \qquad (21)$$

but instead will be a function of the change in refractive index, n_δ. Along a ray through the center of the cylinder and parallel to the direction of propagation of the incident plane wave, the field inside the object becomes a slow (or fast) version of the incident wave or

$$\psi_{\text{object}}(\vec{r}) = Ae^{j(1+n_\delta)\vec{k}_0 \cdot \vec{r}}. \qquad (22)$$

Since the wave is propagating through the object, the phase difference between the incident field and the field inside the object is approximately equal to the integral through the object of the change in refractive index. Therefore, for a cylinder the total phase shift through the object is approximately

$$\text{Phase Change} = 4\pi n_\delta \frac{a}{\lambda}, \qquad (23)$$

where λ is the wavelength of the incident wave. For the first-order Born approximation to be valid, a necessary condition is that the change in phase between the incident field and the wave propagating through the object be less than π. This condition can be expressed mathematically as

$$n_\delta a < \frac{\lambda}{4}. \tag{24}$$

2.3   The Rytov Approximation

The Rytov approximation is valid under slightly less severe restrictions. It is derived by considering the total field to be represented as [15]

$$\psi(\vec{r}) = e^{\phi(\vec{r})}, \tag{25}$$

and rewriting the wave Eq. (1) as

$$(\nabla\phi)^2 + \nabla^2\phi + k_0^2 = -2k_0^2 n_\delta. \tag{26}$$

Expressing the total phase, ϕ, as the sum of the incident phase function ϕ_0 and the scattered complex phase ϕ_s or

$$\phi(\vec{r}) = \phi_0(\vec{r}) + \phi_s(\vec{r}), \tag{27}$$

where

$$\psi_0(\vec{r}) = e^{\phi_0(\vec{r})}, \tag{28}$$

we find that

$$(\nabla\phi_0)^2 + 2\nabla\phi_0\cdot\nabla\phi_s + (\nabla\phi_s)^2 + \nabla^2\phi_0 + \nabla^2\phi_s + k_0^2(1 + 2n_\delta) = 0. \tag{29}$$

As in the Born approximation, it is possible to set the zero perturbation equation equal to zero to find

$$2\nabla\phi_0\cdot\nabla\phi_s + \nabla^2\phi_s = -(\nabla\phi_s)^2 - 2k_0^2 n_\delta. \tag{30}$$

This equation is inhomogeneous and nonlinear but can be linearized by considering the following relation

$$\nabla^2(\psi_0\phi_s) = \nabla^2\psi_0\cdot\phi_s + 2\nabla\psi_0\cdot\nabla\phi_s + \psi_0\nabla^2\phi_s. \tag{31}$$

Recalling that

$$\psi_0 = Ae^{j\vec{k}_0\cdot\vec{r}} = e^{\phi_0(\vec{r})}, \tag{32}$$

we find

$$2\psi_0\nabla\phi_0\cdot\nabla\phi_s + \psi_0\nabla^2\phi_s = \nabla^2(\psi_0\phi_s) + k_0^2\psi_0\phi_s. \tag{33}$$

This result can be substituted into Eq. (30) to find

$$(\nabla^2 + k_0^2)\psi_0\phi_s = -\psi_0[(\nabla\phi_s)^2 + 2k_0^2 n_\delta]. \tag{34}$$

As before, the solution to this differential equation can again be expressed as an integral equation. This becomes

$$\psi_0\phi_s = \int_{V'} G(\vec{r} - \vec{r}')\psi_0[(\nabla\phi_s)^2 + 2k_0^2 n_\delta]dr', \tag{35}$$

where the Green's function is given by (14).

Under the Rytov approximation, it is assumed that the term in brackets in the above equation can be approximated by

$$(\nabla\phi_s)^2 + 2k_0^2 n_\delta \simeq 2k_0^2 n_\delta. \tag{36}$$

When this is done the first-order Rytov approximation to the scattered phase, ϕ_s, becomes

$$\phi_s(\vec{r}) \simeq \frac{2}{\psi_0(\vec{r})} \int_{V'} G(\vec{r} - \vec{r}')\psi_0(\vec{r})k_0^2 n_\delta dr'. \tag{37}$$

Substituting the expression for ψ_s given in Eq. (17) yields

$$\phi_s(\vec{r}) \simeq \frac{\psi_s(\vec{r})}{\psi_0(\vec{r})}. \tag{38}$$

It is important to note that, in spite of the similarity of the Born (17) and the Rytov (37) solutions, the approximations are quite different. As will be seen later, the Born approximation produces a better estimate of the scattered fields for objects small in size with large deviations in the refractive index. On the other hand, the Rytov approximation gives a more accurate estimate of the scattered field for large sized objects with small deviations in refractive index.

When the object is small and the refractive index deviates only slightly from the surrounding media, it is possible to show that the Born and the Rytov approximations produce the same results. Consider our definition of the scattered phase in Eq. (25) and (27). Expanding the scattered phase in the exponential with the Rytov solution to the scattered field, it is seen

$$\psi(\vec{r}) = e^{\phi_0(\vec{r})+\phi_s(\vec{r})} = \psi_0(\vec{r})e^{\exp(-j\vec{k}_0\cdot\vec{r})\psi_s(\vec{r})}. \tag{39}$$

For very small $\psi_s(\vec{r})$, the first exponential can be written in terms of the power series expansion to find

$$\psi(\vec{r}) \simeq \psi_0(\vec{r})[1 + \exp(-j\vec{k}_0\cdot\vec{r})\psi_s(\vec{r})] = \psi_0(\vec{r}) + \psi_s(\vec{r}). \tag{40}$$

Thus when the magnitude of the scattered field is very small the Rytov approximation simplifies to the Born approximation.

The Rytov approximation is valid under a less restrictive set of conditions than the Born approximation [17], [18]. In deriving the Rytov approximation, the assumption was made that

$$(\nabla\phi_s)^2 + 2k_0^2 n_\delta \simeq 2k_0^2 n_\delta. \tag{41}$$

Clearly this is true only when

$$n_\delta \gg \frac{(\nabla\phi_s)^2}{k_0^2}. \tag{42}$$

This can be justified by observing that to a first approximation the scattered phase, ϕ_s, is linearly dependent on n_δ [17]. If n_δ is small, then

$$(\nabla\phi_s)^2 \propto n_\delta^2 \tag{43}$$

will be even smaller and therefore the first term in Eq. (41) above can be safely ignored. *Unlike the Born approximation, the size of the object is not a factor in the Rytov approximation.* The term $\nabla\phi_s$ is the change in the complex scattered phase per unit distance and by substituting $k_0 = 2\pi/\lambda$ we find a necessary condition for the validity of the Rytov approximation is

$$n_\delta \gg \left[\frac{\nabla\phi_s\lambda}{2\pi}\right]^2. \tag{44}$$

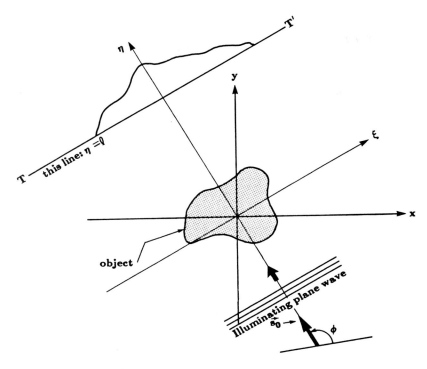

Fig. 2. A typical diffraction tomography experiment.

Therefore in the Rytov approximation, it is the change in scattered phase, ϕ_s, over one wavelength that is important and not the total phase. Thus, because of the ∇ operator the Rytov approximation is valid when the phase change over a single wavelength is small.

2.4   The Scattered Fields

The Fourier Diffraction Theorem relates the Fourier transform of the scattered field, the diffracted projection, to the Fourier transform of the object along a circular arc. While a number of researchers have derived this theory [11], [9], [13], [19] we would like to propose a system theoretic analysis of this result which is fundamental to first order diffraction tomography. This approach is superior not only because it allows the scattering process to be visualized in the Fourier domain but also because it points to efficient FFT-based computer implementations of higher order Born and Rytov algorithms currently under development. Since it appears that the higher order algorithms will be more computationally intensive, any savings in the computing effort involved is potentially important.

Consider the effect of a single plane wave incident on an object. The forward scattered field will be measured at a receiver line as shown in Fig. 2. We will find an expression for the field scattered by the object, $O(\vec{r})$, by analyzing Eq. (17) in the Fourier domain. We will use the plots of Fig. 3 to illustrate the transformations that take place.

The first Born equation for the scattered field (17) can be considered as a convolution of the Green's Function, $G(\vec{r})$, and the product of the object function, $O(\vec{r})$, and the incident field, $\psi_0(\vec{r})$. First we will define the following Fourier transform pairs:

$$O(\vec{r}) \leftrightarrow \tilde{O}(\vec{\Lambda}),$$
$$G(\vec{r}) \leftrightarrow \tilde{G}(\vec{\Lambda}) \tag{45}$$

and

$$\psi(\vec{r}) \leftrightarrow \tilde{\psi}(\vec{\Lambda}),$$

where we have used the relationships

$$\tilde{O}(\vec{\Lambda}) = \int \int O(\vec{r}) e^{-j\vec{\Lambda}\cdot\vec{r}} d\vec{r}, \tag{46}$$

$\vec{\Lambda} = (\alpha, \beta)$ and (α, β) being the spatial angular frequencies along the x and y directions respectively.

The integral solution to the wave Eq. (17) can now be written in terms of these Fourier transforms

$$\tilde{\psi}_s(\vec{\Lambda}) = \tilde{G}(\vec{\Lambda})\{\tilde{O}(\vec{\Lambda}) * \tilde{\psi}_0(\vec{\Lambda})\}, \tag{47}$$

where we have used "$*$" to represent convolution. When the illumination field, ψ_0, consists of a single plane wave

$$\psi_0(\vec{r}) = e^{j\vec{k}_0\cdot\vec{r}} \tag{48}$$

with $\vec{k}_0 = (k_x, k_y)$ satisfying the following relationship

$$k_0^2 = k_x^2 + k_y^2 \tag{49}$$

its Fourier transform is given by

$$\tilde{\psi}_0(\vec{\Lambda}) = 2\pi\delta(\vec{\Lambda} - \vec{k}_0). \tag{50}$$

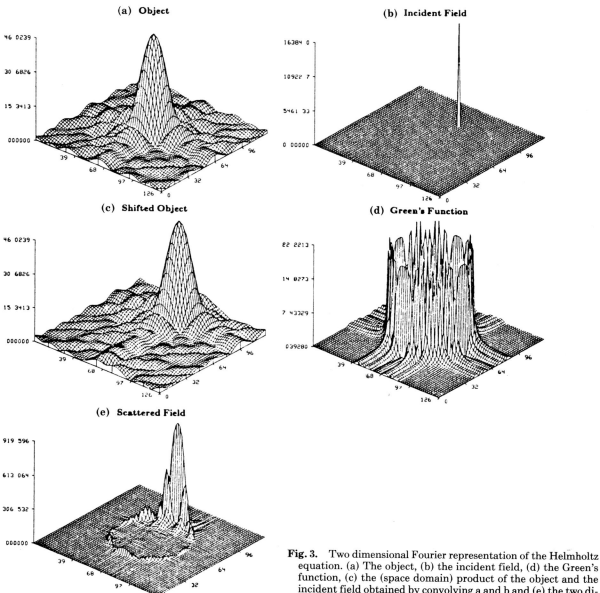

Fig. 3. Two dimensional Fourier representation of the Helmholtz equation. (a) The object, (b) the incident field, (d) the Green's function, (c) the (space domain) product of the object and the incident field obtained by convolving a and b and (e) the two dimensional Fourier transform of the scattered field obtained by multiplication of c and d.

The delta function causes the convolution of Eq. (47) to become a shift in the frequency domain as given by

$$\tilde{O}(\vec{\Lambda}) * \tilde{\psi}_0(\vec{\Lambda}) = 2\pi \tilde{O}(\vec{\Lambda} - \vec{k}_0). \tag{51}$$

This convolution is illustrated in Figs. 3a–c for a plane wave propagating with direction vector, $\vec{k}_0 = (0, k_0)$. Figure 3a shows the Fourier transform of a single cylinder of radius 1λ and Fig. 3b is the Fourier transform of the incident field. The resulting convolution in the frequency domain (or multiplication in the space domain) is shown in Fig. 3c.

To find the Fourier transform of the Green's function, the Fourier transform of Eq. (11) is taken to find

$$(-\Lambda^2 + k_0^2)\tilde{G}(\vec{\Lambda}|\vec{r}') = -e^{-j\vec{\Lambda}\cdot\vec{r}'}, \tag{52}$$

where $\Lambda^2 = \alpha^2 + \beta^2$. Rearranging terms we see that

$$\tilde{G}(\vec{\Lambda}|\vec{r}') = \frac{e^{-j\vec{\Lambda}\cdot\vec{r}'}}{\Lambda^2 - k_0^2}, \tag{53}$$

which has a singularity for all $\vec{\Lambda}$ such that

$$\Lambda^2 = \alpha^2 + \beta^2 = k_0^2. \tag{54}$$

In the space domain the two dimensional Green's function, Eq. (14), has a singularity at the origin so it is necessary to approximate the Green's function by using a two dimensional average of the values of the Green's function near the singularity. An approximation to $\tilde{G}(\vec{\Lambda})$ is shown in Fig. 3d.

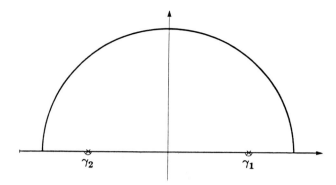

Fig. 4. Integration path in the complex plane for inverting the two dimensional Fourier transform of the scattered field.

The Fourier representation in Eq. (53) is misleading because both point sources and point sinks are valid solutions to Eq. (52). Thus the simple expression of Eq. (53) includes the effects of both waves moving toward and waves moving away from the point at $\vec{r}'$. Later, when we move back from the Fourier domain to the space domain, it will be necessary to choose the proper Fourier components so that only waves traveling away from the point scatterer are retained.

The effect of the convolution shown in Eq. (17) is a multiplication in the frequency domain of the shifted object function, Eq. (51), and the Green's function, Eq. (53), evaluated at $\vec{r}' = 0$. The scattered field is written as

$$\psi_s(\vec{\Lambda}) = 2\pi \frac{\tilde{O}(\vec{\Lambda} - \vec{k}_0)}{\Lambda^2 - k_0^2} \, . \tag{55}$$

This result is shown in Fig. 3e for a plane wave propagating along the y-axis. Since the largest frequency domain components of the Green's function satisfy Eq. (1), the Fourier transform of the scattered field is dominated by a shifted and sampled version of the object's Fourier transform.

We will now derive an expression for the field at the receiver line. For simplicity it will be assumed that the incident field is propagating along the positive y axis or $\vec{k}_0 = (0, k_0)$. The scattered field along the receiver line (x, y = l_0) is simply the inverse Fourier transform of the field in Eq. (55). This is written as

$$\psi_s(x, y = l_0) = \frac{1}{4\pi^2} \int\!\int \psi_s(\vec{\Lambda}) e^{j\vec{\Lambda}\cdot\vec{r}} d\alpha d\beta, \tag{56}$$

which, using Eq. (55), can be expressed as

$$\psi_s(x, y = l_0) = \frac{1}{2\pi} \int\!\int \frac{\tilde{O}(\alpha, \beta - k_0)}{\alpha^2 + \beta^2 - k_0^2} e^{j(\alpha x + \beta l_0)} d\alpha d\beta. \tag{57}$$

We will integrate with respect to β. For a given α, the integral has a singularity at

$$\beta_{1,2} = \pm \sqrt{k_0^2 - \alpha^2} \, . \tag{58}$$

Using contour integration we can close the integration path at infinity and evaluate the integral with respect to β along the path shown in Fig. 4 to find

$$\psi_s(x, l_0) = \int \Gamma_1(\alpha; l_0) e^{j\alpha x} d\alpha + \int \Gamma_2(\alpha; l_0) e^{j\alpha x} d\alpha, \tag{59}$$

where

$$\Gamma_1 = \frac{\tilde{O}(\alpha, \sqrt{k_0^2 - \alpha^2} - k_0)}{j2 \sqrt{k_0^2 - \alpha^2}} e^{j\sqrt{k_0^2 - \alpha^2} l_0} \tag{60}$$

and

$$\Gamma_2 = \frac{\tilde{O}(\alpha, -\sqrt{k_0^2 - \alpha^2} - k_0)}{-j2 \sqrt{k_0^2 - \alpha^2}} e^{j\sqrt{k_0^2 - \alpha^2} l_0}. \tag{61}$$

Examining the above pair of equations, it is seen that Γ_1 represents the solution in terms of plane waves traveling along the positive y axis while Γ_2 represents plane waves traveling in the $-y$ direction. These distinct solutions represent the two solutions to the wave equation for a point discontinuity [see Eq. (53)]. In both cases, as α ranges from $-k_0$ to k_0, Γ represents the Fourier transform of the object along a semicircular arc.

Since we are interested in the forward traveling waves, only the plane waves represented by the Γ_1 solution are valid; and, thus, the scattered field becomes

$$\psi_s(x, l_0) = \int \Gamma_1(\alpha; y) e^{j\alpha x} d\alpha \quad l_0 > \text{object} \tag{62}$$

where we have chosen the value of the square root to lead only to outgoing waves.

Taking the Fourier transform of both sides of Eq. (62) we find that

$$\int \psi_s(x, y = l_0) e^{-j\alpha x} dx = \tilde{\Gamma}(\alpha, l_0). \tag{63}$$

But since $\Gamma(x, l_0)$ is equal to a phase shifted version of the object function, the Fourier transform of the scattered field along the line $y = l_0$ is related to the Fourier transform of the object along a circular arc. The use of the contour integration is further justified by noting that only those waves that satisfy the relationship

$$\alpha^2 + \beta^2 = k_0^2 \tag{64}$$

will be propagated. Thus it is safe to ignore all waves not on the k_0-circle.

This result is diagramed in Fig. 5. The circular arc repre-

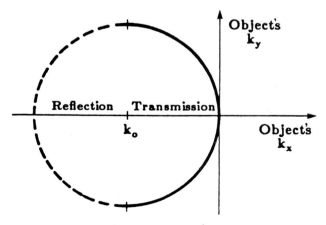

Fig. 5. Estimate of the two dimensional Fourier transform of the object are available along the solid arc for transmission tomography and the dashed arc for reflection tomography.

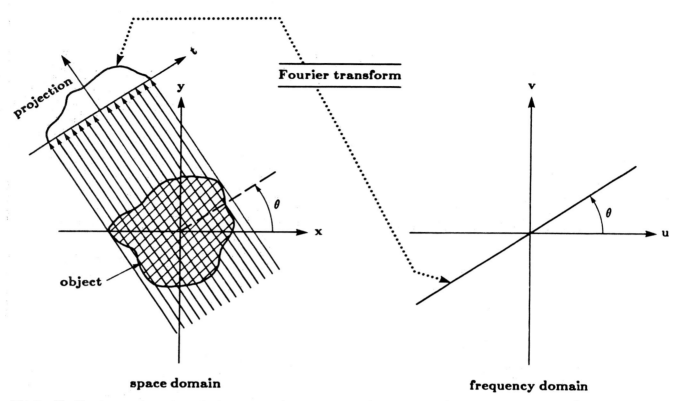

Fig. 6. The Fourier transform of a projection is equal to the two dimensional Fourier transform of the object along a radial line.

sents the locus of all points (α, β) such that $\beta = \pm \sqrt{k_0^2 - \alpha^2}$. The solid line shows the outgoing waves for a receiver line at $y = l_0$ greater than the object. This can be considered transmission tomography. Conversely the dashed line indicates the locus of solutions for $y = l_0$ less than the object or the reflection tomography case.

Straight-ray (i.e., X-ray) tomography is based on the Fourier Slice Theorem [10], [6]

The Fourier transform of a parallel projection of an image f(x,y) taken at an angle θ gives a slice of the 2-D transform, $F(\omega_1, \omega_2)$ subtending an angle θ with the ω_1 axis.

This is diagramed in Fig. 6.

Equation (63) leads us to a similar result for diffraction tomography. Recall that α and β in Eq. (63) are related by

$$\beta = \sqrt{k_0^2 - \alpha^2}. \tag{65}$$

Thus $\tilde{\Gamma}(\alpha)$, the Fourier transform of the received field, is proportional to $\tilde{O}(\alpha, \beta - k_0)$, the Fourier transform of the object along a circular arc. This result has been called the Fourier Diffraction Projection Theorem [13] and is diagramed in Fig. 1.

We have derived an expression, Eq. (63), that relates the scattering by an object to the field received at a line. Within the diffraction limit it is possible to invert this relation to estimate the object scattering distribution based on the received field.

3. THE RECONSTRUCTION PROCESS

The best that can be hoped for in any tomographic experiment is to estimate the Fourier transform of the object for all frequencies within a disk centered at the origin. For objects that do not have any frequency content outside the disc, then the reconstruction procedure is perfect.

There are several different procedures that can be used to estimate the object function from the forward scattered fields. A single plane wave provides exact information (up to a frequency of $\sqrt{2}k_0$) about the Fourier transform of the object along a circular arc. Two of the simplest procedures involve changing the orientation and frequency of the incident plane waves to move the frequency domain arcs to a new position. By appropriately choosing an orientation and a frequency it is possible to estimate the Fourier transform of the object at any given frequency. In addition, it is possible to change the radius of the semicircular arc by varying the frequency of the incident field and thus generating an estimate of the entire Fourier transform of the object. This concept is contained in a companion paper by Farhat elsewhere in this volume.

An important point to notice here is that reflection and transmission tomography provide completely different information about the object (see Fig. 5). A transmission experiment gives information about the object up to a spatial frequency of $\sqrt{2}k_0$. On the other hand, a reflection experiment gives the information for spatial frequencies between

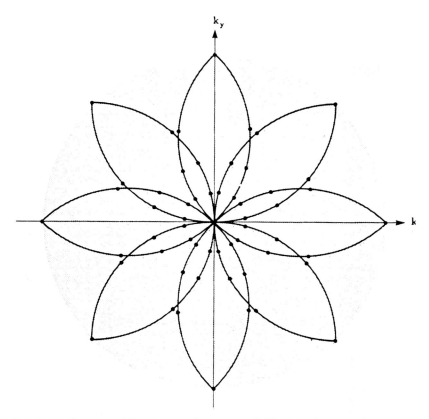

Fig. 7.   Estimates of the object's two dimensional Fourier transform are available along the circular arcs for plane wave illumination.

$\sqrt{2}k_0$ and $2k_0$. In principle it should be possible to combine the two experiments and obtain an estimate of the amplitude of all the spatial frequencies up to $2k_0$.

3.1   Plane Wave Illumination

The most straightforward data collection procedure consists of rotating the object and measuring the scattered field for different orientations. Each orientation will produce an estimate of the object's Fourier transform along a circular arc and these arcs will rotate as the object is rotated. When the object is rotated through a full 360 degrees an estimate of the object will be available for the entire Fourier disk.

The coverage for this method is shown in Fig. 7 for a simple experiment with 8 projections of 9 samples each. Notice that there are two arcs that pass through each point of Fourier space. Generally it will be necessary to choose one estimate as better.

On the other hand, if the reflected data is collected by measuring the field on the same side of the object as the source then estimates of the object are available for frequencies greater than $\sqrt{2}k_0$. This follows from Fig. 5.

The first experimental results for diffraction tomography were presented by Carter and Ho [20], [21], [22]. They used an optical plane wave to illuminate a small glass object and were able to measure the scattered fields using a hologram.

Later a group of researchers at the University of Minnesota carried out the same experiments using ultrasound and gelatine phantoms. Their results are discussed in Ref. [23].

3.2   Synthetic Aperture

Nahamoo and Kak [24], [25] and Devaney [26] have proposed a method that requires only two rotational views of an object. Consider an arbitrary source of waves in the transmitter plane as shown in Fig. 8. The transmitted field, ψ_t, can be represented as a weighted set of plane waves by taking the Fourier transform of the transmitter aperture function [27]. Doing this the transmitted field can be expressed as

$$\psi_t(x) = \frac{1}{4\pi^2} \int_{-\infty}^{\infty} A_t(k_x)e^{jk_xx}dk_x. \qquad (66)$$

Moving the source to a new position, η, the plane wave decomposition of the transmitted field becomes

$$\psi_t(x;\eta) = \frac{1}{4\pi^2} \int_{-\infty}^{\infty} (A_t(k_x)e^{-jk_x\eta})e^{jk_xx}dk_x. \qquad (67)$$

Given the plane wave decomposition, the incident field in the plane follows simply as

$$\psi_i(\eta;x,y) = \int_{-\infty}^{\infty} \left(\frac{1}{4\pi^2} A_t(k_x)e^{-jk_x\eta}\right)e^{j(k_xx+k_yy)}dk_x. \qquad (68)$$

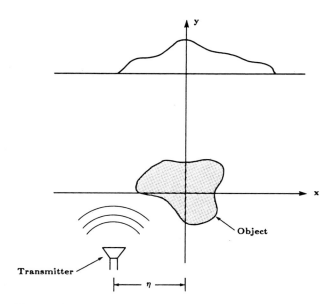

Fig. 8. A typical synthetic aperture tomography experiment.

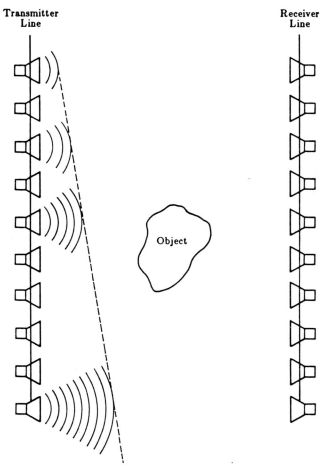

Fig. 9. By adding a phase to the field transmitted from each transmitter any desired plane wave can be synthesized.

Eq. (59) is an equation for the scattered field from a single plane wave. Because of the linearity of the Fourier transform, the effect of each plane wave, $e^{j(k_x x + k_y y)}$, can be weighted by the expression in brackets above and superimposed to find the Fourier transform of the total scattered field due to the incident field $\psi_t(x;\eta)$ as [24]

$$\tilde{\psi}_s(\eta,\alpha) = \int_{-\infty}^{\infty} (A_t(k_x)e^{-jk_x\eta}) \frac{O(\alpha - k_x, \gamma - k_y)}{j2\gamma} e^{j\gamma l_0} dk_x.$$

(69)

The quantity $\tilde{\psi}_s(\eta;\alpha)$ represents the one-dimensional Fourier transform of the field along a receiver line at a distance of l_0 from the origin due to a point source at η. Taking the Fourier transform of both sides with respect to the transmitter position, η, the Fourier transform of the scattered field with respect to both the transmitter and the receiver position is given by

$$\tilde{\psi}_s(k_x;\alpha) = A_t(k_x) \frac{O(\alpha - k_x, \gamma - k_y)}{j2\gamma} e^{j\gamma l_0}.$$

(70)

This approach is named synthetic aperture because a phase is added to the field measured for each transmitter position to synthesize a transmitted plane wave. Thus this method has much in common with the theory of phased arrays. Figure 9 shows that by properly phasing the wave transmitted at each transmitter location a plane wave can be generated that travels in an arbitrary direction. Since the system is linear it doesn't matter whether the phase is added to the transmitted signal or as part of the reconstruction procedure. Thus multiplying the received field for each transmitter position by the pure phase term $e^{-jk_x\eta}$, where η represents the location of the transmitter, is equivalent to an experiment with an incident plane wave with the direction vector $(k_x, \sqrt{k_0^2 - k_x^2})$. The concept is similar to that

of beam steering as discussed by Foti *et al* elsewhere in this volume.

By collecting the scattered field along the receiver line as a function of transmitter position, η, an expression can be written for the scattered field. Like the simpler case with plane wave incidence, the scattered field is related to the Fourier transform of the object along an arc. Unlike the previous case, though, the coverage due to a single view of the object is a pair of circular disks as shown in Fig. 10. Here a single view consists of transmitting from all positions in a line and measuring the scattered field at all positions along the receiver line. By rotating the object by 90 degrees it is possible to generate the complimentary pair of disks and to fill the Fourier domain out to $\pm 2k_0$ along both axes.

The coverage shown in Fig. 10 is constructed by calculating $(\vec{K} - \vec{\Lambda})$ for all vectors $(\vec{K})$ and $(\vec{\Lambda})$ that satisfy the experimental constraints. Not only must each vector satisfy the wave equation, but it is also necessary that only forward traveling plane waves be used. The dashed line in Fig. 10 shows the valid propagation vectors $(-\vec{\Lambda})$ for the transmitted waves. To each possible vector $(-\vec{\Lambda})$ a semicircular set of

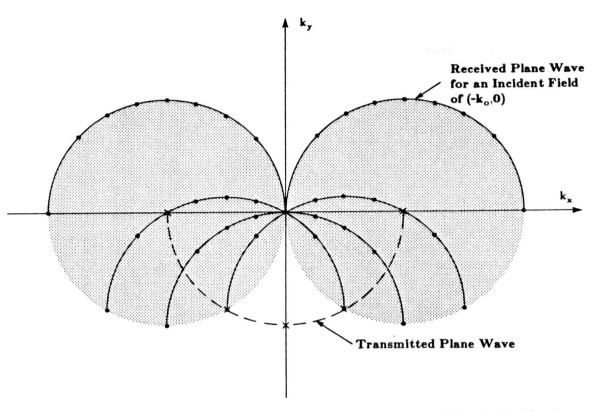

Fig. 10.   Estimates of the Fourier transform of an object in a synthetic aperture experiment are available in the shaded region.

vectors representing each possible received wave can be added. The locus of received plane waves is shown as a solid semi-circle centered at each of the transmitted waves indicated by an "x." The entire coverage for the synthetic aperture approach is shown as the shaded areas.

In addition to the diffraction tomography configurations proposed by Mueller and Nahamoo other approaches have been proposed. In Vertical Seismic Profiling (VSP) [26] the scattering between the surface of the Earth and a borehole is measured. Alternately a broadband incident field can be used to illuminate the object. In both cases, the goal is to estimate the Fourier transform of the object.

In geophysical imaging it is not possible to generate or receive waves from all positions around the object. If it is possible to drill a borehole then it is possible to perform VSP and obtain information about most of the object. A typical experiment is shown in Fig. 11. So as to not damage the borehole, acoustic waves are generated at the surface using acoutic detonators or other methods and the scattered field is measured in the borehole.

The coverage in the frequency domain is similar to the synthetic aperture approach. Plane waves at an arbitrary downward direction are synthesized by appropriately phasing the transmitting transducers. The receivers will receive any waves traveling to the right. The resulting coverage for this method is shown in Fig. 12a. If it can be assumed that the object function is real valued then the sym-

metry of the Fourier transform for real valued functions can be used to obtain the coverage in Fig. 12b.

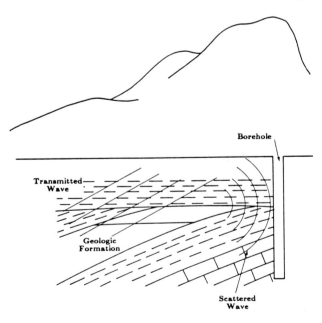

Fig. 11.   A typical Verticla Seismic Profiling (VSP) experiment.

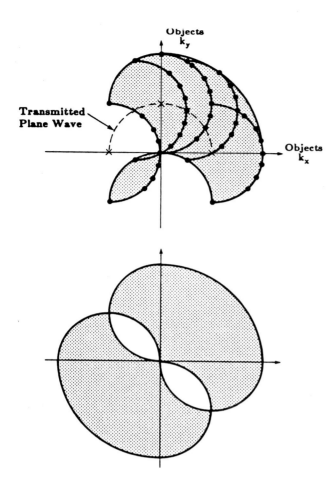

If a plane wave illumination of spatial frequency k_x and a temporal frequency ω leads to the scattered field $u_s(k_x,w;y)$, then the total scattered field is given by a weighted superposition of the scattered fields or

$$\psi_s(k_x;y) = \int_{-\infty}^{\infty} A_t(k_x,\omega)u_s(k_x,\omega;y)d\omega. \qquad (74)$$

For plane wave incidence, the coverage available with this method is shown in Fig. 13a. Figure 13b shows that by doing four experiments at 0, 90, 180, and 270 degrees it is possible to gather information about the entire object.

3.4   Reconstruction by Interpolation

The Fourier Diffraction Theorem as derived in Section 2 shows that when an object is illuminated with a plane wave

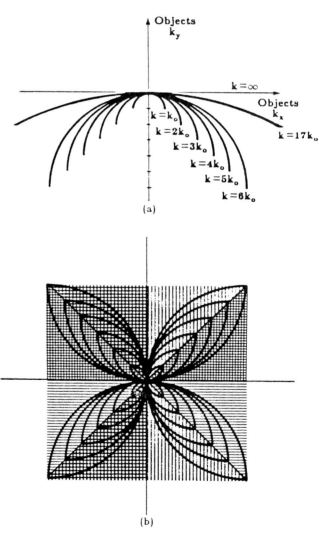

Fig. 12.   Estimate of the Fourier transform of an object are available in the shaded region for a VSP experiment (a). If, in addition, the object is real valued then the symmetry of the Fourier transform can be used to get the coverage shown in (b).

3.3   Broadband Illumination

It is also possible to perform an experiment for broadband illumination [19]. Up until this point only narrow band illumination has been considered; wherein the field at each point can be completely described by its complex amplitude.

Now consider a transducer that illuminates an object with a wave of the form $a_t(k_x,t)$. Taking the Fourier transform in the time domain this wave can be decomposed into a number of experiments. Let

$$A_t(k_x,\omega) = \int_{-\infty}^{\infty} a_t(k_x,t)e^{-j\omega t}dt \qquad (71)$$

where ω is related to k_ω by

$$k_\omega = \frac{c}{\omega}, \qquad (72)$$

c is the speed of propagation in the media and the wavevector (k_x,k_y) satisfies the wave equation

$$k_x{}^2 + k_y{}^2 = k_\omega{}^2. \qquad (73)$$

Fig. 13.   One view of a broadband diffraction tomography experiment will generate estimates of the object along the arcs in (a). With four views of the object complete coverage can be obtained as shown in (b).

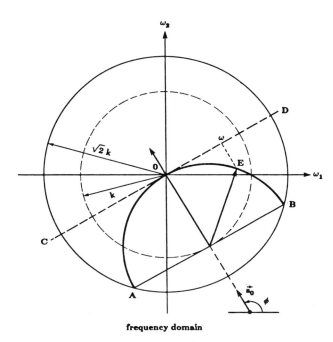

Fig. 14. Each projection is measured using the phi-omega coordinate system shown here.

traveling in the positive y-direction, the Fourier transform of the forward scattered fields gives values of the object's Fourier transform on an arc. Therefore, if an object is illuminated from many different directions it is possible, in principle, to fill up a disc of diameter $\sqrt{2}k_0$ in the frequency domain with samples of the Fourier transform of the object and then reconstruct the object by direct Fourier inversion. Therefore, diffraction tomography, using forward scattered data only, determines the object up to a maximum angular spatial frequency of $\sqrt{2}k_0$. To this extent, the reconstructed object is a low pass version of the original. In practice, the loss of resolution caused by this bandlimiting is negligible, being more influenced by considerations such as the aperture sizes of the transmitting and receiving elements, etc.

The fact that the frequency domain samples are available over circular arcs is a source of computational difficulty in reconstruction algorithms for diffraction tomography since for convenient display it is desired to have samples over a rectangular lattice. It should also be clear that by illuminating the object over 360°, a *double* coverage of the frequency domain is generated; note, however, that this double coverage is uniform. If the illumination is restricted to a portion of 360°, there still will be a complete coverage of the frequency domain; however, in that case, there would be patches in the (ω_1,ω_2)-plane where there would be a double coverage. In reconstructing from circular arc grids to rectangular grids, it is often easier to contend with a uniform double coverage, as opposed to a coverage that is single in most areas and double in patches.

However, for some applications not given to data collection from all possible directions, it is useful to bear in mind that it is not necessary to go completely around an object to get complete coverage of the frequency domain. In principle,

it should be possible to get an equal quality reconstruction when illumination angles are restricted to a 180°-plus interval. The few angles in excess of 180° are required to complete the coverage of the frequency domain.

There are two computational strategies for reconstructing the object given measurements of the scattered field. As pointed out by [28] the two algorithms can be considered as interpolation in the frequency domain and in the space domain and are analogous to the direct Fourier inversion and backprojection algorithms of conventional tomography. Unlike conventional tomography, where backprojection is the preferred approach, the computational expense of space domain interpolation of diffracted projections makes frequency domain interpolation the preferred approach.

The remainder of this section will consist of derivations of the frequency domain interpolation algorithm. The reader is referred to Devaney [9] and Pan [13] for excellent explanations of the space domain interpolation algorithm and [29] for the general case.

In order to discuss the frequency domain interpolation between a circular grid on which the data is generated by diffraction tomography, and a rectangular grid suitable for image reconstruction, parameters for representing each grid must be selected. Then the relationship between the two sets of parameters can be written.

In Section 2, $\mathcal{U}_B(\omega)$ was used to denote the Fourier transform of the transmitted data when an object is illuminated with a plane wave traveling along the positive y direction. Now $\mathcal{U}_{B,\phi}(\omega)$ is used to denote this Fourier transform, where the subscript ϕ indicates the angle of illumination. This angle is measured as shown in Fig. 14. Similarly, $Q(\omega,\phi)$ will be used to indicate the values of $O(\omega_1,\omega_2)$ along a semi-circular arc oriented at an angle ϕ as shown in Fig. 15.

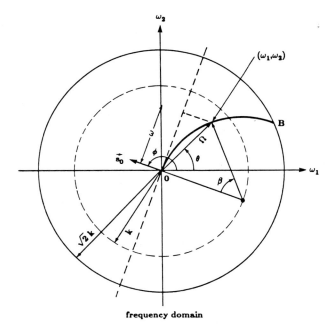

Fig. 15. A second change of variables is used to relate the projection data to the object's Fourier transform.

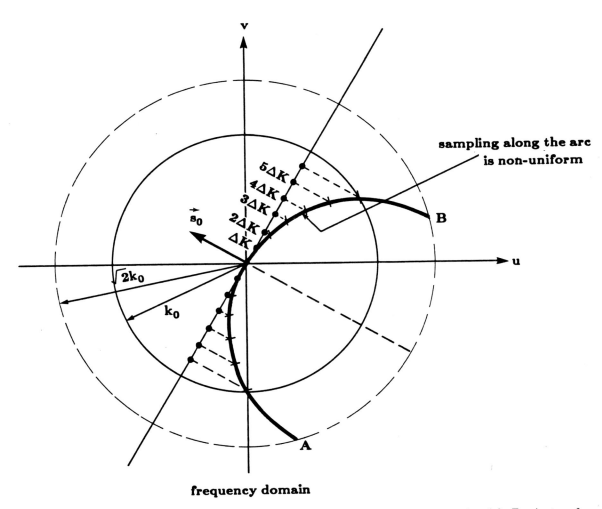

Fig. 16. Uniformly sampling the projection in the space domain leads to uneven spacing of the samples of the Fourier transform of the object along the semi-circular arc.

Therefore, when an illuminating plane wave is incident at angle ϕ, the equality

$$\psi_B(\omega,l_0) = \frac{j}{2\sqrt{k_0^2 - \omega^2}} e^{j\sqrt{k_0^2 - \omega^2}l_0} O(\omega, \sqrt{k_0^2 - \omega^2} - k_0)$$

(75)

can be rewritten as

$$\psi_{B,\phi}(\omega) = \frac{j}{2} \frac{1}{\sqrt{k_0^2 - \omega^2}} \exp[jl_0 \sqrt{k_0^2 - \omega^2}]Q(\omega,\phi)$$

$$\text{for } |\omega| < k_0. \quad (76)$$

In most cases, the transmitted data will be uniformly sampled in space, and a *discrete* Fourier transform of this data will generate uniformly spaced samples of $\psi_{B,\phi}(\omega)$ in the ω domain. Since $Q(\omega,\phi)$ is the Fourier transform of the object along the circular arc AOB in Fig. 14, and since κ is the projection of a point on the circular arc onto the tangent line CD, the uniform samples Q in λ translate into non-uniform samples along the arc AOB as shown in Fig. 16. For this

reason, designate each point on the arc AOB by its (ω,ϕ) parameters. [Note that (ω,ϕ) are *not* the polar coordinates of a point on arc AOB in Fig. 15. Therefore, ω is *not* the radial distance in the (ω_1,ω_2) plane. For point E shown, the parameter ω is obtained by projecting E onto line CD.] The rectangular coordinates in the frequency domain will remain (ω_1,ω_2).

Before the relationships between (ω,ϕ) and (ω_1,ω_2), are presented, it must be mentioned that the points generated by the AO and OB portions of the arc AOB must be considered separately as ϕ is varied from 0 to 2π. This is done because, as mentioned before, the arc AOB generates a double coverage of the frequency domain, as ϕ is varied from 0 to 2π, which is undesirable for discussing a one-to-one transformation between the (ω,ϕ) parameters and the (ω_1,ω_2) coordinates.

Now reserve (ω,ϕ) parameters to denote the arc grid generated by one projection. It is important to note that for this arc grid, ω varies from 0 to k and ϕ from 0 to 2π.

The transformation equations between (ω,ϕ) and (ω_1,ω_2)

will now be presented. This is accomplished in a slightly round-about manner by first defining polar coordinates (Ω,θ) in the (ω_1,ω_2)-plane as shown in Fig. 15. In order to go from (ω_1,ω_2) to (ω,ϕ), first transform from the former coordinates to (Ω,θ) and then from (Ω,θ) to (ω,ϕ). The rectangular coordinates (ω_1,ω_2) are related to the polar coordinates (Ω,θ) by (Fig. 15)

$$\Omega = \sqrt{\omega_1^2 + \omega_2^2} \tag{77}$$

$$\theta = \tan^{-1}\left(\frac{\omega_2}{\omega_1}\right). \tag{78}$$

In order to relate (Ω,θ) to (ω,ϕ), a new angle β, which is the angular position of a point (ω_1,ω_2) on arc OB in Fig. 15, is introduced. Note from the figure that the point characterized by angle β is also characterized by parameter ω. The relationship between ω and β is given by

$$\omega = k_0 \sin\beta. \tag{77}$$

The following relationship exists between the polar coordinates (Ω,θ) on the one hand and the parameters β and ϕ on the other:

$$\beta = 2 \sin^{-1}\frac{\Omega}{2k_0} \tag{79}$$

$$\phi = \theta + \frac{\pi}{2} + \frac{\beta}{2}. \tag{80}$$

By substituting Eq. (79) in (77) and then using (77), ω can be expressed in terms of ω_1 and ω_2. This result is shown below. Similarly, by substituting Eq. (78) in (80), the following expression is obtained for ω and ϕ

$$\omega = \sin\left\{2 \sin^{-1}\left(\frac{\sqrt{\omega_1^2 + \omega_2^2}}{2k_0}\right)\right\} \tag{81}$$

$$\theta = \tan^{-1}\left(\frac{\omega_2}{\omega_1}\right) + \sin^{-1}\left(\frac{\sqrt{\omega_1^2 + \omega_2^2}}{2k_0}\right) + \frac{\pi}{2}. \tag{82}$$

These are the transformation equations for interpolating from the (ω,ϕ) parameters used for data representation to the (ω_1,ω_2) parameters needed for inverse transformation.

To convert a particular rectangular point into (ω,ϕ) domain, substitute its ω_1 and ω_2 values in Eqs. (81) and (82). The resulting values for ω and ϕ may not correspond to any for which $Q(\omega,\phi)$ is known. By virtue of Eq. (76), $Q(\omega,\phi)$ will only be known over a uniformly sampled set of values for ω and ϕ. In order to determine Q at the calculated ω and ϕ, the following procedure is used. Given $N_\omega \times N_\phi$ uniformly located samples, $Q(\omega_i,\phi_j)$, calculate a bilinearly interpolated value of this function at the desired ω and ϕ by using

$$Q(\omega,\phi) = \sum_{i=1}^{N_\omega} \sum_{j=1}^{N_\phi} Q(\omega_i,\phi_j)h_1(\omega - \omega_i)h_2(\phi - \phi_j), \tag{83}$$

where

$$h_1(\omega) = \begin{cases} 1 - \dfrac{|\omega|}{\Delta\omega} & 0 \\ |\omega| \le \Delta\omega & \text{otherwise,} \end{cases} \tag{84}$$

and

$$h_2(\phi) = \begin{cases} 1 - \dfrac{|\phi|}{\Delta\phi} & 0 \\ |\phi| \le \Delta\phi & \text{otherwise;} \end{cases} \tag{85}$$

$\Delta\phi$ and $\Delta\omega$ are the sampling intervals for ϕ and ω, respectively. When expressed in the manner shown above, bilinear interpolation may be interpreted as the output of a filter whose impulse reponse is $h_1(\omega)h_2(\phi)$.

The results obtained with bilinear interpolation can be considerably improved if the sampling density in the (ω,ϕ)-plane is increased by using the computationally efficient method of zero-extending the inverse two-dimensional inverse *Fast Fourier Transform* (FFT) of the $Q(\omega_i,\phi_j)$ matrix. The technique consists of first taking a two-dimensional inverse FFT of the $N_\omega \times N_\phi$ matrix consisting of the $Q(\omega_i,\phi_j)$ values, zero-extending the resulting $N_\omega \times N_\phi$ array of numbers to, perhaps, $mN_\omega \times nM_\phi$ and then taking the FFT of this new array. The result is an mn-fold increase in the density of samples in the (ω,ϕ)-plane. After computing $Q(\omega,\phi)$ at each point of a rectangular grid by the procedure outlined above, the object f(x,y) is obtained by a simple 2-D inverse FFT.

The use of bilinear interpolation and zero padding are both good techniques for resampling a function but they are used here in a non-standard way. Typically interpolation algorithms are derived assuming that the sampled data can be described as nearly linear (when using bilinear interpolation) and frequency limited (when using Fourier domain zero padding) [30], [31], [32]. In this application, when resampling the data from a circular grid to a rectangular grid, the function is assumed to be smooth in the Fourier domain. This assumption is reasonable since the data is assumed to be well behaved.

The interpolation described above, however, is carried out in a rectilinear version of the (ω,ϕ) coordinate system. Thus four points in the (ω,ϕ) space, where data is available, are first assumed to be at the four corners of a rectangle and then the interpolation is calculated for a point in the middle. This is an approximation because the four data points actually define a smooth function that is defined along four points on two circular arcs. As will be seen in the reconstructions, the effect of this approximation is small; but it should be remembered when comparing interpolation schemes.

4. LIMITS OF FIRST ORDER APPROXIMATIONS

In diffraction tomography there are different approximations involved in the forward and inverse directions. In the forward process it is necessary to assume that the object is weakly scattering so that either the Born or the Rytov approximations can be used. Once we arrive at an expression for the scattered field it is necessary to not only measure the scattered fields but then to numerically implement the inversion process.

1.001   **1.01**   **1.10**   **1.20**

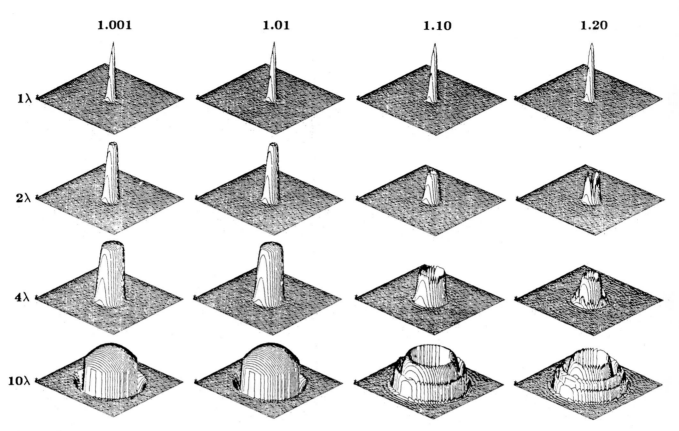

Fig. 17. Simulated reconstructions using the Born approximation for 16 objects with four refractive indices between 1.001 and 1.20 and four radii between 1 and 10λ.

The mathematical and experimental effects limit the reconstruction in different ways. The most severe mathematical limitations are imposed by the Born and the Rytov approximations. These approximations are fundamental to the reconstruction process and limit the range of objects that can be examined. On the other hand, the experimental limitations are caused because it is only possible to collect a finite amount of data. Up to the limit in resolution caused by evanescent waves, it is possible to improve a reconstruction by collecting more data.

By carefully setting up the simulations it is possible to separate the effects of these errors. To study the effects of the Born and Rytov approximations, it is necessary to calculate (or even measure) the exact fields and them make use of the best possible (most exact) reconstruction formulas available. The difference between the reconstruction and the actual object can then be used as a measure of the quality of the approximations.

Only the mathematical limitations will be described here. For a discussion of some of the experimental factors the reader is referred to [33], [34].

4.1 Qualitative Analysis

The exact field for the scattered field from a cylinder as shown by Weeks [35] was calculated for cylinders of various sizes and refractive index. In the simulations that follow a single plane wave was incident on the cylinder and the scattered field was calculated along a line at a distance of 100 wavelengths from the origin.

At the receiver line the received wave was measured at 512 points spaced at 1/2 wavelength intervals. In all cases the rotational symmetry of a single cylinder at the origin was used to reduce the computation time of the simulations.

The simulations were performed for refractive indices that ranged from 0.1% change (refractive index of 1.001) to a 20% change (refractive index of 1.2). For each refractive index, cylinders of size 1, 2, 4, and 10 wavelengths were reconstructed. This gave a range of phase changes across the cylinder [see Eq. (23) above] from 0.004π to 8π. The resulting reconstructions using the Born approximation are shown in Fig. 17.

Clearly, all the cylinders of refractive index 1.001 in Fig. 17 were perfectly reconstructed. As Eq. (24) predicts, the results get worse as the product of refractive index and radius gets larger. The largest refractive index that was successfully reconstructed was for the cylinder in Fig. 17 of radius 1 wavelength and a refractive index that differed by 20% from the surrounding medium.

While it is hard to evaluate the two dimensional reconstructions, it is certainly reasonable to conclude that only cylinders where the phase change across the object was less

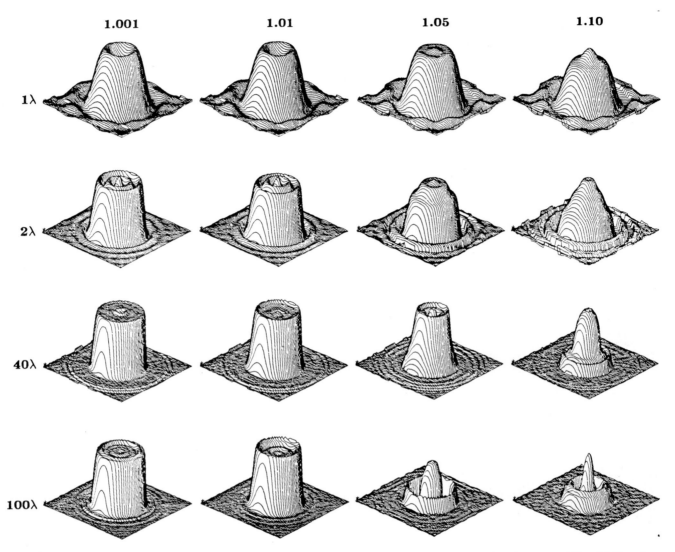

Fig. 18. Simulated reconstructions using the Born approximation for 16 objects with four refractive indices between 1.001 and 1.10 and four radii between 1 and 100λ.

than or equal to 0.8π were adequately reconstructed. In general, the reconstruction for each cylinder where the phase change across the cylinder was greater than π shows severe artifacts near the center. This limitation in the phase change across the cylinder is consistent with the condition expressed in Eq. (24) above.

A similar set of simulations was also done for the Rytov approximation and is shown in Fig. 18. In this case the reconstructions were performed for cylinders of radius 1, 2, 40, and 100 λ and refractive indices of 1.001, 1.01, 1.05, and 1.10. Because of the large variation in cylinder sizes all reconstructions were performed so the estimated object filled half of the reconstruction matrix. While the error in the reconstructions does increase for larger cylinders and higher refractive indices, it is possible to successfully reconstruct larger objects with the Rytov approximation.

4.2 Qualitative Comparison of the Born and Rytov Approximation

Reconstructions using exact scattered data show the similarity of the Born and Rytov approximations for small objects with small changes in the refractive index. For a cylinder of radius 1 wavelength and a refractive index that differs by 1% from the surrounding medium, the resulting reconstructions are shown in Fig. 19. In both cases, the reconstructions are clean and the magnitude of the reconstructed change in refractive index is close to the simulated object.

On the other hand, the reconstructions of objects that are large or have a refractive index that differ by a factor of ca. 20% from unity, illustrate the differences between the Born and the Rytov approximations. Figure 20 shows a simulated

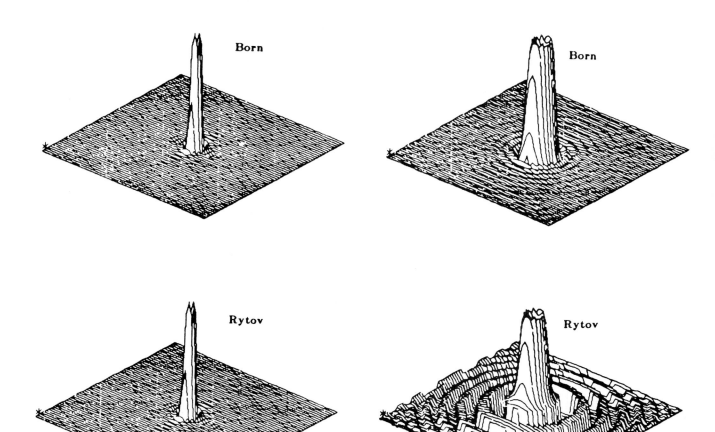

Fig. 19.   Born and Rytov reconstructions of a cylinder of radius 1λ and 1.01 refractive index.

Fig. 20.   Reconstructions of a cylinder of radius 1λ and refractive index 1.20 showing the advantage of the Born over the Rytov.

reconstruction for an object of radius 1 and refractive index 1.20. In this region the Born approximation is superior to the Rytov.

According to Chernov [17] and Keller [18] the Rytov approximation should be much superior to the Born for objects much larger than a wavelength. Reconstructions were done based on the exact scattered wave from a cylinder of radius 40 wavelengths and a refractive index that differed by 1% from the surrounding medium. The reconstructed refractive index is shown in Fig. 21. While the Born approximation has provided a good estimate of the size of the object, the reconstruction near the center is clearly not accurate.

The results in Figs. 20 and 21 are consistent with the regions of validity of the Born and Rytov approximations. The Born approximation is sensitive to the total phase shift in the object. Thus, in the reconstruction of Fig. 21 the Born approximation has done a good job of representing the step change in refractive index; but as the incident field undergoes a phase shift through the object, the reconstruction becomes poor. On the other hand, the Rytov approximation is sensitive to the change in refractive index. Thus the Rytov

reconstruction is accurate near the center of the object but provides a very poor reconstruction near the boundary of the object.

4.3   Quantitative Studies

In addition to the qualitative studies, a quantitative study of the error in the Born and Rytov reconstructions was also performed. As a measure of error we used the relative mean squared error in the reconstruction of the object function integrated over the entire plane. If the actual object function is $O(\bar{r})$ and the reconstructed object function is $O'(\bar{r})$ then the relative Mean Squared Error (MSE) is

$$\mathrm{MSE} = \frac{\iint [O(\bar{r}) - O'(\bar{r})]^2 d\bar{r}}{\iint [O(\bar{r})]^2 d\bar{r}}. \qquad (86)$$

To study the quantitative difference between the Born and the Rytov approximations, several hundred simulated reconstructions were performed. For each simulation the exact scattered field was calculated for a single cylinder with

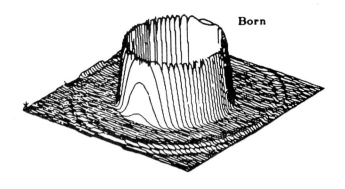

Born

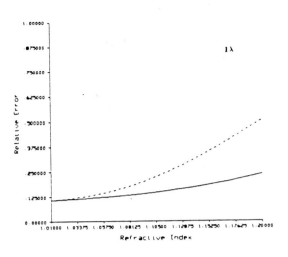

1λ

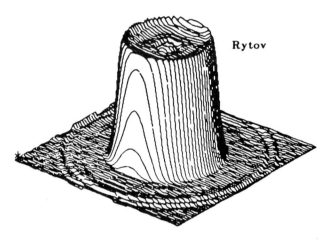

Rytov

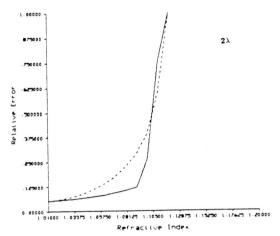

2λ

Fig. 21. Reconstructions of a cylinder of radius 40λ and refractive index 1.01 showing the advantage of the Rytov over the Born.

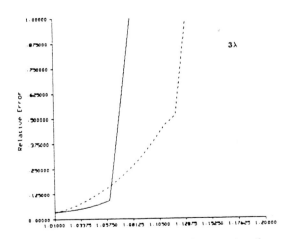

3λ

Fig. 22. The relative mean squared error for reconstructions using the Born (solid line) and Rytov (dashed line) approximations. The error for a total of 60 objects with a radius of 1, 2 and 3 wavelengths are shown.

an arbitrary radius and refractive index. The reconstructions were divided into two sets to highlight the difference between the Born and the Rytov approximations.

The plots of Fig. 22 present a summary of the mean squared error for cylinders of 1, 2, and 3λ in radius and twenty refractive indices between 1.01 and 1.20. In each case the error for the Born approximation is shown as a solid line while the error for the Rytov approximation is shown as a dashed line. The exact scattered fields were calculated at 512 receiver points along a receiver line 10λ from the center of the cylinder.

Only for the 1λ cylinders is the relative mean squared error for the Born approximation always lower than the Rytov. It is interesting to note that while the Rytov approximation shows a steadily increasing error with higher refractive indices the error in the Born reconstruction is relatively constant until a threshold is reached. For the 2λ and the 3λ cylinder, this breakpoint occurs at a phase shift of 0.6 and 0.7π. Thus a criteria for the validity of the Born approximation is that the product of the radius of the cylinder in wavelengths and the change in refractive index must be less than 0.175.

Figure 23 presents a summary of the relative mean squared errors for cylinders of refractive index 1.01, 1.02, and 1.03 and for forty radii between 1 and 40λ. Because the size

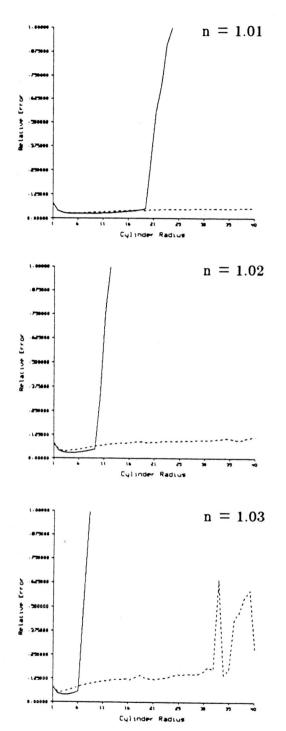

Fig. 23.   The relative mean squared error for reconstructions using the Born (solid line) and Rytov (dashed line) approximations. The error for a total of 120 objects with a refractive index of 1.01, 1.02 and 1.03 are shown.

along a line 2R from the center of the cylinder and spaced at 1/16R intervals.

In each of the simulations the Born approximation is only slightly better than the Rytov approximation until the Born approximation crosses its threshold with a phase shift of 0.7π. Because the error in the Rytov approximation is relatively flat it is clearly superior for large object and small refractive indices. Using simulated data and the Rytov approximation we have successfully reconstructed objects as large as 2000λ in radius.

4.4   Multiple Scattering Effects

The simulations above have only considered reconstructions of simple cylindrical objects with a constant refractive index. While these objects do have the advantage that it is possible to write an expression for the scattered fields they perhaps are not a good model of the real world. Simulations of more complicated objects are needed, but the calculation of the scattered fields are either more difficult or not possible at all. One type of object that can be modeled are those that consist of multiple cylinders.

A major source of difficulty with multi-component* objects is dealing with the interaction between the various components. Depending on the interaction between the components, the total scattered field may or may not bear any resemblance with the simple sum of the scattered fields for each of the components, assuming the others to be absent. A new computational procedure for calculating the inter-component interaction was presented in [36]. With the computer programs developed we are able to generate the scattered fields that are not limited by the first order assumption. Although the results shown will only include the second order fields for a multicomponent object, the computational procedure can easily be generalized for higher order scattering effects.

Since all currently available diffraction tomography algorithms are based on the assumption that the object satisfies the first order scattering assumption, it is interesting to test them under conditions when this assumption is violated. We have used the scattered fields obtained with the new computational procedure to test these algorithms, and shown the resulting artifacts. *Our main conclusion drawn from computer simulation study is that even when object inhomogeneities are as small as 5 percent of the background, multiple scattering can introduce severe distortion in reconstructions of multi-component objects.*

One can, in principle, obtain the exact solution to the wave equation for a multi-component object provided one is able to solve the boundary value problem for the entire object. In practice it is not possible to do so even for two- or three-

of the cylinders varied by a factor of forty, the simulation parameters were adjusted accordingly. For a cylinder of radius R the scattered field was calculated for 512 receivers

---

* By a single component object, we mean one that is composed of a single circular or elliptical cylinder. Analytical expressions are available for the scattered fields of such objects. A multiple-component object has more than one cylinder; for our purposes the cylinders will all be parallel to each other.

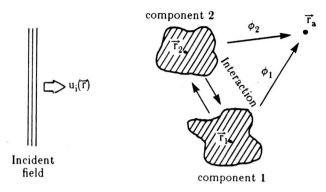

Fig. 24.   A two component phantom illuminated with incident field $u_i(r)$.

component objects, and one must take recourse to computational procedures. We have based our algorithms on Twersky's theory in which the scattered field is expressed as an infinite summation of different order terms [15]. First order fields are obtained by considering the interaction of the original incident field with each component, assuming the others to be absent. First order fields caused by a component when incident on the other components generate the second order fields, and so on.

The basic elements of multiple scattering theory that we will use in our computational modeling will now be explained in detail with the help of a two component object (The approach is easily generalized to more than two components.) Figure 24 shows a two component phantom being illuminated by a field denoted by $u_i(\vec{r})$ (In the absence of the phantom the field everywhere will be $u_i(\vec{r})$.) We will use $u(\vec{r}_a)$ to denote the actual field at a point $(\vec{r}_a)$ as shown in the figure. $u(\vec{r}_a)$ is equal to

$$u(\vec{r}_a) = u_i(\vec{r}_a) + \phi_1(\vec{r}_a) + \phi_2(\vec{r}_a), \quad (87)$$

where $\phi_1(\vec{r}_a)$ and $\phi_2(\vec{r}_a)$ are, respectively, the scattered fields at $\vec{r}_a$ caused by the phantom components 1 and 2.

Let $u_{inc}(\vec{r}_i)$ be the incident field at the center of the ith component, and let $o_s(\vec{r}_a,\vec{r}_i)$ be an operator function such that when it is applied to the fields incident on the scattering component at $\vec{r}_i$, it generates the scattered field at the observation point $\vec{r}_a$. In terms of o_s's, ϕ_1 and ϕ_2 are given by

$$\phi_i(\vec{r}_a) = o_s(\vec{r}_a,\vec{r}_i)u_{inc}(\vec{r}_i) \quad i = 1,2. \quad (88)$$

Substituting (88) in (87), we get

$$u(\vec{r}_a) = u(\vec{r}_a) + o_s(\vec{r}_a|\vec{r}_1)u_{inc}(\vec{r}_1) + o_s(\vec{r}_a|\vec{r}_2)u_{inc}(\vec{r}_2). \quad (89)$$

The field incident at the site of each scatterer may be expressed as

$$u_{inc}(\vec{r}_1) = u_i(\vec{r}_1) + o_s(\vec{r}_1|\vec{r}_2)u_{inc}(\vec{r}_2), \quad (90)$$

$$u_{inc}(\vec{r}_2) = u_i(\vec{r}_2) + o_s(\vec{r}_2|\vec{r}_1)u_{inc}(\vec{r}_1). \quad (91)$$

Substituting (90) and (91) in (89), we get

$$u(\vec{r}_a) = u_i(\vec{r}_a) + o_s(\vec{r}_a|\vec{r}_1)u_i(\vec{r}_1)$$
$$+ o_s(\vec{r}_a|\vec{r}_2)u_i(\vec{r}_2) + o_s(\vec{r}_a|\vec{r}_1)o_s(\vec{r}_1|\vec{r}_2)u_{inc}(\vec{r}_2)$$
$$+ o_s(\vec{r}_a|\vec{r}_2)o_s(\vec{r}_2|\vec{r}_1)u_{inc}(\vec{r}_1). \quad (92)$$

If we again substitute (90) and (91) in (92), we have the following expression

$$u(\vec{r}_a) = u_i(\vec{r}_a) + o_s(\vec{r}_a|\vec{r}_1)u_i(\vec{r}_1) + o_s(\vec{r}_a|\vec{r}_2)u_i(\vec{r}_2)$$
$$+ o_s(\vec{r}_a|\vec{r}_1)o_s(\vec{r}_1|\vec{r}_2)u_i(\vec{r}_2) + o_s(\vec{r}_a|\vec{r}_2)o_s(\vec{r}_2|\vec{r}_1)u_i(\vec{r}_1)$$
$$+ o_s(\vec{r}_a|\vec{r}_1)o_s(\vec{r}_1|\vec{r}_2)o_s(\vec{r}_2|\vec{r}_1)u_{inc}(\vec{r}_1)$$
$$+ o_s(\vec{r}_a|\vec{r}_2)o_s(\vec{r}_2|\vec{r}_1)o_s(\vec{r}_1|\vec{r}_2)u_{inc}(\vec{r}_2). \quad (93)$$

A more compact way to express the above equation is

$$u(\vec{r}_a) = u_i(\vec{r}_a) + u_{first\text{-}order}(\vec{r}_a) + u_{second\text{-}order}(\vec{r}_a)$$
$$+ u_{higher\text{-}order}(\vec{r}_a). \quad (94)$$

The quantities $u_{first\text{-}order}$, $u_{second\text{-}order}$ and $u_{higher\text{-}order}$ represent the first order, second order, and higher order contributions at the observation point $\vec{r}_a$. These are given by

$$u_{first\text{-}order}(\vec{r}_a) = o_s(\vec{r}_a|\vec{r}_1)u_i(\vec{r}_1) + o_s(\vec{r}_a|\vec{r}_2)u_i(\vec{r}_2) \quad (95)$$

$$u_{second\text{-}order}(\vec{r}_a) = o_s(\vec{r}_a|\vec{r}_2)o_s(\vec{r}_2|\vec{r}_1)u_i(\vec{r}_1)$$
$$+ o_s(\vec{r}_a|\vec{r}_1)o_s(\vec{r}_1|\vec{r}_2)u_i(\vec{r}_2) \quad (96)$$

$$u_{higher\text{-}order}(\vec{r}_a) = o_s(\vec{r}_a|\vec{r}_1)o_s(\vec{r}_1|\vec{r}_2)o_s(\vec{r}_2|\vec{r}_1)u_{inc}(\vec{r}_1)$$
$$+ o_s(\vec{r}_a|\vec{r}_2)o_s(\vec{r}_2|\vec{r}_1)o_s(\vec{r}_1|\vec{r}_2)u_{inc}(\vec{r}_2) \quad (97)$$

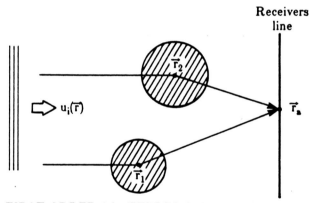

FIRST-ORDER SCATTERED FIELDS AT $\vec{r}_a$

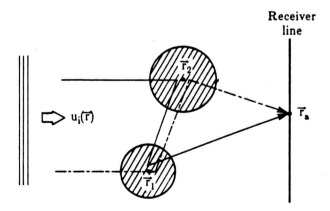

SECOND-ORDER SCATTERED FIELDS AT $\vec{r}_a$

Fig. 25.   This figure depicts the first and second-order scattered fields for a two component object.

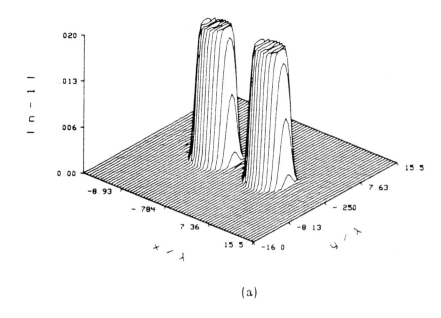

(a)

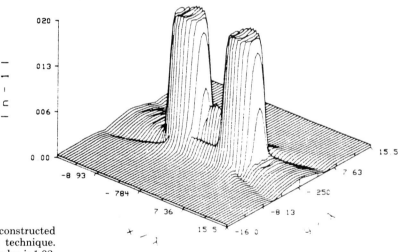

Fig. 26. Cross section of a two-component object reconstructed using the conventional diffraction reconstruction technique. Diameter of the cylinders is 6λ and their refractive index is 1.02. (a) Only the first order scattered fields are used for generating the data for this reconstruction. (b) Doubly scattered fields are included in the projection data for this reconstruction.

(b)

In Fig. 25, we have shown the first and the second order scattering processes for a two component object. For the first order scattering, each component interacts independently with the incident illumination, being oblivious of the existence of the other. To compute the second order scattering terms, each component interacts with the fields sent in its direction by the other component, and so on.

The computing procedure discussed above was used to generate 64 projections over 360° for the object. An interpolation based algorithm was then used to reconstruct the object cross-section; the results follow.

In the reconstructions shown in Figs. 26, 27, 28, and 29 we

have shown the magnitude of the deviation of the reconstructed refractive index from that of the background, which was assumed to be unity. Plots labeled (a) show the reconstruction obtained when the projections were generated by ignoring the second and higher order scattered fields. On the other hand, the plots labeled (b) show the reconstructions when the second order fields are included in the projections.

In Fig. 26 the change in the refractive index of the 6λ cylinders was set to 2%. Although there is some distortion introduced in the direction of the line joining the center of the cylinders, it is negligible. However, when the refractive

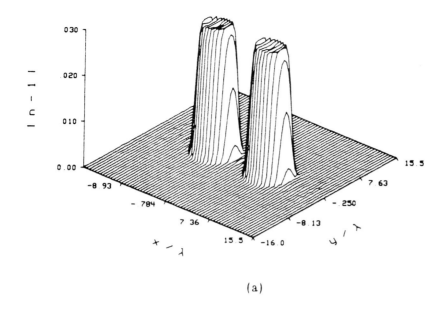

(a)

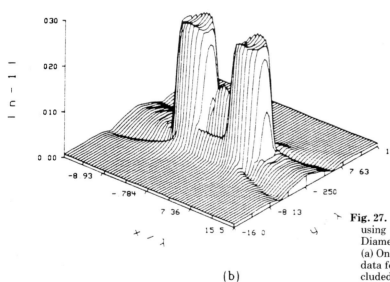

(b)

Fig. 27. Cross section of a two-component object reconstructed using the conventional diffraction reconstruction technique. Diameter of the cylinders is 6λ and their refractive index is 1.03. (a) Only the first order scattered fields are used for generating the data for this reconstruction. (b) Doubly scattered fields are included in the projection data for this reconstruction.

index change is increased in Fig. 27 to 3 percent, the distortion becomes quite noticeable; and in Fig. 28 a 5% change in refractive index is enough to cause the distortion to dominate the reconstruction. When the number of cylinders is increased the distortion is higher as seen in Fig. 29. This is expected because in this case there are more projections affected by second order scattering. It should be mentioned that the computational effort required for generating the projections is enormous. To illustrate, it took three hours of cpu time on the AP120B array processor for computing 64 projections of a three component object.

4.5 The Effect of Attenuation

Although in ultrasonic imaging the role of attenuation is minor, in microwave imaging it can not be ignored. Microwave attenuation rates in water are presented for the attenuation and phase factors at various frequencies in graphical and algebraic form by Foti *et al* elsewhere in this volume. Microwaves at 4 GHz, for example, undergo almost 3 db of attenuation per centimeter of travel in water; thus, the assumption in the Born approximation is quickly violated. In the remainder of this section, the effect of attenu-

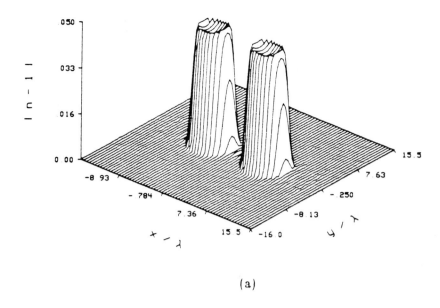

(a)

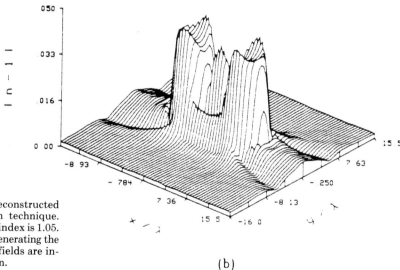

Fig. 28. Cross section of a two-component object reconstructed using the conventional diffraction reconstruction technique. Diameter of the cylinders is 6λ and their refractive index is 1.05. (a) Only the first order scattered fields are used for generating the data for this reconstruction. (b) Doubly scattered fields are included in the projection data for this reconstruction.

(b)

ation will be described as well as its affect on the resolution of the reconstruction.

The angular spectrum of a field on the line $x = x_1$ propagates to line $x = x_2$ according to the following relation

$$A(k_y, x_2) = A(k_y, x_1) e^{j\sqrt{k_0^2 - k_y^2}(x_2 - x_1)}. \qquad (98)$$

Figure 30 shows a plot of the magnitude of the exponential factor in Eq. (98) as a function of both the attenuation of the medium and the spatial frequency of the plane wave in the y direction. We have plotted the magnitude in db as a function of the dimensionless parameter γ, where $k_x = \gamma k_0$. Thus, for $\gamma = 0$ the wave is traveling directly towards the receiver line, while for $\gamma = 1$ the wave is propagating along the direction of the receiver line. In this plot the attenuation

factor and γ have been changed from 0 to 5.0 db/cm and 0 to 1.0, respectively. It is clear that the high frequency components (larger γ) are more attenuated than those components at lower frequencies (those that travel directly towards the receiver line). This means that as the field propagates in an attenuating medium, it loses its high frequency components.

Remembering the Fourier domain coverage of diffraction transmission tomography, one can now associate this phenomenon with a degradation in resolution. This point is illustrated in Fig. 31. As in the case of no attenuation, the inner circle corresponds to transmission tomography, while the outer ring represents the data measured by a reflection tomographic system. The difference made by attenua-

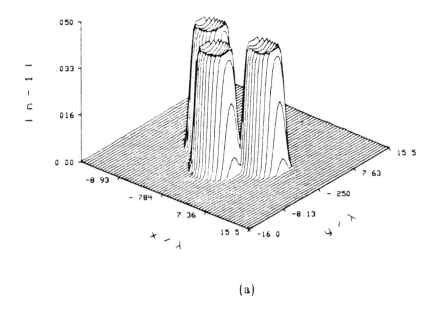

(a)

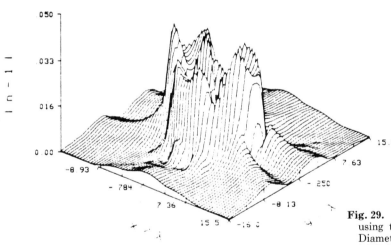

(b)

Fig. 29.   Cross section of a two-component object reconstructed using the conventional diffraction reconstruction technique. Diameter of the cylinders is 6λ and their refractive index is 1.05. (a) Only the first order scattered fields are used for generating the data for this reconstruction. (b) Doubly scattered fields are included in the projection data for this reconstruction.

tion is shown as the shaded area. In this region, the attenuation of the medium reduces the amplitude of the plane wave components below a minimum tolerable Signal to Noise Ratio (SNR). Thus, this region of the object's Fourier transform is unmeasurable.

In the Synthetic Aperture technique, the concept of propagation of the angular spectrum can be applied in an attenuating medium. This results in upper bounds on k_y and β. k_y is the spatial frequency associated with the transmitter, and β is the spatial frequency associated with the receiver. This point is illustrated in Fig. 32. More angular views restore the resolution lost due to the use of an attenuating coupling medium. The coverage in the reflection mode also shrinks as shown in the figure.

5.   CONCLUSIONS

The use of microwave tomographic imaging gives the physician new information about the physiologic status of a patient. X-ray tomography is based on the Fourier Slice Theorem; but, because of the diffraction and refraction of microwaves as they travel through the body, this theory is not useful with microwaves. Instead the Fourier Diffraction Projection Theorem is used to relate the fields scattered by an object illuminated with a plane wave to the Fourier transform of the object. Like the x-ray case, it is possible to measure the scattered fields from a number of different directions and form an estimate of the object's microwave refractive index.

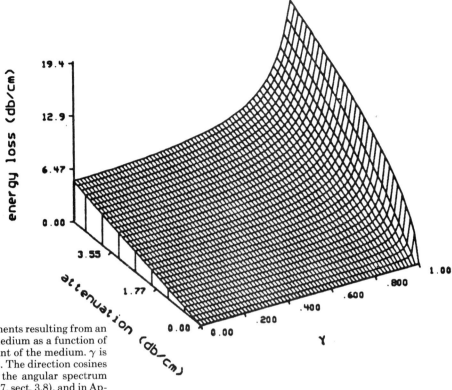

Fig. 30. Energy loss of plane wave components resulting from an angular spectrum expansion in a lossy medium as a function of directional angle and attenuation constant of the medium. γ is the direction cosine of these components. The direction cosines are related to the spatial frequency of the angular spectrum components as described by Goodman (27, sect. 3.8), and in Appendix I of the paper by Farhat elsewhere in this volume.

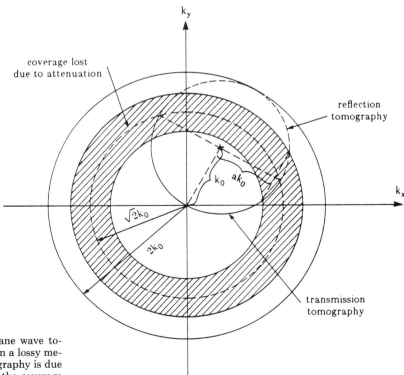

Fig. 31. Object's Fourier domain coverage for plane wave tomography in transmission and reflection modes, in a lossy medium. The lower resolution in transmission tomography is due to the smaller radius ($\sqrt{2}k_0$ compared to $2k_0$) of the coverage circle as shown here. The effect of attenuation is shown in the shaded area.

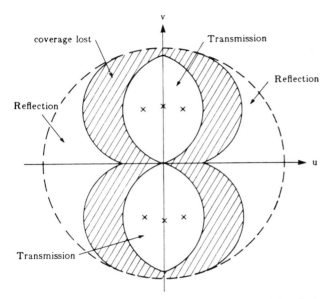

Fig. 32. When the medium is lossy, a larger number of views in the Synthetic Aperature technique can increase the resolution and decrease the distortion in the reconstructed image. The coverage shown corresponds to three rotational views of the object, 60° apart.

In addition, this chapter has presented an alternative derivation of the Fourier Diffraction Projection Theorem. This approach will allow for efficient implementations of

higher order reconstruction techniques on digital computers.

By carefully designing a simulation procedure, we have isolated the effects of the first order Born and Rytov approximations in diffraction imaging. While both procedures can produce excellent reconstructions for small objects with small refractive index changes, they both quickly break down when their assumptions are violated. The assumptions limit the Born approximation to objects where the product of the diameter and the relative refractive index are less than 0.35λ; and the Rytov approximation to objects with a refractive index that differs by less than 2% from the surrounding media, with essentially no constraint on the size of the object.

Two problems need to be solved for microwave imaging to become successful for medical imaging. Foremost, reconstruction algorithms based on higher order approximations to the scattered field will be needed. With 4 GHz microwaves in water, biological structures span tens of wavelengths and often have refractive index variations of more than 100%. In addition, high spatial frequency suffer from large attenuation in water-based microwave systems. An approach that takes into account the attenuation effects of the coupling medium and the target should be studied. Microwave systems for medical imagery must take into account the different Fourier coverage provided by S_{11} and S_{21} data collection system designs and seek high k coupling media with lower loss.

References

1. L. E. Larsen and J. H. Jacobi, "Microwave scattering parameter imagery of an isolated canine kidney," *Med. Phys.*, Vol. 6, pp. 394–403, Sep./Oct., 1979.
2. L. E. Larsen, J. H. Jacobi, and A. K. Krey, "Preliminary observations with an electromagnetic method for the noninvasive analysis of cell suspension physiology and induced pathophysiology," *IEEE Trans. Microwave Theory Tech.*, Vol. MTT-26, pp. 581–595, Aug., 1978.
3. L. E. Larsen and J. H. Jacobi, "Microwave interrogation of dielectric targets: Part 1: By scattering parameters," *Med. Phys.*, Vol. 5, pp. 500–508, Nov./Dec., 1978.
4. O. C. Yue, E. L. Rope, and G. Tricoles, "Two reconstruction methods for microwave imaging of buried dielectric anomaly," *IEEE Trans. Comput.*, Vol. C-24, pp 381–390, Apr., 1975.
5. Loris B, Gregoris and Keigo Iizuka, "Visualization of internal structure by microwave holography," *Proc. IEEE*, Vol. 594, pp. 791–792, May, 1970.
6. A. Rosenfeld and A. C. Kak, *Digital Picture Processing*, Academic Press, Second Edition, 1982.
7. A. H. Andersen and A. C. Kak, "Simultaneous algebraic reconstruction technique (SART): A superior implementation of the ART algorithm," *Ultrason. Imag.*, Vol. 6, pp. 81–94, Jan., 1984.
8. A. H. Andersen and A. C. Kak, "Digital ray tracing in two-dimensional refractive fields," *J. Acoust. Soc. Am.*, Vol. 72, pp. 1593–1606, Nov., 1982.
9. A. J. Devaney, "A filtered backpropagation algorithm for diffraction tomography," *Ultrason. Imag.*, Vol. 4, pp. 336–350, 1982.
10. A. C. Kak, "Tomographic imaging with diffracting and non-diffracting sources," in *Array Signal Processing*, Simon Haykin, ed., Prentice Hall, 1984.
11. R. K. Mueller, M. Kaveh, and G. Wade, "Reconstructive tomography and applications to ultrasonics," *Proc. IEEE*, Vol. 67, pp. 567–587, 1979.
12. E. Wolf, "Three-dimensional structure determination of semi-transparent objects from homographic data," *Opt. Commun.*, Vol. 1, pp. 153–156, 1969.
13. S. X. Pan and A. C. Kak, "A computational study of reconstruction algorithms for diffraction tomography: Interpolation vs. filtered-backpropagation," *IEEE Trans. Acous. Speech Signal Process.*, pp. 1262–1275, Oct., 1983.
14. K. Iwata and R. Nagata, "Calculation of refractive index distribution from interferograms using the Born and Rytov's approximations," *Jap. J. Appl. Phys.*, Vol. 14, pp. 1921–1927, 1975.
15. A. Ishimaru, *Wave Propagation and Scattering in Random Media*, Academic Press, New York, 1978.
16. Philip M. Morse and Herman Feshbach, *Methods of Theoretical Physics*, McGraw Hill Book Company, New York, 1953.
17. L. A. Chernov, *Wave Propagation in a Random Medium*, McGraw Hill Book Company, New York, 1960.
18. J. B. Keller, "Accuracy and validity of the Born and Rytov approximations," *J. Opt. Soc. Am.*, Vol. 59, pp. 1003–1004, 1969.
19. S. K. Kenue and J. F. Greenleaf, "Limited angle multifrequency diffraction tomography," *IEEE Trans. Sonics Ultrason.*, Vol. SU-29, pp. 213–217, July, 1982.
20. William H. Carter, "Computational reconstruction of scattering objects from holograms," *J. Opt. Soc. Am.*, Vol. 60, pp. 306–314, March, 1970.
21. W. H. Carter and P. C. Ho, "Reconstruction of inhomogeneous scattering objects from holograms," *Appl. Opt.*, Vol. 13, pp. 162–172, Jan., 1974.

22. P. L. Carson, T. V. Oughton, and W. R. Hendee, "Ultrasound transaxial tomography by reconstruction," in *Ultrasound in Medicine II*, D. N. White and R. W. Barnes, eds., Plenum Press, pp. 391–400, 1976.

23. M. Kaveh, M. Soumekh, and R. K. Mueller, "Tomographic imaging via wave equation inversion," *ICASSP 82*, pp. 1553–1556, May, 1982.

24. D. Nahamoo and A. C. Kak, *Ultrasonic diffraction imaging*, TR-EE 82-20, School of Electrical Engineering, Purdue University, 1982.

25. D. Nahamoo, S. X. Pan, and A. C. Kak, "Synthetic aperture diffraction tomography and its interpolatioin-free computer implementation," *IEEE Trans. Sonics Ultrason.*, Vol. SU-31, pp. 218–229, July, 1984.

26. A. J. Devaney, "Geophysical diffraction tomography," *IEEE Trans. Geo. Sci., Special Issue Remote Sensing*, Vol. GE-22, pp. 3–13, Jan., 1984.

27. J. W. Goodman, *Introduction to Fourier Optics*, McGraw Hill Book Company, San Francisco, 1968.

28. M. Soumekh, M. Kaveh, and R. K. Mueller, "Fourier domain reconstruction methods with application to diffraction tomography," *Acoust. Imaging*, Vol. 13, pp. 17–30, 1984.

29. M. Soumekh and M. Kaveh, "Image reconstruction from frequency domain data on arbitrary contours," *Int. Conf. Acoust. Speech and Signal Process.*, pp. 12A.2.1–12A.2.4, 1984.

30. S. D. Conte and C. deBoor, *Elementary Numerical Analysis*, McGraw-Hill, New York, 1980.

31. Josef Stoer and Roland Bulirsch, *Introduction to Numerical Analysis*, Springer-Verlag, New York, 1980.

32. F. S. Acton, *Numerical Methods that Work*, Harper & Row, New York, 1970.

33. Malcolm Slaney and A. C. Kak, "Diffraction tomography," *Proc. SPIE*, vol. 413, pp. 2–19, Apr., 1983.

34. Malcolm Slaney and A. C. Kak, *Imaging with Diffraction, Tomography*, TR-EE 85-5, School of Electrical Engineering, Purdue University, March, 1985.

35. W. L. Weeks, *Electromagnetic Theory for Engineering Applications*, John Wiley and Sons, Inc., New York, 1964.

36. M. Azimi and A. C. Kak, "Distortion in diffraction imaging caused by multiple scattering," *IEEE Trans. Med. Imag.*, Vol. MI-2, pp. 176–195, Dec., 1983.

Inverse Methods in Electromagnetic Imaging

Wolfgang-M. Boerner and Chung-Yee Chan

An attempt is being made here to scan some of the pertinent review literature on electromagnetic inverse problems for the purpose of isolating those inverse methods which have already been or most likely may be applied to medical imaging and remote diagnostics of exterior and interior body sections. Specifically we consider those inverse techniques which have been developed in many seemingly unrelated fields of physical sciences where the characteristic descriptors of a medium are estimated from experimental data, obtained from measurements made usually at a distance from the medium, utilizing the laws that relate these characteristic parameters to the experimental data in a given situation. Particular emphasis will be placed on the description of relevant radar target imaging techniques. No mathematical formulas are presented and we refer to available publications in the open literature.

1. INTRODUCTION

It is the objective of this succinct review to highlight the major contributions made in electromagnetic inverse scattering over the past three decades placing major emphasis on radar target imaging. As such, the Chapters by Slaney, *et al.*, Guo *et al.*, as well as that by Kim & Webster are directly pertinent.

The complete description of electromagnetic scattering processes requires polarization information to recover the descriptive parameters of the scatterer from the measured field, knowing the incident field. The problem can be defined in terms of the target scattering matrices and their particular properties. In radar target discrimination, identification, classification and imaging use of the entire spatial frequency domain of the radar cross section must be made for obtaining a reliable or sufficiently unique solution leading to various approximate approaches.

In this Introductory Section we first introduce general definitions of the electromagnetic inverse problem, discuss relationships with associated medical imaging techniques, and conclude with comments on the selection methods chosen for presentation in this review.

1.1 The Electromagnetic Inverse Problem of Scattering

Various descriptions on the nature of the Inverse Problem are given in the literature (see, e.g. [1]–[10]) and solutions to inverse problems have become of increasing importance in aeronomy, geo-physical exploration, remote sensing,

radar/sonar/ladar target imaging, material and medical diagnostics, etc. An elaborate survey of inversion techniques as applied to a variety of different physical disciplines is compiled in a NASA Technical Memorandum [1] and more recently a comprehensive State of the Art Review of Electromagnetic Inverse Scattering was attempted in [6], [7] containing an extensive list of pertinent references.

It is far beyond the scope of this succinct review to provide a complete treatment of this topic also for the reason that this new scientific discipline is in a state of rapid and dynamic development. Instead, it will be one of our main objectives to scrutinize those methods and techniques developed in electromagnetic (radar) imaging which look promising to be applicable to some extent to medical microwave imaging and profile inversion.

The inverse problem of electromagnetic scattering may be described as follows:

Whereas on the one hand, in the direct problem of electromagnetic diffraction, total a priori information on the size, shape and material constituents of a scatterer, the incident field vector and its orientation with respect to the fixed scatterer coordiante system are given, and the scattered field is to be calculated everywhere over the total frequency domain; on the other hand, in the inverse problem we need to recover the size, shape, and the constitutive characteristics of an *a priori* unknown scattering target with the knowledge of the incident field and the resulting scattered field data.

1.2 Electromagnetic Inverse Methods Applicable to Medical Imaging

Although electromagnetic waves span a wide spectrum of frequencies also including the high-energetic x-ray region,

Communications Laboratory, Department of Information Engineering, University of Illinois at Chicago Circle, P.O. Box 4348, SEO-1104, Chicago, IL, 60680.

in the following discussion we shall concentrate mainly on the m-to-mm wavelength region of the electromagnetic spectrum, i.e., we will mainly be concerned with microwave imaging. It is worthwhile mentioning that in medical imaging [11] we will have available in addition to conventional roentgenology and its recent extension to x-ray Computer Assisted Tomographic Scanning, three additional methods: ultrasonic methods based on either forward scattering or interferometric techniques, Zeugmatagraphy based on nuclear magnetic resonance effects, and water immersed microwave imaging. All of these three principally different methods will interrogate with different material tissue parameters, i.e., they will image different tissue properties. Thus, the ultimate medical imaging system of highest overall image resolution may be comprised of a set of four major imaging subsystems (x-ray CAT-scan, SONAR TT-scan, Zeugmatographic scan, and microwave time-delay spectroscopy-scan). In this paper we will be concerned mainly with the latter technique and refer for literature reviews on other topics primarily to the medical-applications oriented papers of Lin, Lauterbur, Burdette, Barrett and Meyers, Jacobi and Larsen published in these Proceedings.

As it had been clearly established in the companion papers by Larsen and Jacobi and Larsen (see also [12]–[14]), polarimetric properties of various tissue layers make themselves very pronounced and therefore aspects of polarization utilization in electromagnetic imaging [15], [16] will be given specific attention in the selection of the material chosen for presentation.

1.3   Selection of Relevant Inverse Scattering Methods

Although the inverse problem of electromagnetic scattering encompasses that of x-ray and optical scattering, we shall be mainly concerned with macroscopic wave interaction phenomena and microscopic material properties and associated wave interrogation will not be considered here. We will not be concerned with the polarizability of matter and tissue layers as reviewed most recently in [17], [18], but rather with properties of the coherent radar scattering matrix in relation to the problem of vector diffraction from radar targets [15]. This problem is also of immediate interest to microwave scattering parameter imagery [14] and therefore increased importance is placed on the peculiar properties of the coherent radar scattering matrix, its associated optimal polarizations, i.e., the use of the cross/co-polarization null pairs and of the associated polarization transformation invariant are considered in detail. It is the objective to introduce novel methods developed in radar imaging which should also be of considerable use in advancing the state of the art of inverse scattering in electromagnetic dosimetric imaging.

The material chosen for this succinct review is presented in three parts, the first dealing with basic measurement principles and mathematical theories; the second deals with the inverse problem of the wave equation and far field in-

version methods; and the third part with vector extension of existing electromagnetic inverse techniques. The presentation does however not deal with any direct medical applications and/or the interpretation of existing profile inversion theories developed in biomedical diagnostics [19].

PART I: BASIC PRINCIPLES & THEORIES

2.   MEASUREMENT INPUT AND OUTPUT

Since it is the main objective to review inverse scattering methods, which may be defined as the problem of deducing the characteristic parameters of a scatterer from measurement data obtained by interrogating the scatterer with a probing medium, it is appropriate to analyze some fundamental aspects of the probing medium, the input and output of the experiment, and the completeness or limitedness of the resulting measurement data.

2.1   The Probing Medium and Complete Measurements

The complete description of the geometrical and material properties of an unknown target will in general require a multitude of different probing techniques using different probing media such as acoustic waves, seismic p and s-waves, electromagnetic waves, etc. Each of the specific probing media displays specific distinct wave interaction properties and the design of a complete, unique measurement requires that these specific wave interaction properties are completely satisfied. For example, in pure sonar scattering using acoustic waves propagating in a fluid, a purely scalar treatment is satisfactory to completely describe the interrogation phenomenon [8]; whereas at microwave frequencies we do require the complete measurement of the scattering matrix per aspect and frequency [15] to obtain unambiguous, complete input data for inversion algorithm. This latter case applies also to water-immersed microwave scattering parameter imagery. For completeness' sake, it should be noted that in the case of vertical seismic profiling [9] complete information for every stack on both p- and s-wave scattering parameters is required to obtain a complete unambiguous set of input data for profile inversion.

In conclusion, we need to emphasize that depending on the nature of the interrogating medium as well as the resulting scattering phenomenon, a complete unambiguous set of measurement data per aspect and frequency is to be obtained for a self-consistent solution by use of an appropriate vector-extended profile inversion algorithm.

2.2   Measurement Limitations

In practice, the measured field data are limited in format (phase, amplitude, doppler, polarization), aspect (monostatic and/or bistatic over a limited measurement aperture; discrete and/or sparse) and frequency (discrete multifrequency, band-limited) and various limited approximate techniques may have to be applied depending upon the degree of completeness of data, as well as, on the amount of a priori knowledge available on the nature of the scatterer under interrogation.

Since measurement data in practice are available over limited frequency bands, profile inversion commonly is facilitated by taking recourse to appropriate approximate inverse theories applicable to the specific spatial frequency region of a scatterer's cross section, for which scattered field data are available. In Section 5 various inverse theories applicable only within such limited regions of the spatial frequency domain will be discussed and limitations on accuracy and well-posedness of solution in dependence of the format and completeness of the available data will be clarified.

2.3 Incompleteness and Data Limitedness, *A Priori* Knowledge and Self Consistency

Due to the very nature of compiling measurement data in a noisy environment and due to the nonavailability of measurement data over the total unit sphere of directions, it is essential that every bit of available a priori information be incorporated in the total profile inversion algorithm.

For example, in the case of radar target identification and/or imaging, it can be assumed that a rather extensive amount of a priori knowledge on both target and clutter properties is available so that a trade-off on the amount of measured data is possible. Various aspects of these limitations are discussed in detail in [7], where we refer specifically to Section 7.1.

In essence, we find that whenever a self-consistent, unique solution to an inverse problem is to be found given overall limited measurement data, an equal amount of complementary a priori information on the nature of the scatterer equivalent to the unavailable information is required. It should be noted here that the solution to the general data-limited problem given some a priori information is best facilitated by utilizing Radon's projection space analysis [20] yielding a self-consistent though not necessarily unique solution [7].

3. MATHEMATICAL INVERSE THEORIES

It has become evident during the discussion of complete measurement input data, that the profile inversion problem is highly mathematical in nature and we require some background knowledge of at least the more pertinent mathematical inverse theories. It would be totally beyond the scope of this succinct review to provide even a more or less complete overview on existing inverse theories and, therefore, we have chosen only four major topics which are of direct interest to applications of electromagnetic inverse

problems to dosimetric imagery. We, however, refer the reader to [1]–[10], [20]–[28] and [221]–[230] for further references.

3.1 General Inverses and the Backus-Gilbert Method

The mathematics of generalized inverses was developed side by side with the mathematics of profile inversion [29], and it has its roots essentially in the context of so-called "ill posed" linear problems [28], i.e., for which the questions of numerical stability and sensitivity are an issue. These include problems in which one either specifies too much information, or too little which is generally the case in most of the numerical profile inversion methods where the characteristics of a medium need to be recovered from remote, incomplete measurements [30]. Using the notion of "least-squares solution of minimal norm" it can be shown that a "unique solution" may be found, though these problems cannot be solved in the sense of a solution of a non-singular problem [31]. An excellent treatment of the theory of generalized inverses and its applications was given in [28] in which particular emphasis is placed on the Moore-Penrose inverse which is of great importance not only to optimization and pattern recognition, but also to numerical profile inversion techniques [1], antenna synthesis and electromagnetic inverse problem [32] (also see [227]).

Another procedure that can be used to solve ill-posed problems is the method of regularization [29]–[36]. An excellent tutorial exposition on the application of regularization methods to electromagnetic inverse problems is given in [32], where several regularization methods have been compared and discussed. Regularization methods are not a cure-all, e.g., in such cases for which insufficient data are available and no meaningful guess can stabilize the solution [5], [30], [37], [38] (also see [227]).

A rather startling extension of the regularization method may be seen in the optimum strategy method applicable to inversion of limited data also known as the Backus-Gilbert method [39], [40] which was further developed in [41]–[43] where it is shown that this method is well suited to treat a class of nonlinear problems of the solitary wave type [22] (also see [230]).

3.2 The Minkowski Inverse Problem in Differential Geometry

In differential geometry there exist two classical inverse problems concerning closed shaped surfaces which can be related to the radar problem of profile inversion of closed convex-shaped scatterers, known as the Minkowski [44] and the Christoffel-Hurwitz [45] inverse problems. These problems are based on Minkowski's basic paper on "Volumen und Oberfläche (1903)" in which he introduces a transform procedure to map the topological image of a closed convex surface onto the unit sphere of directions in terms

(i) the Gaussian curvature, being the Minkowski problem, or (ii) the sum of the principal radii of curvature, being the Christoffel-Hurwitz problem [46]. The basic information contained in the resulting theories is essential for establishing proofs of uniqueness and selfconsistency in the geometrical and/or electromagnetic inverse problem [47]–[50].

3.3　Radon's Projection Theory and Object Shape Density Profile Reconstruction from Projections

In a wide variety of profile reconstruction techniques a need often arises to deduce the two- or three-dimensional distribution of different physical quantities from their projections, e.g., in radio astronomy [51], [52], in structural biology [53]–[56], in X-ray photography [57]–[59], in geophysics [60], [61], in 3-dimensional quantum inverse scattering [63], [64]. Investigations of such image reconstruction problems in various specialized areas resulted in the establishment of the new interdisciplinary subject known as "Reconstruction from Projections" [65]–[70]. By far the most spectacular application of these techniques to date has been the development of Computer-Assisted-Tomographic Scanners (CAT-Scan) which have brought about a revolution in medical radiology (Hounsfield and MacCormack, 1979). Basically, these scanners (Hounsfield, 1973) take a discrete number of X-ray shadowgraphs at different angles and digitally process them using an algorithm of "reconstruction from projections" to produce a sharp image of any selected cross-sectional plane. The underlying theory was first described in lecture notes by Lorentz [71] and then rigorously formulated by Radon [64]–[70], [72]–[74]. Since this theory [221] is and will be of increasing importance to many different imaging methods in various physical fields and particularly in various schemes of dosimetric electromagnetic imaging, it is essential that a complete researcher in this field will have to attain a high standard of proficiency in the theories and techniques developed from Radon's theory. Specifically we refer for thorough studies to references [51]–[74] placing specific importance to [70], [73], [69] and [65]–[67], and we note that resulting techniques are very important to ultrasonic transmission tomography, zeugmatography, microwave scattering parameter imagery, and of course, X-ray CAT-scanning (also see [221]).

It should be noted here that in ultrasonic transmission tomography as well as in water immersed microwave scattering parameter imagery, it is essential to generalize Radon's theory such that ray path curvature and multi-path problems may be incorporated in to the reconstruction-from-projection algorithm [75], [5].

3.4　Pattern Classification Theory

Another peripheral discipline is strongly interwoven into the new science and technique of profile inversion and must be mentioned here, namely that of Pattern Classification Theory [76]–[80]. Following [77], the concept is based on the fact that scattering of electromagnetic waves from conducting bodies is a linear phenomenon and can be mathematically specified, for any incident pulse shape, by means of the impulse response of the target. In this sense, as for all linear systems, the scattered wave produced by an incident pulse is the Unique [77] electromagnetic characterization of the target. Proposing a classification system based on this idea is not practical because construction of short pulse, long-range radars is beyond the current state-of-the-art in radar technology. Since the impulse response is, however, equivalent to the frequency response at all frequencies, a classification scheme using measurements at many appropriately chosen harmonic frequencies should be considered. To optimize the choice of selected frequencies, it seems recommendable to choose one sampling frequency each in the Rayleigh, the resonant, the physical optics, and geometrical optics spectral regions of the radar cross-section as is explained in great deal in [76]–[80] and also in [68]. We strongly recommend the review papers [77] and [79] for further details on how methods of detection, estimation and modulation theory, the theory of testing statistical hypotheses, the theory of pattern classification, and scene analysis are used in conjunction with specific classification schemes as, e.g., the nearest neighbor method [76], [79], [80].

PART II:　SCALAR WAVE APPROACHES

In problems of remote sensing and probing we are essentially dealing with the inversion of the wave equation. Although in the following we are dealing exclusively with linear problems, here in the introduction we should mention that there exist many non-linear wave phenomena which are canonical and allow a solution in terms of solitary waves or solitons [81]–[90] and the reader is referred to the extensive list of readily available references provided. Although solitary wave theory is of great importance to the general inverse problem of scattering as was shown in the excellent tutorial exposition [91], in the following we are treating classical linear wave propagation phenomena only.

4.　INVERSE PROBLEM OF THE LINEAR WAVE EQUATIONS

In many remote sensing and profile inversion problems we are dealing with the inversion of the wave equation for non-homogeneous media with discontinuous stratifications from reflection and transmission data, as opposed to the determination of shapes of obstacles [1]. There exist various asymptotic techniques for the determination of the refractive index $n(\bar{r})$ associated with the wave equation and excellent recent reviews are given for example in [5], [8]–[10], [21]. Various solutions to this problem have been treated in radiowave propagation for many years and we refer to [10], [92]–[97] for a broad treatment of the problem. Of particular importance is the case of a slowly varying index profile n($\bar{r}$), in which case the classical WKBJ method [98] is often applied making use of the "phase memory" concept developed in geometrical optics [99]. The inversion of the wave equation has become one of the central issues in electromagnetic imaging of penetrable distributed inhomogeneous scatterers, and in the following a few of the more pertinent methods are reviewed (also see [227], [229]), and in the paper by Slaney *et al.* elsewhere in this volume.

4.1 Passive Sounding of the Propagation Path

A detailed review on passive and active atmospheric sounding is given in [1, Sects. 1 and 2] which usually requires the solution of ill-posed integro-differential equations. Since aspects of these problems have been considered in Section 3.1, we shall here concentrate on the Born and Rytov approximations of the wave equation [10], [92], [95], [229].

For the case in which ($n^2 - 1$) is small in the wave equation, it is possible to recover n($\bar{r}$) using the Born and Rytov approximations [10], [92], [93], [95] given measurements of the scattered outside the domain of support at a fixed frequency and for different angles of plane wave incidence. The Rytov approximation presents a significant improvement on the Born approximation providing that scattering is mainly in forward direction [94] as was shown earlier in [100]. In a recent study [75], [101] we have examined the Rytov approximation for smoothly varying media with the assumption that refraction predominates over reflection; and its significance for acoustic remote sensing was demonstrated for the case of slight ray bending. The extension of this problem in conjunction with the problem of reconstructing the profile from projections using an extended Radon transform approach is urgently required [5], [7] (see also [221], [229]).

4.2 Geophysical Inverse Problems

Today the most widely used geophysical method is the seismic exploration technique which is well reviewed in [102] making extensive use of spectral analysis [102]–[106], [8]. Exploration seismology is divided into two main disciplines that of reflection seismology and that of refraction seismology [21], where the time of arrival of seismic pulses at various locations, generated by sources at different locations, i.e., the "wave equation migration method" [107] is used to formulate the problem in terms of non linear functional equations. In practice the problem is linearized by treating n($\bar{r}$) as a linear combination of an unknown perturbed and a prescribed unperturbed part as described in detail in [9], [21], [102]–[107]. Particularly, the powerful signal versus clutter suppression methods developed in geophysics should be of considerable use to profile inversion and image reconstruction methods such as have been proposed in [12]–[14]. A vector extension of the above two methods is Vertical Seismic Profiling (VSP) in which use is made of both p- and

s-wave reflection and refraction (also see [230]). As was shown in [9], VSP permits inverse scattering in elastic media such that the complex multi-wave field as it is observed on the surface is split into its various wave types which can so be identified and allow determination of the cross-section with which the wave types are associated. An extension of this method utilizing optimal polarization information is feasible and should be rather useful in further developing dosimetric imaging methods of sub-aquadic microwave scattering parameter mapping.

4.3 Profile Inversion in Quantum Mechanics

In the framework of quantum mechanics major attention has been given to the case of non-relativistic particle interaction via potentials, with a scattering center or equivalently, that of two particles interacting through a potential depending on their relative distance [1], [4]. It is well known in quantum mechanics that the scattering of particles by a potential field is completely determined by the asymptotic form of the wave function at infinity (see also [227]).

In accordance with Heisenberg's postulate [108], it is precisely the asymptotic behavior of the wave functions that has a physical meaning [109]. The question therefore arises naturally as to whether it is possible to reconstruct the potential from the knowledge of the wave function at infinity [4], [108]–[110]. One of the most commonly used techniques in quantum mechanical inverse problems is the Gelfand-Levitan-Marchenko (GLM) procedure [4], next to the WKBJ and the Newton-Sabatier methods [1], [4], which has been developed to a very high state of perfection [227].

Of particular interest to this succinct review for advancing dosimetric electromagnetic imagery is the extension of the GLM method to the electromagnetic case which was considered for example in [111]–[114] for the one- and the two-dimensional cases. An extension to the three-dimensional case was attempted in [110], [115] and more recently in [116] where a Radon transform approach was introduced. It should be noted here that in the excellent tutorial exposition [91] the existing relations between time evolutionary, the inverse, and the solitary wave problems have been established which requires further interpretation for the direction application to electromagnetic dosimetric imagery.

4.4 Inverse Electromagnetic Radiation Problem

The inverse problem of the Helmholtz equation relates to the problem of determining scattering parameters from far field data which in essence becomes an inverse source problem [27] or that of reconstructing the near field from far field data [117]. This latter problem has been considered first in [118], [119], where it is shown that the near field can be recovered from the far field down to the minimum sphere enclosing the sources, i.e., the Wilcox-Müller sphere. For further details, we refer to the pertinent monographs presented in [27] and we shall not consider further the inverse source problem here in any detail.

5. SHAPE RECONSTRUCTION FROM FAR-FIELD DATA

A unique radar measurable of far field scattering data is the radar cross-section [24]–[26] which, in general, is highly frequency, aspect, and polarization dependent. There exist various asymptotic scattering theories valid only within narrow spatial frequency regimes of the monostatic radar cross section [5], [7], [15]; and the monostatic radar cross-section of an ideally conducting sphere usually is used as a reference cross section (see, e.g. [24]–[26] or [120].) Of particular interest are inverse scattering techniques developed from asymptotic or special wave solutions in i) the low frequency, ii) the resonant, iii) the Physical Optics, and iv) the Geometrical Optics spectral regions which are being considered next.

5.1 Low Frequency Inverse Scattering

In the Rayleigh-Gans or low frequency region the scattering properties of any finite 3-dimensional object provide interesting and useful reactions on the impulse and transient response [12], [122]. At sufficiently low frequencies [ka ≪ 1), the phasor response G(s) of a scattering object can be expanded in a Taylor series about the origin s = 0 and one can show [123]–[125] that according to the Rayleigh-Gans theory the first two expansion coefficients vanish, whereas the third is proportional to the volume of the scatterer times a polarization-dependent multiplication factor; and for symmetrical targets the fourth coefficient is also vanishing. Consequently, at very low frequencies the radar cross-section is proportional to the squared volume and the fourth power in frequency which establishes one of the most celebrated inverse scattering identities [122].

5.2 Natural Mode Excitation in Mie-Lorentz Region

In the Mie-Lorentz [126]–[128] the low and/or the high frequency asymptotic approximations fail and one has to resort to computational procedures [24]–[26]. One of the most efficient and direct methods for predicting scattering properties of an object in the resonance region has been the modal expansion [29]–[131] which is closely related to the singularity expansion method reviewed most recently in [132], [133]. The basic concept of the SEM method is to express the electromagnetic behavior in terms of the singularities in the complex frequency plane, which is a well known problem used in many physical problems of transient signature analysis [133]. The extension of the method to characterize the electromagnetic response in terms of its complex

frequencies as a function of the two-sided Laplace transform was developed in [129]–[132] and further expanded in [135], [136]. Since the Laplace-transform of a damped sinusoid corresponds to one pole or a pair of poles in the complex frequency plane, the scattering object may be expected to have a large response at frequencies near such poles. A broadband pulse excites these poles which may be referred to as the natural frequencies of the object. It should be noted that the distribution of the (infinite) set of eigen frequencies is independent of excitation and aspect; however, depending on excitation, polarization and aspect only a finite set of modes may be excited to produce an observable modal signature return. The basic methods of a target's natural frequencies concept were primarily developed at the ElectroScience Laboratory, Ohio State University [129], [130], [137], [138] and are well documented in [139] and in the paper by Young & Peters elsewhere in this volume where the biomedical implications are presented. It should be emphasized that the entirety of observable eigen modes for one aspect is obtained if and only if the broadband properties of all non-identical components of the radar scattering matrix are obtained [5], [7], [15]. Of particular importance to this resonant mode concept is the concept of pulse shaping for the purpose of direct mode extraction [140], [141] which should prove to become a very useful method also in various other schemes of dosimetric interrogation.

5.3 Physical Optics (PO) Inverse Methods

The PO inverse methods are basically derived from the PO approximation of Kirchhoff (see e.g. [99], [142]–[145] which provides a solution of reliable accuracy only for flat, or, surfaces with large radii of curvature excluding current regions close to or within the shadow regions. Furthermore, PO fails to satisfy reciprocity everywhere except in the direction of a specular return and it does not account for depolarization effects [146], [147].

(a) *Fourier Transform Method of Physical Optics.* It can be shown that there exists a linear relationship between the far scattered fields and the currents excited on the "equivalent surface" of a perfect conductor in terms of a Fourier transform pair [148]–[153]. This important Fourier transform identity [5], [7] between the target surface properties and the far scattered field have been utilized in many fields of remote sensing and target mapping [153]–[156] and in essence it is one of the basic underlying principles of High Frequency Inverse Scattering. We refer to [5], [7], [15], [221] for further details.

(b) *The Target Ramp Response Methods.* Employing the PO approximation and evaluating the diffraction integral over the illuminated side of the target, it is shown in [157] that the target's ramp response can be related to the cross-sectional projected area function of a target. This method was then applied to reconstruct the shape of rotationally symmetric, perfectly conducting targets for nose-on incidence [158], and in [159] a three-look angle approach was introduced to obtain the approximate shape of three-dimensional targets of more complicated shape. It should be noted that this method is not unique and that one would require to use projection area data for aspect angles covering the total unit sphere of directions to obtain a complete 3-dimensional reconstructed image. Such a method was developed in [160]–[162] where use is made of the Radon transform approach discussed in Section 3.3 of this review paper.

(c) *The $\bar{\kappa}$ Frequency-Space Formulation of POFFIS.* As was shown in [160]–[162], a straight forward application of an inverse Fourier-Radon transform to the ramp response identity leads to the $\bar{\kappa}$-frequency space formulation of Physical Optics Far Field Inverse Scattering [163] which was first reported in [164] and is known as the Bojarski identity [165]. It should be emphasized that the Bojarski identity represents a milestone in the development of high frequency inverse scattering techniques and many papers have been written on this subject matter as was reviewed in [5], [7] and also applied to the geophysical inverse problem more recently [166]. It would be beyond the scope of this succinct review paper to go into any further details which are basically all reported in [5], [7] and in [222] for the ultrasonic/seismic case.

5.4 Geometrical Optics (GO) Inverse Scattering

Whereas PO Inverse Scattering Theories are mainly broadband techniques, GO Inverse Scattering Methods primarily make use of narrowband radar data [167–169]. There exist basically two principal methods, that of equivalent curvature comparison and that of scattering center imagery based upon a Fourier transform method.

(a) *G.O. Equivalent Curvature Comparison Method.* This method is based on the G.O. asymptotic expansion of the radar cross-section which is primarily dependent on the local radius of curvature at the specular point [170], [171], and it is related closely to the Minkowski problem of differential geometry reviewed briefly in Section 3.2 [47], [50]. This behavior in the backscattering direction, is the foundation of a "system synthesis approach" introduced in [49] for the two-dimensional, and in [50] for the three-dimensional cases, respectively. This iterative averaging method compares the averaged magnitude of the backscattered fields, given off by the unknown, with that resulting from a known equivalent-curvature model scatterer. It should be noted that in the three-dimensional general case, we require total polarization information [5], [6], [16] to implement this approach.

(b) *Scattering Center Imagery, Kell's Monostatic Bistatic Equivalence Theorem.* The scattering center discrimination technique [172]–[174] draws heavily from the monostatic-bistatic equivalence theorem derived in [172] and extended to the general polarization-sensitive case in [175]. This basic technique was recently further generalized in [176] and is also one of the underlying principles in satellite far field imagery [177]. The principle if extended to a relatively narrow band multiple harmonic system allows

shape portrayal of rotating targets utilizing doppler information [178]–[180].

PART III: VECTOR INVERSE THEORIES

The complete description of electromagnetic scattering processes implies polarization and since an electromagnetic scatter acts like a polarization transformer, we require measurements for the complete description of the target scattering matrices so that the descriptive parameters of a scatterer can be uniquely recovered from the measured field data. For the purpose of demonstrating the concept of polarization utilization in practice the radar case is chosen and generalization to include other electromagnetic imaging techniques are briefly mentioned. Mainly for historical reasons of having had available amplitude data only, in most cases the approximations used in deriving inverse scattering theories had been simplified to purely scalar nature, i.e., polarization-dependent properties were discarded, and the resulting theories may no longer be valid or even unique. It is the main objective to demonstrate in Part III that due to the vector nature of electromagnetic waves, electromagnetic inverse scattering theories if applicable in practice require incorporation of complete polarization information into their formulation [16]. By applying this approach to existing theories, it is shown that remarkable improvements in fidelity and quality of the reconstructed images are obtained and that indeed there is ample justification for continuing the efforts in developing methods and theories of inverse scattering applicable to all those fields of physical sciences where information on the characteristic parameters of a scattering process is to be drawn from remote measurement be it the electromagnetic vector case or the even more complicated seismic case of s- and p-wave interactions in elastic media. In the following a succinct summary on important aspects of polarization utilization are presented.

6. ELECTROMAGNETIC VECTOR INVERSE PROBLEMS

Because it is the major objective of this Section to analyze the specific vector nature of electromagnetic inverse theories applicable in the m to mm wavelength regions, the quantity of primary interest in describing the far scattered field properties is the normalized radar cross-section which relates the far scattered fields to the incident fields [120]. In case the scatterer shape is not spherical but of general asymmetric shape and the scatterer possesses partially conducting surface properties, the simplified scalar description is no longer valid since depolarization occurs whenever the local curvatures differ [16]. In this more general vector case the target polarization transformation properties must be expressed [181] in terms of the target scattering matrix [S] which is introduced in Section 6.1. In particular, we will show that the polarization transformation properties of [S] establish some very important fundamental optimal polarization phenomena [182]–[185] which in the past have been overlooked in the proper formulation of vector inverse scattering theories [5], [7], [15], [16], [186] as is briefly summarized in Section 6.2. We also briefly describe properties of vector holography in Section 6.3.

Throughout the presentation of this material we shall make use of the succinct review paper [186] on this subject matter and only absolutely necessary symbolic formulae are presented here.

6.1 Scattering Matrix Description of Polarization

In propagation studies of electromagnetic waves it is common usage to decompose any complex polarization vector **h** into its horizontal (parallel to earth's surface) and vertical components so that $\mathbf{h} = \mathbf{h}_h + \mathbf{h}_v$. There exist various descriptions of the state of the polarization vector in terms of its ellipse or its unique presentation on the Poincaré polarization sphere which was introduced for the radar case in [182] as is assumed in [15], [16], [186]. Further development of this topic is presented in the paper by Larsen & Jacobi elsewhere in this volume. To overcome limitations of the conventional radar range equation [120], Sinclair [181] initiated a systematic approach to the analysis of radar target detection in clutter using polarization and defining the scattering matrix [S] which for two general orthogonal elliptical polarization base vectors $\mathbf{h}_A$ and $\mathbf{h}_B$ such that $\mathbf{h}_A \cdot \mathbf{h}_{\bar{B}} = 0$ may be defined as

$$[S] = \begin{bmatrix} S_{AA} & S_{AB} \\ S_{BA} & S_{BB} \end{bmatrix}, \text{ monostatic case } S_{AB} = S_{BA} = |S_{AB}|$$

This scattering matrix [S] can be related to the 4 × 4 Mueller matrix [M] as is shown, e.g., in [15] and in more detail in [10], and its properties are also of importance here in case incoherent scatter properties need to be analyzed.

Returning to [S], it is obvious that there exist an infinite number of general elliptical orthogonal polarization base vector pairs ($\mathbf{h}_A$ and $\mathbf{h}_B$) and an infinite number of possible invariant transformations which can be best performed on the Poincaré polarization sphere of radius p = $|S_{AA}|^2 + |S_{AB}|^2 + |S_{BA}|^2 + |S_{BB}|^2$ [182], [183]. It was Kennaugh [185] who first observed that associated with [S] are two optimal polarization pairs, the minimal polarization (CO-POL nulls) for which the diagonal terms S_{AA} and S_{BB} vanish, and the maximal polarization pairs (X-POl nulls) for which the off-diagonal cross (X)-polarized terms in [S] vanish. The important property that the four polarization nulls lie on a main circle on associated polarization sphere of radius p such that the X-POl nulls are antipodal and the CO-POl nulls are bisected by the line joining the X-POl null locations was utilized first with definite success by Huynen [184] and its unique importance was more recently strongly advocated by Poelman [185] as is summarized in detail in [186]. It should be noted here that under certain conditions of particularly pronounced target symmetry the two distinct CO-POL nulls may degenerate into one double solution being identical to one of the X-POL nulls [183].

We need to emphasize here that each specific type of target and/or clutter produces its own characteristic CO-POL null distribution within bounded regions on the Poincaré sphere [15]. In fact, it has been demonstrated that the co-polarization null distribution of clutter is strongly stable if no target is present but it is strongly affected in the presence of a target and thus the copolarization null concept should serve as a useful target versus clutter detection as well as imaging discriminant [185]. Due to the limited space available, the very interesting properties of the associated Mueller matrix [M] are not discussed here and we refer to [10] for some details.

6.2 Polarization Correction of Electromagnetic Inverse Scattering Theories

Various inverse theories in Optics have recently been considered in [27] and aspects of polarization utilization in electromagnetic imaging are reviewed by this author in [27]. Therefore, we shall be very brief and only highlight the most important applications of optimal polarization concepts introduced above to the most relevant inverse scattering theories in the various major frequency regimes.

(a) *Low Frequency Regime: Mapping of Hydrometeors.* As was pointed out in Section 5.1 the coefficient a_2 in a low frequency Taylor series expansion of the phasor response, depends upon shape, orientation and on constitutive parameters of the scatterer [122] and very strongly on polarization [125]. Although radar targets do not usually fall within the Rayleight-Gans region (target dimension l smaller than wavelength: $kl < 1$), other scatterers such as hydrometeors (precipitation particles) do, and recently considerable efforts have been expanded on using polarization in radar meteorology [186]–[191]. Of particular interest should be the observation that properly used polarization information is very effective in analyzing differential state changes of hydrometeors and also will uniquely allow us to identify the boundaries in between hydrometeor ensembles of different species. It must be stated here that decisive improvements [192] still can be made by fully exploiting the optimal polarization concepts discussed in Section 6.1.

(b) *Resonant Region: Identification of a Target's Natural Frequencies.* As was discussed in Section 5.2 the basic idea of identifying a target by a limited set of its infinite number of natural frequencies is that the location of the eigen frequencies in the complex frequency plane is independent of aspect and excitation [136]. Yet, depending upon characteristics of the probing pulse, it is possible to excite only one or the other pulse [140] and pulse-shaping is also very dependent on polarization [186]. There still exist a number of unresolved questions relating to the self-consistency of this identification method in terms of a target's natural frequencies. However, results [137]–[139], [193]–[196] clearly indicate that this method will play an essential role in developing reliable techniques of target identification and classification.

It is pertinent here to mention the properties of non-linear reradiating effects from metal junctions which were utilized in the design of the METRRA system [197], [198], [78]. In this *METal Reradiating RAdar* system a non-linear metal junction effect is utilized which cases odd 3rd, 5th, 7th, 9th, . . .) higher order harmonics to reradiate [199]. A combination of METRRA and excitation of specific high frequencies utilizing the method of pulse shaping [140], [141] may yield a self-consistent target classification technique.

(c) *Polarization Correction of HF-Inverse Scattering Methods*. As was shown before, in the high frequency region there exist three major inverse scattering theories: (i) POFFIS, (ii) ECIM, and (iii) GOIS which all, in principle, are derived from polarization insensitive approximations. Since extensive analyses and reviews of these problems are presented in [5], [7], [15], [16], [186] utilizing vector correction of the P.O. approximation first presented in [200] and the optimal polarization concept, no further details are presented here.

We note, however, that it is essential to utilize total polarization information otherwise ambiguity in the reconstruction of the unique solution may result.

6.3   Polarization Microwave Imagery

Although some scattered reports on utilizing polarization information in microwave holography exist [201]–[203], two major recent reviews on microwave holography [204] and radar imaging of terrain structures [205] neglected to treat this important aspect. A first systematic analysis on polarization diversity was presented in [206], based on experiments described in [207]. In these studies it is shown how the incoherent super-position of the images reconstructed from microwave hologram recordings of the three different monostatic elements $|S_{HH}|^2$, $|S_{HV}|^2$ and $|S_{VV}|^2$ of [S] yield the highest image resolution and image fidelity [15]. This finding is consistent with the properties of the polarization transformation invariant p = span{[S]} = {$|S_{HH}|^2$ + $|S_{HV}|^2$ + $|S_{VH}|^2$ + $|S_{VV}|^2$} = {$|S_{AA}|^2$ + $|S_{AB}|^2$ + $|S_{BA}|^2$ + $|S_{BB}|^2$}, where A,B denote any pair of general orthogonal elliptical polarizations. Hence the optimal reconstructed image is obtained whenever an incoherent super-position of the, in general, four images reconstructed according to the law of polarization transformation invariance is performed. We note that any general orthogonal elliptical pair of polarization base vectors may be employed [186]. It should be noted that incorporation of the optimal polarization concept may add another dimension to microwave imaging with the specific aim at reducing the inherent clutter return [207].

6.4   Vector Holography: The Concept of Inverse Boundary Conditions

In direct problems of scattering and diffraction, the shape and the material constituents of the scatterer, all of which

are assumed to be known *a priori* together with the prespecified incident fields may be incorporated into the boundary conditions. On the other hand, in the formulation of the true inverse problem, in general, no specific information about the scatterer may be assumed. Therefore, in the inverse case such *inverse boundary contions* must be sought, which neither depend on the shape nor the material properties of a scattering body, but allow to specify those characteristic parameters uniquely from the near field which need to be recovered from the far field [208]–[211].

The problem of recovering the near field from far field measurements is still an open problem [212], [213], although remarkable progress has been made recently [5], [7], [15], [16], [214], [215]. Thus, assuming that the electric and magnetic near fields can be recovered from the measured far field data, the question arises as to how many and which characteristic parameters must be defined to determine uniquely or at least self-consistently the shape, the size, and the material constituents of an unknown scatterer, given the fields everywhere within and in its neighborhood [210].

The formulation of such inverse boundary conditions was initiated under the guidance of V. H. Weston [216], [217], [208], [209] and is formulated to some detail in [210], and verified in [211]. The resulting necessary but not sufficient conditions, resembling the character of interference relations commonly recorded on a hologram as was reviewed in [15], can be used to specify both the correct scattering surface and its local material surface properties provided one reference point and its associated local reference impedance is given at one point on the proper surface locus [15].

The concept of necessary but not sufficient inverse boundary conditions will require further extended studies in connection with developing systematic methods of discriminating the proper locus from the improper pseudo loci of either surface location or surface impedance. To accomplish this goal will require highly improved and more accurate polarization measurement techniques. One such method which rends itself useful for this purpose is the new unified theory of near-field analysis and measurement developed by Wacker [214], [215]. This unified nearfield theory was instrumental for establishing one of the most excellent near field measurement facilities hitherto available (at the EM Division, NBS, Boulder [219]).

6.5   Measurement Requirements in Electromagnetic Inverse Scattering

It was one of the main objectives of this presentation to emphasize the fact that in electromagnetic inverse problems we are dealing with a vector holographic problem. It is the common trend to shy away from vector or polarization measurement problems and rather implement a simplified scalar treatment which more often does not apply in practice. Although the phenomenon of polarization seems to be well understood in classical optics, the rather important optimal polarization concept developed from a radar scattering matrix approach have been overlooked. Therefore, the very useful concepts of co/cross polarization nulls have been stressed, since they will undoubtedly become instrumental in radar target versus background image discrimination, and in recovering the useful coherent target signal in polarization microwave holography from clutter perturbed data. Similarly, the very important properties of the polarization transformation invariant $p = \Sigma_i \Sigma_j |S_{ij}|^2$, which states that the local scattering matrix power must be recovered, otherwise information essential to completely describe the image properties of a scatterer are lost, is of particular importance to radar target mapping and holographic imaging in the mm-to-m wave range being considered here. Finally we have indicated that further development of polarization microwave holography may enable us to map both the image and material properties of the surface enclosing a scatterer.

7. CONCLUSIONS

It was one of the main objectives of this contribution to the Workshop on Electromagnetic Dosimetric Imaging to emphasize the fact that the problems of target and density profile imagery and mapping are an integral part of the newly developing multidisciplinary science of "Inverse Scattering," or "General Profile Inverse," or "Remote Sensing," etc.

The choice of the presented material was dictated mainly by the objective to assist researchers engaged in studies related to Electromagnetic Dosimetric Imagery and provide them with rudimentary reference material (some major 230 references) to analyze inverse problems appearing in their work.

7.1   Summary on State of the Art

It was shown that inverse problems appear in many different fields of science and that, in principle, the underlying basic problems can be solved with the aid of unique mathematical tools which may be named for example "General Inverses Theory of Profile Inversion," "Radon's Projection Theory," "The Maximum Entropy Method," etc. The overall state of the art is still in the development phase, and we require many state of the art reviews, monographs, workshop proceedings and special conference sessions with subsequent special issues on the subject matter. Yet, it is evident that several inverse theories have had already a upshaking effect on the state of the art of electromagnetic dosimetric imagery such as for example Radon's projection theory to CAT-scan, or to zeugmatography or to aquadically matched microwave imagery. In summary, we expect that there will be many more effective and upshaking applications of inverse theories to electromagnetic dosimetric imagery.

7.2   Omissions and Limitations

In this contribution we have not reviewed the total pertinent literature on inverse scattering theories due to (i) lack of space, (ii) priorities and applicability, and (iii) the current state of rapid changes and development of particular fields. However, the choice and selection of contributions made by the coordinators of this worthwhile workshop should ensure that many aspects which should have been mentioned in this paper are presented in companion papers as for example microwave deep sounding (Peters, et al.), zeugmatography (Lauterbur), Microwave Holography (Farhat).

Furthermore, there have appeared several State-of-the-Art review monographs which are referenced and should be consulted for further reading (e.g. [227]).

7.3   Recommendations

The Symposium on Electromagnetic Dosimetric Imagery clearly showed that in addition to conventional Roentgenology and X-ray CAT-scanning, ultrasonic imagery, the two more recently developed methods of zeugmatography and of aqua immersed microwave imagery add two new very important techniques to medical imaging. Of specific importance to advancing the technique of water immersed microwave imaging are properties of the extended Radon-Fourier transform theory, of polarization information utilization and methods for correcting raypathblending and diffraction effects. Recent advances in electromagnetic inverse scattering theory should be of great use to perfect the method.

The invitation by Dr. L. E. Larsen and Mr. John H. Jacobi for participating in such a stimulating workshop is highly appreciated and I thank them for this opportunity. Similarly, I wish to thank my peer colleague Dr. Arthur K. Jordan for many stimulating discussions and his encouragement.

The research reported was supported, in parts, by the National Science and Engineering Research Council of Canada (Grant No. A7240), The U.S. Army Office of Research, the U.S. Office of Naval Research, a continuing Humboldt fellowship, a Nato Research Grant (No. 1405) and with clerical assistance of the Department of Information Engineering, UICC, Chicago, as well as of the German Space Research Administration (DFVLR) at Oberpfaffenhofen, Bavaria, and the HF-Institute, Technical Faculty, University of Nürnberg in Erlangen, FRG. Last not least I wish to acknowledge the financial assistance by the Walter Reed Army Institute of Research for participating in the Symposium.

References

1. L. Colin, "Mathematics of Profile Inversion," Proc. of a Workshop, NASA Technical Memorandum X-62.150, Ames Research Center, Moffet Field, Ca., 1972, (also see additional refs. [2.221]–[2.230].
2. J. B. Keller, *Amer. Math. Monthly*, **83,** pp. 107–118, 1976.
3. R. L. Parker, *Ann. Res. Earth Planet Sci.*, **5,** pp. 35–64, 1977.
4. K. Chadan and P. C. Sabatier, *Inverse Problems in Quantum Scattering Theory*, Springer, New York, Heidelberg, Berlin, 1977.

5. W. M. Boerner, "State of the Art Review on Polarization Utilization in Electro-magnetic Inverse Scattering," Rept. No. 78-3, Communications Laboratory, University of Illinois, Chicago, 1978.

6. C. L. Bennett, "Inverse Scattering: Time-Domain Solutions Via Integral Equations," Nato Advanced Study Institute on Theoretical Methods for Determining the Interaction of Electro-magnetic Waves with Structures, J. K. Skwircynski, Ed., Sijphoff & Noordhoff Intern. Publ. Co., Amsterdam, Netherlands, 1980, in press.

7. W-M. Boerner, "Polarization Utilization in Electromagnetic Inverse Scattering," Chapter 7 in *Inverse Scattering Problems in Optics*, Vol. II, H. P. Bates, Ed., Topics in Current Physics, Vol. 20, Springer Verlag, July 1980.

8. M. T. Silvia and E. A. Robinson, *Deconvolution of Geophysical Time Series in the Exploration for Oil and Natural Gas.* Developments in Petroleum Science 10, Elsevier, 1979.

9. E. I. Gal'perin, *"Vertical Seismic Profiling,"* translated by Alfred J. Hermont, J. E. White, Ed., Society of Exploration Geophysicists, Special Publication No. 12, 1974.

10. A. Ishimaru, *Wave Propagation and Scattering in Random Media,* Vol. I: Single Scattering and Transport Theory: Vol. II: Multiple Scattering, Turbulence, Rough Surfaces, and Remote Sensing, Academic Press, New York, 1978.

11. K. Preston, Jr., K. J. W. Taylor, S. A. Johnson, and W. R. Ayers, (Eds.), *Medical Imaging Techniques—A Comparison,* Plenum Press, New York, 1979.

12. L. E. Larsen and J. H. Jacobi, "Microwave interrogation of dielectric targets. PART I: By Scattering Parameters"; "Part II: By microwave time delay spectroscopy," *Med. Phys.* 5(6), pp. 500–513, Nov/Dec., 1979.

13. L. E. Larsen, J. H. Jacobi and A. K. Krey, "Preliminary observations with an e.m. method for the noninvasive analysis of all suspension physiology and induced pathophysiology," *IEEE Trans. Microwave Theory Tech.,* 26(8), pp. 581–595, Aug. 1978.

14. J. H. Jacobi, L. E. Larsen, and C. T. Hast, "Water-immersed microwave antennas and their applications to microwave interrogation of biological targets," IEEE Trans. Microwave Theory Tech., 27(1), pp. 70–78, Jan. 1979.

15. W-M. Boerner, "Polarization Microwave Holography: An extension of scalar to vector holography (INVITED)," 1980 Intern. Opt. Computing Conference, SPIE's Tech. Symposium East, Wash., D.C., April 9, 1980, Sess. 3B, Paper No. 231-23, 1980.

16. W-M. Boerner, "Use of *Polarization* Information in Electromagnetic Inverse Scattering," Inverse Scattering Sess. III, Paper No. 1, 1980 Int. URSI-Comm. B, Wave Propation Sym., Munch. GRG, Aug. 1980.

17. J. M. Bennett and H. E. Bennett, "Polarization," in *Handbook of Optics*, W. G. Criscoll and W. Vaughan, Eds., McGraw-Hill, New York, 1978, Chap. 10.

18. J. C. Lin, "Microwave Biophysics," in *Microwave Bioeffects and Radiation Safety*, M. Stuchly, Ed., IMPI, Edmonten, Alta., pp. 15–54, 1978.

19. L. E. Larsen and J. H. Jacobi, "Microwave scattering parameter imagery of an isolated carine bridney," *Med Phys.*, 6(5) pp. 394–403, Sept./Oct. 1979.

20. W-M. Boerner and C-M Ho, "Physical Optics Far Field Inverse Scattering for Space Aspect and Limited Aperture Data," *Wave Propagation*, 4, Sept. 1980, in print.

21. J. F. Claerbout, "Fundamentals of geophysical data processing, with applications to petroleum prospecting," Interim-Series in Earth & Planetary Sciences, McGraw-Hill, New York, 1976.

22. V. H. Weston: "Electro-magnetic Inverse Scattering," in *Electromagnetic Scattering*, P. L. E. Uslenghi, Ed., Academic Press, New York, 1978, pp. 289–313.

23. Y. T. Lin and A. A. Ksienski Radio & Electr. Eng., **46**, pp. 472–486, 1976. (also see: A. A. Ksienski: "Inverse Scattering as a Target Identification Problem," in *International Symposium on Recent Developments in Classical Wave Scattering—Focus on the T-Matrix Approach*, Ohio State University, Columbus, Ohio, June 25–27, 1979, V. K. Varadan and V. V. Varadan, Eds., Pergamon Press, New York, 1980, in press).

24. J. J. Bowman, T. B. A. Senior, and P. L. E. Uslenghi *Electromagnetic and Acoustic Scattering by Simple Shapes*, North-Holland, Amsterdam, 1969.

25. J. W. Crispin, Jr. and K. M. Siegel *Methods of Radar Cross Section Analysis*, Academic Press, New York, 1968.

26. G. T. Ruck, D. E. Barrick, W. D. Stuart, and C. K. Krichbaum, *Radar Cross Section Handbook*, Vols. I/II, Plenum Press, New York, 1970.

27. H. P. Baltes, *Inverse Source Problems in Optics*, Topics in Current Physics, Vol. 9, Springer-Verlag, New York, 1978. (Vol. 20, 1980.)

28. M. Z. Nashed, *Generalized Inverses and Applications*, Proc. of Adv. Seminar, Univ. of Wisconsin, Madison, Wisc., Oct. 8–10, 1973, Academic Press, New York, 1976.

29. M. M. Laverentiev, *Some improperly posed problems of mathematical physics*, Springer Verlag, New York, 1967.

30. Z. S. Agranowich and V. A. Marchenko, *The Inverse Problem of Scattering Theory*, (transl. by B. D. Seckler), Gordon & Breach Science Publishers, New York, 1963.

31. S. Twomey, "The application of numerical filtering to the solution of integral equations encountered in indirect sensing measurements," *J. Franklin Inst.*, **279**(2), pp. 95–109, 1964.

32. G. A. Deschamps and H. S. Cabayan, "Antenna synthesis and solution of inverse problems by regularization methods," *IEEE Trans. Antennas Propag.*, **20**(3), pp. 168–174, 1972.

33. B. Van der Pol and H. Bremmer, *Operational Calculus* (based on the two-sided Laplace Transform), Cambridge Univ. Press, 1955.

34. A. Tikhonov et V. Arsénine, "Méthodes de résolution de problémes mal posés," *Dokl. Nauk* SSSR, **39**, 1944.

35. A. N. Tikhonov, "Regularization of mathematically incorrectly posed problems," *Sov. Math.*, **4**, pp. 1624–1627, 1963.

36. H. S. Cabayan, R. C. Murphy, and R. J. F. Pavlasek, "Numerical stability and near field reconstruction," *IEEE Trans. Antenna Propag.*, **21**(3), pp. 346–351, 1973.

37. M. Bertero, C. DeMol, and G. A. Viano, "Restoration of optical objects using regularization, *Opt. Lett.*, 1978, in press.

38. W. L. Perry, "Approximate solution of inverse problems with piecewise continuous solutions," *Radio Sci.*, **12**(5), pp. 637–642, 1977.

39. G. Backus and F. Gilbert, "The resolving power of gross earth data," *Geophys. J. R. Soc.*, **16**, pp. 169–205, 1968.

40. G. Backus and G. Gilbert, "Uniqueness in the inversion of inaccurate gross earth data," *Philos. Trans. R. Soc.*, **A266**, pp. 123–192, 1970.

41. R. J. Banks, "Geomagnetic variations and the electrical conductivity of the upper mantle," *Geophys. J. R. Astron. Soc.*, **17**, pp. 457–487, 1979.

42. R. C. Bailey, "Inversion of the geomagnetic induction problem," *Proc. R. Soc.*, **315**, pp. 185–194, 1970.

43. R. L. Parker, "The inverse problem of elec. conductivity in the mantle," *Geophys. J. R. Astron. Soc.*, **22**, pp. 121–138, 1970, [also see: 7.21–7.28 in (Colin, 1971): The Backus-Gilbert method and its application to the electr. conductivity problem].

44. H. Minkowski, "Volumen und Oberflache," *Math. Ann.*, **57**, pp. 447–495, 1903.

45. L. Nirenberg, *Commun. Pure Appl. Math.*, **6**, pp. 337–394, 1953.

46. J. J. Stoker, *Commun. Pure Appl. Math.*, **3**, pp. 231–257, 1950.

47. P. C. Waterman and M. R. Weiss, "Inverse Scattering and the Minkowski Problem," Proc. GISAT II Symp., Bedford, Ma., Oct. 2–4, 1967, Vol. 2, Pt. I, pp. 371–376, Mitre Corp.

48. W. T. Payne, "Determination of Shape and Size of Non-Axisymmetric Conducting Targets by Geometrical Optics," Proc. GISAT II Symp., Bedford, Ma., Oct. 2–4, 1967, Vol. 2 Pt. I, pp.

303–313, Mitre Corp.

49. F. H. Vanderberghe and W-M. Boerner, *Radio Sci.*, **6,** pp. 1163–1171, 1971.

50. S. K. Chaudhuri and W. M. Boerner, "A monistatic inverse scattering model based on polarization utilization." *Appl. Phys.* (Springer), 11(4), pp. 337–350, 1976. [Also see: S. K. Chaudhuri and W. M. Boerner, "Polarization utilization in profile inversion of a perfectly conducting prolate spheroid," *IEEE Trans. Antennas Propag.*, 25(4), pp. 505–511, 1977a.]

51. R. M. Bracewell, "Strip integration in radioastronomy," *Austr. J. Phys.*, **9,** pp. 198–217, 1956. (See also: "Two-dim. aerial smoothing in radio astronomy," *Austr. J. Phys.*, **9,** pp. 297–314, 1956); Chapt 2: Image reconstruction in radio astronomy, in *Image Reconstruction from Projection*, G. T. Herman, Ed., Topics in Applied Physics, 32, Springer, 1979.

52. R. M. Bracewell and A. C. Riddle, "Inversion of fan-beam scans in radio-astronomy," *Astrophys. J.*, **150,** pp. 427–434, 1967.

53. R. N. Bracewell and S. J. Wernecke, "Image reconstruction over a finite field of view," *J. Opt. Soc. Am.*, Vol. 65, No. 11, pp. 1342–1347, 1975.

54. R. A. Crowther, D. J. DeRosier, and A. Klug, "The reconstruction of a three-dimensional structure from projections and its applications to electron microscopy," *Proc. R. Soc. London Ser. A*, **317,** pp. 319–340, 1970.

55. R. B. Gordon, Ed., Proc. of Topical Meeting on Image Processing for 2-D and 3-D Reconstruction from Projections: Theory and Practice in Medicine and the Physical Sciences, Aug. 4–7, 1975, Stanford University, Stanford, 1975; R. B. Gordon (1978), "A Treatise on Reconstruction from Projections and Computer Tomography," Golem Press, in preparation; R. B. Gordon and G. T. Herman, "Three-dimensional reconstruction from projections: A review of algorithms," Intern. Rev. of Cytology, Acad. Press, 38, pp. 111–151, 1974: R. Gordon, G. T. Herman and S. A. Johnson, "Image reconstruction from projections," Scientific American. Oct., pp. 56–68, 1975.

56. A. M. Cormack, "Representation of a function by its line integrals, with some radiological applications," *J. Appl. Phys.*, **34,** pp. 2722–2727, 1963. (Also see: A. M. Cormack, "Representation of a function by its line integrals, with some radiological applications, II," *J. Appl. Phys.*, **35,** pp. 2908–2913, 1964.)

57. G. N. Hounsfield, "Computerized transverse axial scanning (Tomography): Part I, Description of the system," British J. of Radiology, **46,** pp. 1016–22, 1973. (Also see: Patent specification 1283915, Patent Office, London, 1972.)

58. D. G. Grant, "Tomosynthesis: a three-dimensional radiographic imaging technique," IEEE Trans. Biomed. Eng., 19(1), pp. 20–28, 1972. (Also see: M. Kock and U. Tiemens, "Tomosynthesis: A holographic method for variable depth display," *Opt. Commun.*, 7(3), pp. 260–265, 1973.)

59. B. K. P. Horn, "Fan-beam reconstruction methods," *Proc. IEEE*, 67(12), pp. 1616–1623, Dec., 1979.

60. W. Munk and C. Wunsch, "Ocean acoustic tomography: a scheme for large scale monitoring," *Deep-Sea-Res.*, **26A,** pp. 123–161, 1978.

61. K. Aki, A. Christoffersen, and E. S. Huxxbye, "Determination of the three-dimensional seismic structure of the lithosphere," *J. Geophys. Res.*, **82,** pp. 277–296, 197X.

62. D. L. Lager and R. J. Lytle, "Computer algorithms useful for determining a subsurface electrical profile via HF probing," Lawrence Livermore Laboratory, UCRL-51748/TID-4500/UC-11, Feb., 1975.

63. R. G. Newton, "Scattering Theory in the Mixed (Radon Space) Representation," in *1979 Conf. on Mathematical Methods and Applications of Scattering Theory*, Catholic University Washington, D.C. USA, May 21–25; J. A. DeSanto, A. W. Saenz and W. W. Zachari, Eds., Lecture Notes in Physics, Springer Verlag, Berlin, Heidelberg, New York, 1980, in press.

64. Y. Das and W. M. Boerner, "Application of algorithms for 3-D image reconstruction from 2-D projections to electromagnetic inverse scattering," USNC/URSI Annual Meeting, Boulder, Col., 20–23 Oct., 1975, Session III-7-7, p. 184. (Also see: Y. Das and W. M. Boerner, "An interdisciplinary approach to electromagnetic inverse scattering using Radon transform theory." USNC/URSI-Comm B, Sess. B-4, Paper No. B-4-3, IEEE/URSE Int. Symp. 1978, Washington, D.C. May 15–19, 1978. Y. Das and W. M. Boerner, "On radar target shape estimation using algorithms for reconstruction from projections," *IEEE Trans. Antennas Propag.*, 26(2), pp. 274–279, 1978. Y. Das, "Application of concepts of image reconstruction from projections and Radon transform theory to radar target identification," Ph.D. dissertation, Fac. of Grad. Studies, U. of Manitoba, Winnipeg, MB, Nov., 1977.)

65. L. A. Shepp and J. B. Kruskal, "Computerized tomography: the new medical x-ray technology," *Amer. Math. Monthly*, **85,** pp. 420–439, 1978.

66. H. T. Scudder, "Introduction to computer-aided tomography," *Proc. IEEE*, 66(6), pp. 628–637, 1978.

67. G. B. Folland, "Introduction to partial differential equations," Math. Notes, Princeton Univ. Press, pp. 233–238, 1976.

68. B. F. Logan and L. A. Shepp, "Optimal reconstruction of a function from its projections," *Duke Math J.*, **42,** pp. 645–659, 1975.

69. R. H. T. Bates and T. M. Peters, "Towards improvements in tomography," New Zealand J. Sci., Vol. 14, No. 4, pp. 883–896, 1971.

70. K. T. Smith, D. C. Solomon, and S. L. Wagner, "Practical and Mathematical aspects of the problem of reconstructing objects from radiographs," *Bull. Am. Math. Soc.*, 83(6), pp. 1227–1270, 1977.

71. G. E. Uhlenbeck, "Over een stelling van Lorentz en haar uitbreiding voor meer dimensionale ruimten," Physica, Nederlands Tijdschrift voor Natuurkunde, **5,** pp. 423–428, 1925. (Also see: H. B. A. Bockwinkel, Versl. Kon. Akad. Wis., XIV, 2, 636, 1906.)

72. J. Radon, "Über die Bestimmung von Funktionen durch ihre Integralwerte längs gewisser Mannigfaltigkeiten" Ber. Verh. Sächs. Akad. Wiss. Leipzig," Math.-Nat. K1, **69,** pp. 262–277, 1917. (Also see: P. Mader, "Uber die Darstellung von Punktfunktionen im n-dim. euklidischen Raum durch Ebenenintegrale, Berichte der math-phys. Kl. der sächs," Gesellschaft der Wissenschaften, 1926, 646–652.)

73. D. Ludwig, "The Radon transform on Euclidean spaces," *Commun. Pure Appl. Math.*, Vol. XIX, pp. 49–81, 1966.

74. B. K. Vainshtein, "Finding the structure of objects from projections," *Kristallografiya*, **15,** pp. 894–902, 1970 (transl. in *Sov. Phys. Crystallogr.*, **15,** pp. 781–787, 1971). "Synthesis of projecting functions," *Dokl. Akad. Nauk*, *SSSR*, **196,** pp. 1072–1075, 1971 (translated in Sov. Phys. Dokl., **16,** pp. 66–69, 1971).

75. G. R. Dunlop, "Ultrasonic Transmission Imaging," PhD. Thesis, Univ. of Canterbury, Christchurch, New Zealand. (Also: W. M. Boerner, G. R. Dunlop, and R. H. T. Bates, "An extended Rytov approximation and its significance for remote sensing and inver scattering." *Opt. Commun.* 18(4), pp. 421–423 1976.)

76. J. P. Toomey, R. M. Hieronymus, and C. L. Bennett, "Multiple Frequency Classification Technique," Sperry Rand Recentre, Sudberry, MA, Final Rept. RADC-TR-76-59, March, 1976, AD-A023-204.

77. J. L. Mesla and D. L. Cohn, *Decision and Estimation Theory* McGraw Hill, New York, 1978.

78. J. K. von Schlachta, "A contribution to radar target classification," Radar 1977, IEE Conf. Publ. 155 (IEEE, Savoy Place, London, England, 1977) pp. 135–139.

79. C. Brindley, "Target recognition," Space/Aeronautics, pp. 62–68, June, 1965.

80. A. G. Repjar, "The linear separability of multiple-frequency radar returns, with applications to target identification, ESL-OSU Tech. Rept. 2768-5, 16 Nov., 1970, AD 717198.

81. S. M. Ulam, *Problems in Modern Mathematics*, J. Wiley, N.Y.,

1964.

82. W. F. Ames, *Nonlinear ordinary differential equations in transport processes*, Acad Press, New York, 1968. (Also see: W. F. Ames, ed., *Nonlinear partial differential equations: A symposium on methods of solution*, Acad. Press, N.Y., 1967. W. F. Ames, *Nonlinear partial differential equations in Engineering*. Acad. Press, N.Y., 1965.)

83. P. P. Zabreyko, M. A. Koshelev, S. G. Mikhin, L. S. Rakoreshchik, and V. Ya. Ste'senko, *Integral Equations—A reference text*. Noordhoff, Leyden, 1975.

84. V. E. Zakharov and L. D. Faddayev, "Korteweg-de Vries equation: A completely integrable Hamiltonian system," *Funct. Anal. Appl.*, **5**, pp. 280–287, 1971.

85. R. M. Miura, "Korteweg-de Vries equation and generalizations. I. A remarkable explicit nonlinear transformation," *J. Math. Phys.*, **9**, pp. 1202–1204, 1968, (also see 1204–1209).

86. G. B. Whitham, *Linear and Nonlinear Waves*, Pure & Applied Mathematics, Wiley-Interscience Series, J. Wiley & Sons, New York 1974.

87. V. I. Karpman and F. F. Cap, *Nonlinear waves in dispersive media*, (transl. from Russian by S. M. Hamberger), Pergamon Press, N.Y., 1975.

88. V. G. Makhankov, "Dynamics of classical solitons in non-integrable systems," Physics Reports, A review of *Phys. Lett. C*, North-Holland Publ. Co., **35**(1), pp. 1–128, Jan. 1978.

89. A. C. Scott, F. Y. F. Chu, and D. W. McLaughlin, "The Soliton: A new concept in applied science," *Proc. III*, **61**(10), pp. 1443–1483, 1973.

90. G. Backus and I. Gilbert, *Philos. Trans. R. Soc., Ser. A*, **266**, pp. 123–192, 1970, (also see: *Geophys. J. Res. Astr. Soc.*, **13**, pp. 247–276, 1967).

91. P. D. Lax, "Integrals of nonlinear equations of evolution and solitary waves," *Commun. Pure Appl. Math.*, **21**, pp. 467–490 1968.

92. L. Chernov, *Wave Propagation in a Random Medium*, Dover, N.Y., 1967.

93. V. Tatarski, "The effect of the turbulent atmosphere on wave propagation," Springfield, Va. U.S. Dept. Commerce, 1971 (also see: "Wave propagation in a turbulent medium," Dover, New York, 1967).

94. R. L. Fante, "Electromagnetic beam propagation in turbulent media," *Proc. IEEE*, **63**(12), pp. 1669–1692, 1975.

95. J. W. Strohbehn, Ed., *Laser Beam Propagation in the Atmosphere*, Topics in Applied Physics, Vol. 25, Springer, Berlin, Heidelberg, New York, 1978.

96. K. G. Budden, *Radio Waves in the Ionosphere. The mathematical theory of the reflection of radio waves from stratified ionized layers*, Cambridge, University Press, 1961.

97. B. B. Baker and E. T. Copson, *The Mathematical Theory of Huygens Principle*, Oxf. Univ. Press, 2nd ed., 1950.

98. Sir H. Jeffreys and Lady B. Jeffreys, *Methods of Mathematical Physics*, 3rd ed., Cambridge University Press, 1956.

99. M. Kline and I. W. Kay, *Electromagnetic Theory and Geometrical Optics*, Pure & Appl. Math., Vol. XIII, J. Wiley, Intersc. Publ., New York, 1965. (Also see: M. Kline, "The asymptotic solution of Maxwell's equations," *Commun. Pure Appl. Math.*, **4**, pp. 225–262, 1951. "An asymptotic solution of Maxwell's equation, in *The Theory of EM Waves*, New York, Intersc., 1951; "Electromagnetic theory and geometrical optics," in *Electromagnetic Waves*, R. E. Langer, Ed., Madison, Univ. of Wisconsin Press, 1962.

100. J. B. Keller, "Accuracy and validity of the Born and Rytov approximations," *J. Opt. Soc. Am.*, **59**, pp. 1003–1004, 1969.

101. R. H. T. Bates, W. M. Boerner, and G. R. Dunlop, "An extended Rytov approximation and its significance for remote sensing and inverse scatterings," Optics Commun., 18(4), pp. 421–423, 1976.

102. Bäth, M., *Mathematical Aspects of Seismology*, Elsevier, Amsterdam, 1968.

103. Bäth, M., *Introduction to Seismology*, Birkhäuser, Basel and Stuttgart, 1973.

104. Bäth, M., *Spectral Analysis in Geophysics*, Elsevier, Amsterdam, 1974.

105. Dobrin, M. B. *Introduction to Geophysical Prospecting*, McGraw-Hill, New York, N.Y., 3rd ed., 1976.

106. G. M. Jenkins, and D. G. Watts, *Spectral Analysis and its Applications*, Holden-Day, San Francisco, Calif. 1968.

107. F. S. Grant and G. F. West, *Interpretation Theory in Applied Geophysics*, McGraw-Hill, New York, 1965.

108. W. Heisenberg, "Über den anschaulichen Inhalt der quanten-theoretischen Kinematik und Mechanik," *Z. Phys.*, **43**, pp. 172–198, 1927.

109. R. G. Newton, "A New Representation for Quantum Mechanics," *Physica*, **96A**, pp. 271–279, 1979.

110. L. D. Faddeyev, "Itogi naukiitechniki (Vsesoyuznyi institut nauchnoi i toknicheskoi informatsi)," Sorremennye Problemy Mathematika, **3**, pp. 98–180, 1974.

111. I. Kay and H. E. Moses, "Reflectionless transmission through dielectrics and scattering potentials," *J. Appl. Phys.*, **27**, pp. 1503–08, 1956, (also see: *Nuovo Cimento*, **2**, pp. 917–961, 1955, *Nuovo Cimento*, **3**, pp. 66–84, 1956. Determination of the scattering potential from the spectral measured function. I. Kay and H. E. Moses, "The Gel'fand-Levitan equation for the three-dimensional scattering problem," *Nuovo Cimento*, **22**, pp. 689–705, 1961; see also: *Commun. Pure & Appl. Math.*, **14**, p. 435, 1961).

112. A. K. Jordan and S. Ahn, "Profile reconstruction for inhomogeneous regions from the response to a delta-function input: Reflection coefficient with two poles and one zero," 1978 URSI Intern. Electromagnetic Wave-theory Symposium, Stanford Univ., June 23, 1977 (Proc. 148–149). A. K. Jordan and S. Ahn, "Profile Inversion of Simple Plasmas and Non-uniform Regions: Three-Pole Reflection Coefficients," *IEEE Trans. Antennas Propag.*, **24**, pp. 879–882, 1916.

113. V. H. Weston and R. J. Krueger, "On the inverse problem for a hyperbolic dispersive P. D. E. II," *J. Math. Phys.*, **14**, pp. 400–408, 1973.

114. R. J. Krueger, "An inverse problem for a dissipative hyperbolic equation with discontinuous coefficients." *Quart. Appl. Math.*, **23**(2), pp. 129– 147, 1976. (Note: Several additional papers in print and in progress.)

115. R. G. Newton, *Scattering theory of waves and particles*, McGraw-Hill, New York, 1966, (also see: The *comples-j-plane*, Benjamin, New York, 1964).

116. R. G. Newton, "Inverse problems in physics," *SIAM Rev.*, **12**, 346–356, 1970.

117. P. C. Clemmov, *The plane wave spectrum representation of electromagnetic fields*, Pergamon, New York, 1966.

118. C. Müller, "Electromagnetic radiation patterns and sources," *IRE Trans. Antennas Propag.*, **4**(3), pp. 224–232, 1956. (See also: Radiation patterns and radiation fields, *J. Rational Mech. Anal.*, **4**, pp. 235–246, 1955. *Grundprobleme der mathematischen Theorie elektromagnetischer Schwingungen*, Springer, Berlin, pp. 140–142, 1957.)

119. C. H. Wilcox, "An expansion theorem for electromagnetic fields," *Commun. Pure Appl. Math.*, pp. 115–134, 1956. (Also see for corrections: P. Hartmann and C. Wilcox, "On solutions of the Helmholtz equation in exterior domains," Math. 2, **75**, pp. 228–255, 1961.)

120. M. I. Skolnik, *Introduction to Radar Systems*, McGraw-Hill, New York, 1962. (Also see: M. K. Skolnik, *Radar Handbook*, McGraw-Hill, New York, 1978.)

121. J. S. Asvestas and R. E. Kleinman, "Low Frequency Electromagnetic Scattering," in *Electro-magnetic Scattering*, P. L. E. Uslenghi, Ed., Academic Press, New York, 1978.

122. P. J. Wyatt, Appl. Pt. 7, pp. 1879, 1896, 1968.

123. J. W. Strutt, and Lord Rayleigh, *Collected Scientific Papers*, Dover, New York, 1964.

124. H. C. van de Hulst, *Light Scattering by Small Particles*, John Wiley & Sons, New York, 1957. (Also see: D. McIntyre and F. Gormick, Eds., *Light scattering from dilute polymer solutions*, Gordon & Smead, New York, 1964.)

125. N. A. Logan, "Survey of some early studies of the scattering of plane waves by sphere," *Proc. IEEE*, 53, Special Issue on Radar Reflectivity, pp. 773–785, Aug., 1965.

126. G. Mie, "Beiträge zur Optik trüber Medien, speziell kolloidaler Metallösungen," *Ann Phys.*, **25**, pp. 377–422, 1908.

127. G. Kortüm, *Reflectance Spectroscopy*, Springer, Berlin, Heidelberg, New York, 1969.

128. J. C. Dainty, Ed., *Laser Speckle, and Related Phenomena*, Topics in Applied Physics, 9, Springer, New York, 1975.

129. R. J. Garbacz, "Modal expansion for resonance scattering phenomena," *IEEE Proc.*, 53(8). Special Issue on Radar Reflectivity, pp. 856–864, Aug. 1965. (Also see: "A general expansion for radiated and scattered fields," Ph.D. thesis, OSU, Columbus, OH, 1968.)

130. J. H. Richmond, "Radiation and scattering by thin wire structures in the frequency domain." NASA GR-2396, May 1974, also see: *IEEE Trans. Antennas Propag.*, 18(6), pp. 820–821, 1970 *IEEE Trans. Antennas Propag.*, 14(6), pp. 782–786, 1966.

131. R. F. Harrington, "Characteristic Modes for Antennas and Scatterers," in *Numerical and Asymptotic Techniques in Electro-magnetics*, R. Mittra, Ed. Topis in Applied Physics, Vol. 3 Springer, New York, Heidelberg, Berlin, pp. 51–87, 1975.

132. C. E. Baum, *Proc. IEEE*, **64**, pp. 1598–1616, 1976.

133. F. M. Tesche, *IEEE, Trans. Antennas Propag.*, **21**, pp. 53–62. 1973.

134. L. Marin, *IEEE Trans. Antennas Propag.*, **21**, pp. 266–274, 1974.

135. L. Marin, *IEEE Trans. Antennas Propag.*, **21**, pp. 809–818, 1973.

136. A. J. Berni, *IEEE Trans. Aerosp. Electron. Sys.*, **11**, pp. 147–154, 1975.

137. D. L. Moffatt and R. K. Mains, *IEEE Trans. Antennas Propag.*, **23**, pp. 358–367, 1975.

138. C. W. Chuang and D. L. Moffatt, 1975 URSI NRC Meeting. UIUC, Urbana, Il., June 3–5, 1975, pp. 67–68 (Abstracts).

139. Electro Science Lab—Ohio State University, Class Notes: *Radar Target Identification I/II*, Columbus, Ohio, Sept., 1976.

140. E. M. Kennaugh, "The K-Pulse Technique," ONR-N00014-78-C0049, 1980.

141. K. M. Chen, "Radar Waveform Synthesis Method—A Radar Detection Scheme," NAU-AIR-0019-79C-0385, 1980.

142. E. T. Copson, *Asymptotic expansions*, Cambridge racts in Maths. & Math. Physics, Vol. 55, 1965, Cambridge Univ. Press.

143. M. Born and E. Wolf, *Principles of Optics*, Pergamon Press, New York, 1959.

144. H. Hönl, A. W. Maue, and K. Westpfahl, *Theorie der Beugung*, Handbuch der Physik, Vol. XXV-I, Springer, New York, Heidelberg, Berlin, 1961.

145. V. A. Borovikov and B. Ye. Kinber, "Some problems in the asymptotic theory of diffraction," *Proc. IEEE*, 62(11), pp. 1416–1437, 1974.

146. R. G. Koujoumjian, *Proc. IEEE*, 53, pp. 864–876, 1965.

147. P. H. Pathak and R. G. Kouyoumjian, "An analysis of the radiation from apertures in curved surfaces by GTD," *Proc. IEEE*, 62(11), pp. 1438–1447, 1974.

148. P. M. Woodward, *J. IEEE*, **94-11A**, p. 1554, 1943. (See also: *IEE*, **95-111A**, pp. 363–370, 1948; D. R. Rhodes, *Proc. IEEE*, **53**, pp. 1013–1021, 1965; *IEEE Trans. Antennas Propag.*, **19**, pp. 162–166, 1971; *IEEE Trans. Antennas Propag.*, **20**, pp. 143–145, 1972. R. A. Hurd, *Proc. IEEE*, **121**, pp. 32–48, 1974.)

149. A. Freedman, "The portrayal of body shape by a sonar or radar system." Radio Electron. *Eng.*, **25**, pp. 51–64, 1963.

150. J. B. Keller, "The inverse scattering problem in geometrical optics and the design of reflectors," *IEEE Trans. Antennas Propag.*, 7, p. 146, 1959. (Also see: J. B. Keller. "On the use of a short-pulse broad-band radar for target identification," RCA,

Moorestown, N.J. Feb. 17, 1965.)

151. N. M. Tomljanovich, H. S. Ostrowsky, and J. F. A. Ormsby, "*Narrowband interferometer imaging*," Mitre Corp. Prop. 4966, Nov., 1968, AD 679–208.

152. J. L. Altman, R. H. T. Bates, and E. N. Fowle, *Introductory notes relating to electromagnetic inverse scattering*, Mitre Corp.-SR-121, Sept., 1964.

153. R. M. Bracewell, *The Fourier Transform and its Applications*. McGraw-Hill, New York, 1965.

154. J. W. Goodman, *Introduction to Fourier Optics*, McGraw-Hill, New York, 1968.

155. A. Papoulis, *The Fourier Integral and its Applications*. McGraw-Hill, New York, 1962. (Also see: A. Papoulis, *Probability, Random Variables and Stochastic Processes*, McGraw-Hill, N.Y., 1965; A. Papoulis, *Systems and transforms with applications to optics*, McGraw-Hill, 1968; A. Papoulis, *Signal Analysis*, McGraw-Hill Book Co., New York, 1977.)

156. R. O. Harger, *Synthetic Aperture Radar Systems: Theory and Design*, Acad. Press, New York.

157. E. M. Kennaugh and R. L. Cosgriff, "The use of impulse response in electromagnetic scattering problems," IRE Nat. Convention Record, Part I, pp. 72–77, 1958.

158. E. M. Kennaugh and D. L. Moffatt, "Transient and impulse response approximation," *IEEE Proc.*, 53(8), pp. 893–901, 1965.

159. J. D. Young, "Target imaging from multiple-frequency radar returns," ESL-OSU, Columbus, Ohio, Techn. Rept. 2768-6, Jan., 1971 (AD728 235), (also see: "Radar imaging from ramp response signatures," *IEEE Trans. Antennas Propag.*, 24, pp. 276–282, 1976).

160. W-M. Boerner and C. M. Ho, "The importance of RADON's projection reconstruction theory in radar target scattering," Kleinheubacher Berichte, Vol. 30, FTZ, Deutsche Bundespost, Darmstadt, FRG, West Germany, 1980.

161. C. M. Ho, "Development of Physical Optics inverse scattering using RADON transform theory and polarization utilization," M.Sc. thesis, University of Illinois at Chicago Circle, Chicago, USA, Jan., 1980.

162. Y. Das and W. M. Boerner, On radar target shape estimation using algorithms for reconstruction from projections," *IEEE Trans. Antennas Propag.*, 26(2), pp. 274–279, 1978a. (Also see: Y. Das, "Application of concepts of image reconstruction from projections and Radon transform theory to radar target identification," Ph.D. dissertation, Fac. of Grad. Studies, U. of Manitoba, Winipeg, Manitoba, Nov., 1977.)

163. N. Bleistein and J. K. Cohen, "A survey of recent progress in inverse problems," University of Denver, Rept. MS-R-7806, 1978 (prepared for ONR). (Also see Chpt. in [2.222].)

164. N. N. Bojarski, "A survey of electromagnetic inverse scattering," Syracuse Univ. Res. Corp., Special Projects Lab. Rept., Oct. 1967, DDC AD-813-581; N. N. Bojarski, "Inverse Scattering, Company Rep N 00019-73-C-0312/F prepared for NASD, AD-775-235/5, 1964.

165. R. M. Lewis, "Physical Optics inverse diffraction," *IRE Trans. Antennas Propag.*, 17, pp. 308–314, 1969.

166. R. O. Mager and N. N. Bleistein, "An examination of the limited aperture problem of physical optics inverse scattering," *IEEE Trans. Antennas Propag.* (5), pp. 695–698, Sept., 1978.

167. J. A. Hammer, "Method to determine the scattering centers from the backscatter pattern of a body," Proc. GISAT II Symp. Vol. II, Part I, 223–235, Oct. 2–4, 1967, Mitre Corp., Bedford, Mass. AD8397000.

168. W. B. Goggins, Jr., "Identification of radar targets by pattern recognition," Ph.D. thesis, USAF Institute of Technology, Air University, Dayton, OH, June, 1973.

169. E. C. Burt and R. F. Wallenberg, "Aircraft targetmodeling for radar image simulation," SRC-TN-78-344, March, 1979 (SRC).

170. J. B. Keller and R. M. Lewis, "Asymptotic solutions of some

diffraction problems," *Commun. Pure Appl. Math.*, **9**, pp. 207–265, 1956.

171. J. B. Keller and R. M. Lewis, "Inverse scattering problems for bodies of revolution," RCA Report, 1961. (Also see: J. B. Keller, *IEEE Trans. Antennas Propag.*, **7**, p. 146, 1959).

172. R. E. Kell, "On the deviation of bistatic RCS from monostatic measurements," *Proc. IEEE*, **53**, pp. 983–988, 1965.

173. M. E. Bechtel and R. A. Ross, "Radar Scattering Analysis," CAL Rept. ER IRIS-10, August 1966.

174. S. H. Bickel, "Polarization extensions to the monostatic-bistatic equivalence theorem," The MITRE Corp., Bedford, Mass., Internal Rept., 1965. (Also see: S. H. Bickel, "Some invariant properties of the polarization scattering matrix," *Proc. IEEE.* **53**(8), pp. 1070–72, 1965. Also see MITRE, Rept. Contract No. AF19(628)-2390, April, 1965.)

175. G. A. Dike, E. C. Burt and R. F. Wallenberg, "An application of GTD and PO to object identification through inverse scattering," 1980 Interm. Symp. IEEE-APS/URSI-B, Quëbec, Special Session III on Inverse Scattering.

176. Gniss, H. and K. Magura, *mm-Wave of Ground Based Objects*, Rept. 1-78-153, FHP-FGAN, Wachtberg-Werthoven, FRG, Jan., 1978.

177. G. Graf, *IEEE Trans. Antennas Propag.*, **24**, pp. 378–381, 1976.

178. K. Magura, "Probleme bei der holographischen Abbildung im Mikrowellenbereich," Tech. Rpt. 6-72-11, HFP-FGAN Wachtberg Werthoven, 1972.

179. R. Karg, *Arch. Elektrotech. Übertr.*, **31**, pp. 150–156, 1976.

180. J. Detlefsen, Nachrichtentechn. Zeitschrift 30(9), pp. 723–725, 1977, (also see: *IEEE Trans. Antennas Propag.*, **28**, July, 1980).

181. G. Sinclair: "Modification of the Radar Range Equation for Arbitrary Targets and Arbitrary Polarization," Rept. 302–19, Antenna Laboratory, Electro-Science Lab, Ohio State University, Columbus, 1948.

182. V. H. Rumsey, G. A. Deschamps, M. L. Kales, and J. I. Bohnert, *Proc. IRE*, **39**, pp. 535–553, 1951.

183. E. M. Kennaugh, "Effects of Type of Polarization on Echo Signals," Tech. Rpt. 389-9, Antenna Lab. Columbs, 1951, Correction to Rpt. (1978). (Also see: E. M. Kennaugh, "Polarization Properties of Target Reflection," Tech. Rpt. 389-2, Griffis AFB, 1952.)

184. J. R. Huynen, "Radar Target Sorting Based Upon Polarization Signature Analysis," Repts. 28-82-16, 1960, and AD 318597, 1972, Lockheed Missiles & Space Division. (Also see: *Proc. IEEE*, **53**, pp. 936–946, 1965; "Phenomenological Theory of Radar Targets," Ph.D. Thesis, Techn. Univ. Delft, Druckkerij Bronder-Offset N.V., Rotterdam, 1970; "Radar Target Phenomenology in Electro-magnetic Scattering," Selected papers of a NH Conf. on *Electro-magnetic Scattering*, UICC, Chicago, Il., June 15–18, 1976, P.L.E. Uslenghi, Ed., Academic Press, New York.

185. A. J. Poelman, Tijdschrift van het Nederlands Electronica en Radiogenootschap, **44**, pp. 93–106, 1979. (Also see: *IEEE Trans. Aerosp. Electron. Sys.*, **11**, pp. 660–662, 1975; *IEEE Trans. Aerosp. Electron. Sys.*, **12**, pp. 674–682, 1976. *Electronics Lett.*, **13**, pp. 533–534, 1977.)

186. W-M. Boerner and C-Y. Chen, "Polarization Dependence in Electromagnetic Inverse Scattering," *IEEE Trans. Antennas Propag.*, **29**, Special Issue on Inverse Problems in Electromagnetics, March, 1981.

187. S. Weisbrod and L. A. Morgan, "RCS matrix studies of sea clutter," NAVAIR Systems Command, Rpt. R2-79, January, 1979, Teledyne Micronetics, San Diego, Ca.

188. R. Rosien, D. Hammers, G. Ioannidis, J. Bell and J. Nemit: "Implementation techniques for polarization control in ECCM," RADC-TR-79-4, Feb., 1979. Griffis AFB, Rome, N.Y.

189. R. J. Doviak and D. Sirmans, "Doppler Radar with Polarization Diversity," *J. Atmos. Sci.*, **30**(4), pp. 737–738. May, 1973. (Also see: R. J. Doviak, D. S. Zrnic, and D. S. Sirmans, "Doppler Weather Radar," *Proc. IEEE*, **67**(11), pp. 1522–1553, Nov. 1979.)

190. L. E. Allan and G. C. McCormick, "Measurements of backscattered matrix of dielectric spheroids," *IEEE Trans. Antennas Propag.*, **26**(4), pp. 579, 587. July, 1978. (See also: A. Hendry, G. C. McCormick, and B. L. Barge, "The degree of common orientation of hydrometeors observed by polarization diversity radars," *J. Appl. Meterol*, **15**(6), pp. 633–640, June 1976.

191. T. A. Seliga, V. N. Bringi, and H. H. Al-Khatib, "Differential reflectivity measurements of rainfall rate: raingauge comparison," 19th Conf. on Radar Meterology, 15–18 April, 1980, Miami Beach, Fl. (*Am. Met. Soc. Proc.*).

192. S. M. Cherry, J. W. F. Goodard, and M. P. M. Hall, "Examination of rain dron sizes using a dual-polarization radar," 19th Conf. on Radar Meterology, 15–18 April, 1980, Miami Beach, Fl.

193. J. I. Metcalf and J. D. Echard, "Coherent polarization-diversity radar techniques in meteorology," *J. Atmos. Sci.*, **35**, pp. 2010–2019, 1978.

194. L. C. Chan, D. L. Moffatt, and L. Peters, Jr., "A Characterization of Subsurface Radar Targets," *Proc. IEEE*, July, 1979.

195. L. C. Chan, "Subsurface Electromagnetic Target Characterization and Identification," Ph.D. Dissertation, 1979, The Ohio State University, Department of Electrical Engineering.

196. K. A. Shubert, J. D. Young, and D. L. Moffatt, "Synthetic Radar Imagery," *IEEE Trans. Antennas Propag.*, **25**, No. 4, pp. 477–483, July, 1977.

196. J. N. Brittingham, E. K. Miller, and J. L. Willows, "Pole Extraction from Real Frequency Information," Proceedings of the IEEE, Vol. 68, No. 2 February, 1980.

197. R. F. Elsner, "Vehicular Variable Parameter METRRA Systems Final Report, III." Inst. of Techn. Rept. No. E. 6224, May, 1974 (AD782214).

198. G. L. Opitz, "Metal-detecting radar rejects clutter naturally," *Microwaves*, **15**(8), pp. 12–14, 1976.

199. R. Elsner and M. Frazier, "Engineering study for electrical hull interference, III. Inst. Tech. Research Center Rept. No. 11R1-56013-14 (AD462970), 1965.

200. C. L. Bennett, A. M. Auckenthaler, R. S. Smith, and J. D. DeLorenzo, "Space time integral equation approach to the large body scattering problem," Sperry Res. Centre, Sudbury, Ma., Rept. No. SCRCR-Cr-73-1, 1973.

201. N. Farhat, "High resolution microwave holography and the imaging of remote objects, *Opt. Eng.*, **14**, pp. 499–505, 1975.

202. K. Iizuka, *Proc. IEEE*, **57**, pp. 812–814, 1969.

203. W. E. Kock, Real-time detection of metallic objects using liquid crystal microwave holograms," *Proc. IEEE.* **60**, p. 1104, 1972.

204. G. Tricoles, and N. H. Farhat, "Microwave Holography: Applications and Techniques," *Proc. IEEE*, **65**(1), pp. 108–121. 1977, (in Special Issue on *Optical Computing*).

205. J. A. Stiles and J. C. Holtzman, Eds., "*Radar Backscatter from Terrain*," Tech. Rpt. RSLR374-2, US-AE Topogr. Lab., Fort Belvoir, Remote Sensing Lab. Kansas, 1979.

206. R. W. Larson, F. Smith, R. Lawson, and M. L. Brian, "Multispectral Microwave Imaging Radar for Remote Sensing Applications," in *Proc. URSI Special Meeting on Microwave Scattering and Emission from the Earth*, E. Schanda, Ed., pp. 305–315, U. of Bern, 1974.

207. S. Marder, "Synthetic Aperture Radar," in *Atmospheric Effects on Radar Target Identification and Imaging*. H. E. G. Jeske, Ed., Proc. NATO-ASI, Goslar, 1975. (D. Reidel Publ. Co., Dordrecht-Holland, 1976).

208. V. H. Weston, and W. M. Boerner, "Inverse Scattering Investigations," Tech. Rpt. 8575, Univ. of Michigan Radiation Lab., 1968.

209. V. H. Weston, and W. M. Boerner, "An inverse scattering technique for electromagnetic bistatic scattering," *Can. J. Phys.*, **47**, pp. 1177–1184, 1969.

210. W. M. Boerner and H. P. S. Ahluwalia, On a set of continuous wave electromagnetic inverse scattering boundary conditions," *Can. J. Phys.*, **50**(10), pp. 3023–3061, May, 1972.
211. H. P. S. Ahluwalia and W-M. Boerner, *IEEE Trans Antennas Propag.*, **21**, pp. 673–672, 1973. H. P. S. Ahluwalia and W-M. Boerner, *IEEE Trans. Antennas Propag.*, **22**, pp. 663–682, 1974.
212. W. M. Boerner and F. H. Vandenberghe, "Determination of the electrical radius ka of a special scatterer from the scattered field." *Can. J. Phys.*, **49**, pp. 1507–1535. (Also see: *Can. J. Phys.*, **49**, pp. 804–819, 1971; *Can. J. Phys.*, **50**, pp. 754–759, 1972; *Can. J. Phys.*, **50**, pp. 1987–1992, 1972.)
213. W. M. Boerner and O. A. Abou-Atta, *Utilitas Math.*, **3**, pp. 163–273, 1973.
214. P. F. Wacker: "Non-planar near field measurements spherical scanning," Tech. Rpt. NSRIR 75-809, National Bureau of Standards, Boulder, Colorado, 1975.
215. P. F. Wacker, "A Qualitative Survey of Near Field Analysis and Measurement," Tech. Rpt. NBSIR-79-1602. National Bureau of Standards Boulder, Colorado, 1979.
216. V. H. Weston and J. J. Bowman, "The plane wave representation and the inverse scattering problem," Proc. GISAT II Symp. Vol. II, Part I, pp. 289–301, Oct. 2–4, 1967, Mitre Corp., Bedford, Mass, AD 839700.
217. V. H. Weston, J. J. Bowman, E. Ar. On the inverse electromagnetic scattering problem, Arch. National Mech. Anal. *31* (1968), 199–213. (Also see: Inverse Scattering Investigations, Final Report, U of Mich., (RCS) Radiation Lab., Contract AF-19 628-4884, 1966.)
218. P. J. Wood, The prediction of antenna characteristics from spherical near field measurements, MARCONI REVIEW XL, 1977, 42–68 (Part I), 117–155 (Part II).
219. A. C. Newell, R. C. Baird, P. F. Wacker: IEEE Trans. AP-21, 418–431 (1973).
220. J. Brown, E. V. Jull: Proc. IEE *1088*, 635–644 (1961).

Recent Monographs

221. G. T. Herman, Ed., *Image Reconstruction from Projections*, Implementation and Applications, Topics in Applied Physics, **32**, Springer, 1979.
222. J. A. De Santo, *Ocean Acoustics*, Topics in Current Physics, **8**, Springer, 1979.
223. V. M. Babiĉ and N. Y. Kirpiĉnikova, *The boundary-layer method in diffraction problems*, Springer Series in Electrophysics, **3**, Springer, 1979.
224. V. M. Babiĉ, *Mathematical questions in the theory of wave diffraction and propagation*, STEKLO/114-LC. 74-2363; ISBN 0-8218-3015-5, AMS Catalogue, 1979.
225. D. Cassanent, *Optical Data Processing*, Topics in Applied Physics, **23**, Springer, 1978.
226. C. van Schooneveld, *Image formation from coherence functions in astronomy*, D. Reidel Publ. Co., Dorebrecht, 1979.
227. P. C. Sabatier, *Applied Inverse Problems*, Lecture Notes in Physics, **85**, Springer, 1978. (See also: K. Chadon and P. C. Sabatier, "Inverse Problems in Quantum Mechanics," Texts and Monographs in Physics, Springer, 1977.)
228. S. Haykin, *Non-linear methods of spectral analysis*, Topics in Current Physics, **34**, Springer, 1979.
229. A. Deepak, *Inversion methods in atmospheric remote sensing*, NASA CP-004, 1977.
230. F. Fuchs, G. Müller, Ed., Proc. of XI Int. Symposium on Mathematical Geophysics at Seeheim/Odenwald, FRG, 18–27 Aug., 1976, *J. Geophys.*, **43**, (I), 1977, Special Issue.

Editors' Biographies

Lawrence E. Larsen (M'81–SM'82) was born in Denver, CO. He attended the University of Colorado at Boulder and the School of Medicine in Denver. He received the M.D. degree (*magna cum laude*) in 1968. He pursued post-doctoral training in biophysics at the University of California, Los Angeles, from 1969 to 1970 (under NIH sponsorship).

He entered active duty at the Walter Reed Army Institute of Research, Washington, DC, as a Research Physiologist in the Department of Microwave Research as a Medical Corps Captain. He was promoted to Major in 1972. He left the Army Medical Research and Development Command in 1973 to accept a faculty position (Computer Science and Physiology) at the Baylor College of Medicine, Waco, TX, where he developed digital x-ray and ultrasound imaging systems. He returned to the Walter Reed Army Institute of Research in 1975 as Associate Chief for Biophysics in the Department of Microwave Research. He was named the Chief of that department in 1977 and promoted first to Lt. Col. in 1979 and to Col. in 1984. His major research areas were in biomedical hazards and medical applications of microwave radiation. In 1985 he started the Medical Microwave Research Corporation in Silver Spring, MD, and serves as its President.

Dr. Larsen is a member of the IEEE Microwave and Techniques Society. In 1984 he was honored with the U.S. Army Research and Development Achievement Award. He is the author of over 100 contributions to the scientific literature, contributor to four books, and author of the first book to be published on microwave methods for medical imagery. He also holds seven patents in microwave medical technology.

John H. Jacobi (M'70–M'77–SM'81) received the B.S.E.E. degree from Rose-Hulman Institute of Technology, Terre Haute, IN, in 1959 and the M.S.E.E. degree from the University of Maryland, College Park, in 1969.

From 1974 to 1983, he was employed at the Department of Microwave Research, Walter Reed Army Institute of Research, where he held the position of Associate Chief for Engineering. In that capacity, he was a co-investigator with Dr. Lawrence E. Larsen in the design and development of microwave imaging systems used for interrogation of biological targets. He is currently Division Manager of the Radio Frequency Systems Division at Information Development and Applications, Inc. (IDEAS) in Beltsville, MD. He is the author of numerous technical papers, has given many conference presentations, and is the co-inventor on seven patents pertaining to microwave imaging and other microwave-related technologies.

Mr. Jacobi is a member of the Executive Committee of the Washington/Northern Virginia Chapter of the IEEE Microwave Theory and Techniques Society and is a past Chairman of the Chapter. He was on the Steering Committee of the 1980 International Microwave Symposium and has served as Professional Activities Chairman for the Washington/Northern Virginia Section of the IEEE.